New Diagnostic and Therapeutic Approaches in Diabetic Microvascular Complications

New Diagnostic and Therapeutic Approaches in Diabetic Microvascular Complications

Guest Editors

Ana Maria Dascalu
Dragos Serban

MDPI

Basel • Beijing • Wuhan • Barcelona • Belgrade • Novi Sad • Cluj • Manchester

Guest Editors
Ana Maria Dascalu
"Carol Davila" University of Medicine and Pharmacy
Bucharest
Romania

Dragos Serban
"Carol Davila" University of Medicine and Pharmacy
Bucharest
Romania

Editorial Office
MDPI AG
Grosspeteranlage 5
4052 Basel, Switzerland

This is a reprint of the Special Issue, published open access by the journal *Biomedicines* (ISSN 2227-9059), freely accessible at: https://www.mdpi.com/journal/biomedicines/special_issues/1RZSW3OS96.

For citation purposes, cite each article independently as indicated on the article page online and as indicated below:

Lastname, A.A.; Lastname, B.B. Article Title. *Journal Name* **Year**, *Volume Number*, Page Range.

ISBN 978-3-7258-2499-1 (Hbk)
ISBN 978-3-7258-2500-4 (PDF)
https://doi.org/10.3390/books978-3-7258-2500-4

Contents

About the Editors

Ana Maria Dascalu

Ana Maria Dascalu is a habilitated doctor of ophthalmology and Associate Professor of Ophthalmology, Faculty of Medicine, "Carol Davila" University of Medicine and Pharmacy Bucharest (Romania). Dr. Dascalu is a member of the Romanian Society of Ophthalmology, Romanian Glaucoma Society, and the Global Burden Disease (GBD) 2019 Blindness and Vision Impairment Collaborators. Her areas of scientific expertise include glaucoma, diabetic retinopathy, age-related macular degeneration, and the role of systemic inflammation in ophthalmologic cases. She is the author of over 90 original articles, 15 monographs and book chapters, and over 50 abstracts presented during international and Romanian professional meetings. Furthermore, she has collaborated as a reviewer or Guest Editor with more than 40 medical journals. Dr. Dascalu also has over ten years of experience in ophthalmic surgery, with special interests in cataracts and anterior segment surgery.

Dragos Serban

Dragos Serban is a habilitated Professor of General Surgery, Faculty of Medicine, "Carol Davila" University of Medicine and Pharmacy Bucharest (Romania), and Chief of the Discipline Surgery IV SUUB (Emergency University Hospital Bucharest). He is a member of the Romanian Society of Surgery, Sigma Xi Scientific Research Honor Society, and the Global Burden of Diseases (GBD) research group. His special research interests are improving quality of care, early biomarkers, laparoscopic surgery, inflammation, emergency surgery, colorectal cancer, and diabetic foot issues. He is an author and co-author of 93 scientific papers presented at important national and international congresses and conferences, and has another 148 articles published in medical journals, of which 117 are included in the Web Of Science Core Collection. He has a Hirsch Index of 25 on Web of Science, and is a reviewer of 23 prestigious international medical publications. Prof. Dr. Serban has designed two OSIM patents (registered at national institutions) for surgical devices. He is the main author of the monograph "Postbulbar Duodenal Ulcer", and the author and co-author of chapters from 13 other medical books, some of which are for students and young medical trainees. He has participated as the chairman or invited speaker in numerous seminars and symposia on surgical topics with Romanian and foreign teachers. Furthermore, he is a senior surgeon, with more than 20 years' experience, currently the Chief of the Fourth General Surgery Department, Emergency University Hospital Bucharest.

biomedicines

MDPI

Editorial

New Diagnostic and Therapeutic Approaches in Diabetic Microvascular Complications

Dragos Serban [1,2] and Ana Maria Dascalu [1,3,*]

1 Faculty of Medicine, "Carol Davila" University of Medicine and Pharmacy, 020021 Bucharest, Romania; dragos.serban@umfcd.ro
2 Fourth Surgery Department, Emergency University Hospital Bucharest, 050098 Bucharest, Romania
3 Ophthalmology Department, Emergency University Hospital Bucharest, 050098 Bucharest, Romania
* Correspondence: ana.dascalu@umfcd.ro

Diabetes mellitus is a major global health problem with an ascendant trend that makes it expected to reach up to 700 million cases by 2045 [1]. Population aging, obesity, and sedentarism, as well as the recent COVID-19 pandemic, are all risk factors that contribute to the increased incidence of this chronic, debilitating disease [2–5].

The microvascular triad unique to diabetes includes diabetic retinopathy, nephropathy, and neuropathy. Diabetic nephropathy is one of the most common causes of end-stage renal failure, and it is associated with having a severe impact on the patient's quality of life [6,7]. Despite significant achievements in early diagnosis and therapy, diabetic retinopathy remains the leading cause of blindness in the working-age population [8]. Diabetic peripheral neuropathy affects nearly 50% of adults who have diabetes during their lifetime and represents a major risk factor for diabetic foot ulcer (DFU), the most common cause of non-traumatic amputations worldwide [9]. Classically, the duration of diabetes, the level of hyperglycemia, and the co-existence with arterial hypertension, as well as dyslipidemia, are related to the onset and progression of these invalidating diabetic complications. However, various training specialties and regular screening visits are required for the early diagnosis of these complications, placing a significant burden on healthcare systems worldwide. There is still a significant unmet need for novel diagnostic tools for early diagnosis, which is based on non-invasive, cost-effective screening methods that can provide personalized management for diabetic patients [10].

Novel research has found multiple molecular pathways that may interfere with vascular dysfunction, ischemia, and tissue damage. In this Special Issue, we showcase nine original articles and two comprehensive reviews that provide new data on novel biomarkers, early diagnosis, pathology, molecular mechanisms, and new therapies in the fields of diabetic retinopathy, nephropathy, neuropathy, and diabetic foot ulcers.

Postprandial plasma glucose (PPG), traditionally measured 1–2 h after a main meal, is a well-recognized clinical tool used not only in diabetes diagnosis but also in evaluating the risk of micro- and macrovascular diabetic complications, as well as therapeutic control of the disease [11]. In the first contribution to this Special Issue, Dr. Yutang Wang and col. examined a large number of patients included in the NHANES III study, which aimed to assess whether predicted PPG after 4–7.9 h could be used to diagnose diabetes. A multivariate prediction model considering 30 possible risk factors was developed to calculate predicted PPG after 4–7.9, and demonstrated a high diagnostic accuracy at 87.3%. The authors consider this parameter to be less influenced by diet or other confounders and suggest that PPG after 4–7.9 could be a promising diagnostic indicator for diabetes.

Fedulovs et al., in the second article, and Hayden et al., in the third, provide new insights regarding the pathological changes in metabolic syndrome (MS) and diabetes. Meanwhile, in the study conducted by Fedulovs and col. (our ninth contribution), they performed for the first time a comparative analysis of the LPS, LBP, EndoCAb IgM, EndoCAb

Citation: Serban, D.; Dascalu, A.M. New Diagnostic and Therapeutic Approaches in Diabetic Microvascular Complications. *Biomedicines* **2024**, *12*, 1858. https://doi.org/10.3390/biomedicines12081858

Received: 27 July 2024
Accepted: 13 August 2024
Published: 15 August 2024

IgG, and fecal calprotectin in patients with T1D, as well as in controls, followed by a stratification of the results according to MS status. The authors evidence higher endotoxemia in T1DM patients with MS and reinforce the need for screening and treating MS in these patients. Dr. Melvin R Hayden and col. (contribution 3) provide a comprehensive review regarding the role of the perivascular unit, a novel concept developed by Troili and col. [12], in the development of neuroinflammation, cerebrovascular disease, and neurodegeneration encountered in metabolic syndrome, obesity, and T2DM.

In the fourth paper, Dr. Ngema and Col. present in a narrative review, with evidence based on the Developmental Origins of Health and Disease (DOHaD) theory that T2DM during pregnancy could not only impact fetal development but also induce permanent impairment of hypothalamic–pituitary–adrenal (HPA) axis activity into adulthood.

Skin autofluorescence (SAF) is a novel non-invasive biomarker on and invasive marker of advanced glycation end products [13]. The level of SAF describes the AGE accumulation in the body and is associated with its increased production, decreased degradation, and renal clearance. Several comparative studies have found that SAF values are increased in diabetic patients versus normal subjects and that skin autofluorescence is even higher when microvascular complications are present. However, there are still many challenges to be resolved, as SAF values appear to be influenced by skin pigmentation and the use of skin products; more studies on a larger number of patients are needed. In this context, we welcome the research of Dr. Reurean-Pintilei and Col. (contribution 5) on a large group of 885 T2DM patients. Their study found that SAF levels were correlated with HbA1c, diabetic kidney disease, and cardiovascular risk. A cut-off value of 2.36 was associated with very high CV risk, especially after adjusting for age, gender, and HbA1c level, and could be useful in selecting candidates for revascularization procedures.

Another approach to this challenging topic is that of Dr. Hiroki Yamagami and Col., whose study is featured in the sixth contribution to this Special Issue. A 1-year prospective study on 350 Japanese people with T2DM was carried out to comparatively evaluate the predictive power of SAF for the progression of diabetic kidney disease, by both urine albumin-to-creatinine ratio (uACR), as a biomarker of glomerular injury, and urine liver-type fatty acid-binding protein (uLFABP)-to-creatinine ratio (uL-FABPCR), as a biomarker of tubular injury. The authors found that SAF correlated with uL-FABP, but not with uACR, in people with T2D, suggesting that SAF can serve as a novel predictor for the development of diabetic tubular injury.

Angiopoietins and angiopoietin-like proteins are essential factors in angiogenesis, endothelium maturation, and inflammation [14,15]. Recently, these were investigated for their potential role in the endothelial dysfunction and inflammation processes encountered in diabetic nephropathy. Dr. Eman Alshawaf et al. (in contribution 7) quantified circulating ANGPTL3, ANGPTL4, ANGPTL8, Ang1, and Ang2 in the fasting plasma of patients with T2DM with and without diabetic nephropathy. They found a close association between Ang2 and ANGPTL8 in a population with DN, suggesting that they can function as DN risk predictors. In line with previous experimental and clinical evidence regarding the effects of Ang2 expression on podocytes through paracrine signaling, leading to glomerular EC destabilization and impaired filtration, this study highlights the importance of Ang2 as an early marker of tubular damage before the appearance of clinical symptoms like microalbuminuria.

A significant portion of this Special Issue is dedicated to novel insights into diabetic retinopathy. In the context of there being few specialists and an increased overload in ophthalmology departments due to screening visits, there is an increased need to develop simple screening tools that may be used by primary healthcare and diabetology departments to prioritize the patients who need urgent referrals to retina specialists. Serum inflammatory biomarkers based on a simple complete blood cell counts were investigated as potential predictive tools. There is solid evidence that neutrophil-to-lymphocyte ratio (NLR) and platelet-to-lymphocyte ratio (PLR) increase in T2DM, compared to normal subjects, and the values found in this regard are well correlated with the presence and the severity of diabetic complication [16–18]. These serum inflammatory biomarkers were thought to be correlated with chronic hyperglycemia and systemic inflammation levels. However, implementing those biomarkers in clinical practice is still challenging, due to the wide variation in means and cut-off values encountered in previous studies, and there is still a need for more evidence on the topic. The study conducted by Dr. Ana Dascalu et al. (this Special Issue's eighth contribution) found that no statistical differences were noticed between NDR and NPDR groups for any of the white cell inflammatory biomarkers. However, significantly higher values for NLR, MLR, SII, and MPV were found in the PDR group when compared with the NDR and NPDR groups. This finding may signify that the level of systemic inflammation is higher in the advanced stage of DR associated with neovascularization. Age, sex, race, and ethnicity, as well as a large array of physiologic and pathological conditions, could impact serum inflammatory biomarkers. However, serum inflammatory biomarkers could be useful when they are integrated into comprehensive risk prediction models.

Optical coherence angiography (OCTA) is a novel imagistic approach that offers valuable information regarding macular vessels in a repeatable, non-invasive manner. Dr. Irini Chatziralli and col. (contribution 9) investigated changes in macular microvasculature using optical coherence tomography angiography (OCTA) in association with functional changes in patients with proliferative diabetic retinopathy (PDR) treated with pan-retinal photocoagulation (PRP) with a 12-month follow-up. The study results showed that at months 6 and 12 after PRP, foveal and parafoveal vessel density (VD) at the superficial capillary plexus (SCP) significantly increased compared to baseline, while the foveal avascular zone (FAZ) area significantly decreased and FAZ became more circular. The improvement of the choroidal circulation at the macula could be explained by the PRP-induced inflammatory response and the redistribution of choroidal flow from obliterated peripheral capillaries to the macula [19,20]. However, the authors could not find an improvement in BCVA in laser-treated patients.

Dr. Tanasie et al., in contribution 10, offer an interesting perspective on diabetic retinopathy as a neurodegenerative disease. In a comparative evaluation of the electrophysiological findings between patients with and without DR, the results indicate a substantial dysfunction of the neural retina in all stages of DR, with significant differences in response delays among study groups.

Intravitreal anti-VEGF is currently the mainstay therapy in diabetic retinopathy associated with macular edema. However, up to 30% of patients have a suboptimal response, suggesting that other molecules are involved in increased vascular permeability and capillary occlusion [21,22]. The valuable contribution of Dr. Lietuvninkas and Col. (contribution 11) is an experimental study that aims to analyze the "in vitro" properties of a novel agent: the (transforming growth factor beta) TGFβ receptor inhibitor RepSox (RS). The results of this study are encouraging, as RS performed better compared to anti-VEGF agents, not only in preventing barrier relaxation but also in promoting the closing of the relaxed barrier, induced by VEGF and TNF α. Thus, RS could be a more potent agent in treating diabetic macular edema by inhibiting multiple pathways.

Finally, we would like to take the opportunity to acknowledge all of these contributors for sharing their valuable work, taking a step forward in unveiling significant biological and molecular changes, and characterizing potential new tools for improving the diagnosis and management of diabetic microvascular complications.

Author Contributions: Conceptualization, D.S. and A.M.D.; methodology, D.S.; software, A.M.D.; validation, D.S. and A.M.D.; formal analysis, D.S.; investigation, A.M.D.; resources, D.S.; data curation, A.M.D.; writing—original draft preparation, A.M.D.; writing—review and editing, D.S.; visualization, A.M.D.; supervision, D.S.; project administration, A.M.D.; funding acquisition, D.S. All authors have read and agreed to the published version of the manuscript.

Funding: This research received no external funding.

Acknowledgments: We offer our sincere thanks to the authors in this Special Issue for their valuable scientific contributions.

Conflicts of Interest: The authors declare no conflicts of interest.

List of Contributions

1. Wang, Y.; Fang, Y.; Aberson, C.L.; Charchar, F.J.; Ceriello, A. Postprandial Plasma Glucose between 4 and 7.9 h May Be a Potential Diagnostic Marker for Diabetes. *Biomedicines* **2024**, *12*, 1313. https://doi.org/10.3390/biomedicines12061313.
2. Fedulovs, A.; Pahirko, L.; Jekabsons, K.; Kunrade, L.; Valeinis, J.; Riekstina, U.; Pīrāgs, V.; Sokolovska, J. Association of Endotoxemia with Low-Grade Inflammation, Metabolic Syndrome and Distinct Response to Lipopolysaccharide in Type 1 Diabetes. *Biomedicines* **2023**, *11*, 3269. https://doi.org/10.3390/biomedicines11123269.
3. Hayden, M.R. A Closer Look at the Perivascular Unit in the Development of Enlarged Perivascular Spaces in Obesity, Metabolic Syndrome, and Type 2 Diabetes Mellitus. *Biomedicines* **2024**, *12*, 96. https://doi.org/10.3390/biomedicines12010096.
4. Ngema, M.; Xulu, N.D.; Ngubane, P.S.; Khathi, A. A Review of Fetal Development in Pregnancies with Maternal Type 2 Diabetes Mellitus (T2DM)-Associated Hypothalamic–Pituitary–Adrenal (HPA) Axis Dysregulation: Possible Links to Pregestational Prediabetes. *Biomedicines* **2024**, *12*, 1372. https://doi.org/10.3390/biomedicines12061372.
5. Reurean-Pintilei, D.; Pantea Stoian, A.; Salmen, T.; Stoica, R.-A.; Mititelu-Tartau, L.; Lazăr, S.; Timar, B. Associations between Skin Autofluorescence Levels with Cardiovascular Risk and Diabetes Complications in Patients with Type 2 Diabetes. *Biomedicines* **2024**, *12*, 890. https://doi.org/10.3390/biomedicines12040890.
6. Yamagami, H.; Hara, T.; Yasui, S.; Hosoki, M.; Hori, T.; Kaneko, Y.; Mitsui, Y.; Kurahashi, K.; Harada, T.; Yoshida, S.; et al. Cross-Sectional and Longitudinal Associations between Skin Autofluorescence and Tubular Injury Defined by Urinary Excretion of Liver-Type Fatty Acid-Binding Protein in People with Type 2 Diabetes. *Biomedicines* **2023**, *11*, 3020. https://doi.org/10.3390/biomedicines11113020.
7. Alshawaf, E.; Abu-Farha, M.; Mohammad, A.; Devarajan, S.; Al-Khairi, I.; Cherian, P.; Ali, H.; Al-Matrouk, H.; Al-Mulla, F.; Al Attar, A.; et al. Angiopoietin-2 and Angiopoietin-like Proteins with a Prospective Role in Predicting Diabetic Nephropathy. *Biomedicines* **2024**, *12*, 949. https://doi.org/10.3390/biomedicines12050949.
8. Dascalu, A.M.; Serban, D.; Tanasescu, D.; Vancea, G.; Cristea, B.M.; Stana, D.; Nicolae, V.A.; Serboiu, C.; Tribus, L.C.; Tudor, C.; et al. The Value of White Cell Inflammatory Biomarkers as Potential Predictors for Diabetic Retinopathy in Type 2 Diabetes Mellitus (T2DM). *Biomedicines* **2023**, *11*, 2106. https://doi.org/10.3390/biomedicines11082106.
9. Chatziralli, I.; Dimitriou, E.; Agapitou, C.; Kazantzis, D.; Kapsis, P.; Morogiannis, N.; Kandarakis, S.; Theodossiadis, G.; Theodossiadis, P. Optical Coherence Tomography Angiography Changes in Macular Area in Patients with Proliferative Diabetic Retinopathy Treated with Panretinal Photocoagulation. *Biomedicines* **2023**, *11*, 3146. https://doi.org/10.3390/biomedicines11123146.
10. Tănasie, C.A.; Dan, A.O.; Ică, O.M.; Mercuț, M.F.; Mitroi, G.; Taisescu, C.-I.; Sfredel, V.; Corbeanu, R.I.; Mocanu, C.L.; Danielescu, C. Retinal Functional Impairment in Diabetic Retinopathy. *Biomedicines* **2024**, *12*, 44. https://doi.org/10.3390/biomedicines12010044.
11. Lietuvninkas, L.; Baccouche, B.; Kazlauskas, A. The Multi-Kinase Inhibitor RepSox Enforces Barrier Function in the Face of Both VEGF and Cytokines. *Biomedicines* **2023**, *11*, 2431. https://doi.org/10.3390/biomedicines11092431.

References

1. Saeedi, P.; Petersohn, I.; Salpea, P.; Malanda, B.; Karuranga, S.; Unwin, N.; Colagiuri, S.; Guariguata, L.; Motala, A.A.; Ogurtsova, K.; et al. Global and regional diabetes prevalence estimates for 2019 and projections for 2030 and 2045: Results from the International Diabetes Federation Diabetes Atlas, 9th edition. *Diabetes Res. Clin. Pract.* **2019**, *157*, 107843. [CrossRef] [PubMed]
2. Klein, S.; Gastaldelli, A.; Yki-Järvinen, H.; Scherer, P.E. Why does obesity cause diabetes? *Cell Metab.* **2022**, *34*, 11–20. [CrossRef] [PubMed]
3. Wong, R.; Lam, E.; Bramante, C.T.; Johnson, S.G.; Reusch, J.; Wilkins, K.J.; Yeh, H.-C. Does COVID-19 Infection Increase the Risk of Diabetes? Current Evidence. *Curr. Diabetes Rep.* **2023**, *23*, 207–216. [CrossRef] [PubMed]
4. Naveed, Z.; García, H.A.V.; Wong, S.; Wilton, J.; McKee, G.; Mahmood, B.; Binka, M.; Rasali, D.; Janjua, N.Z. Association of COVID-19 Infection with Incident Diabetes. *JAMA Netw. Open* **2023**, *6*, e238866. [CrossRef] [PubMed]
5. Deng, Y.; Li, N.; Wu, Y.; Wang, M.; Yang, S.; Zheng, Y.; Deng, X.; Xiang, D.; Zhu, Y.; Xu, P.; et al. Global, Regional, and National Burden of Diabetes-Related Chronic Kidney Disease From 1990 to 2019. *Front. Endocrinol.* **2021**, *12*, 672350. [CrossRef] [PubMed]

6. Kumar, M.; Dev, S.; Khalid, M.U.; Siddenthi, S.M.; Noman, M.; John, C.; Akubuiro, C.; Haider, A.; Rani, R.; Kashif, M.; et al. The Bidirectional Link Between Diabetes and Kidney Disease: Mechanisms and Management. *Cureus* **2023**, *15*, e45615. [CrossRef]
7. Thomas, M.C.; Brownlee, M.; Susztak, K.; Sharma, K.; Jandeleit-Dahm, K.A.M.; Zoungas, S.; Rossing, P.; Groop, P.-H.; Cooper, M.E. Diabetic kidney disease. *Nat. Rev. Dis. Prim.* **2015**, *1*, 1–20. [CrossRef]
8. Kropp, M.; Golubnitschaja, O.; Mazurakova, A.; Koklesova, L.; Sargheini, N.; Vo, T.-T.K.S.; de Clerck, E.; Polivka, J.; Potuznik, P.; Stetkarova, I.; et al. Diabetic retinopathy as the leading cause of blindness and early predictor of cascading complications—Risks and mitigation. *EPMA J.* **2023**, *14*, 21–42. [CrossRef]
9. Lu, Q.; Wang, J.; Wei, X.; Wang, G.; Xu, Y. Risk Factors for Major Amputation in Diabetic Foot Ulcer Patients. *Diabetes Metab. Syndr. Obes. Targets Ther.* **2021**, *14*, 2019–2027. [CrossRef]
10. Ortiz-Martínez, M.; González-González, M.; Martagón, A.J.; Hlavinka, V.; Willson, R.C.; Rito-Palomares, M. Recent Developments in Biomarkers for Diagnosis and Screening of Type 2 Diabetes Mellitus. *Curr. Diabetes Rep.* **2022**, *22*, 95–115. [CrossRef]
11. Cheng, P.-C.; Kao, C.-H. Postprandial plasma glucose excursion is associated with an atherogenic lipid profile in individuals with type 2 diabetes mellitus: A cross-sectional study. *PLoS ONE* **2021**, *16*, e0258771. [CrossRef]
12. Troili, F.; Cipollini, V.; Moci, M.; Morena, E.; Palotai, M.; Rinaldi, V.; Romano, C.; Ristori, G.; Giubilei, F.; Salvetti, M.; et al. Perivascular Unit: This Must Be the Place. The Anatomical Crossroad between the Immune, Vascular and Nervous System. *Front. Neuroanat.* **2020**, *14*, 17.
13. Semedo, C.D.M.; Webb, M.; Waller, H.; Khunti, K.; Davies, M. Skin autofluorescence, a non-invasive marker of advanced glycation end products: Clinical relevance and limitations. *Postgrad. Med. J.* **2017**, *93*, 289–294. [CrossRef]
14. Li, M.; Popovic, Z.; Chu, C.; Reichetzeder, C.; Pommer, W.; Krämer, B.K.; Hocher, B. Impact of Angiopoietin-2 on Kidney Diseases. *Kidney Dis.* **2023**, *9*, 143–156. [CrossRef] [PubMed]
15. Szymczak, A.; Kusztal, M.; Gołębiowski, T.; Letachowicz, K.; Goździk, A.; Kościelska-Kasprzak, K.; Tukiendorf, A.; Krajewska, M. High Plasma Angiopoietin-2 Levels Predict the Need to Initiate Dialysis within Two Years in Patients with Chronic Kidney Disease. *Int. J. Mol. Sci.* **2023**, *24*, 10036. [CrossRef] [PubMed]
16. Chen, W.; Chen, K.; Xu, Z.; Hu, Y.; Liu, Y.; Liu, W.; Hu, X.; Ye, T.; Hong, J.; Zhu, H.; et al. Neutrophil-to-Lymphocyte Ratio and Platelet-to-Lymphocyte Ratio Predict Mortality in Patients with Diabetic Foot Ulcers Undergoing Amputations. *Diabetes Metab. Syndr. Obes. Targets Ther.* **2021**, *14*, 821–829. [CrossRef]
17. Zeng, J.; Chen, M.; Feng, Q.; Wan, H.; Wang, J.; Yang, F.; Cao, H. The Platelet-to-Lymphocyte Ratio Predicts Diabetic Retinopathy in Type 2 Diabetes Mellitus. *Diabetes Metab. Syndr. Obes. Targets Ther.* **2022**, *15*, 3617–3626. [CrossRef]
18. Serban, D.; Papanas, N.; Dascalu, A.M.; Kempler, P.; Raz, I.; Rizvi, A.A.; Rizzo, M.; Tudor, C.; Tudosie, M.S.; Tanasescu, D.; et al. Significance of Neutrophil to Lymphocyte Ratio (NLR) and Platelet Lymphocyte Ratio (PLR) in Diabetic Foot Ulcer and Potential New Therapeutic Targets. *Int. J. Low. Extremity Wounds* **2024**, *23*, 205–216. [CrossRef]
19. Abdelhalim, A.S.; Abdelkader, M.F.S.O.; Mahmoud, M.S.E.-D.; Mohamed, A.A.M. Macular vessel density before and after panretinal photocoagulation in patients with proliferative diabetic retinopathy. *Int. J. Retin. Vitr.* **2022**, *8*, 1–8. [CrossRef]
20. Faghihi, H.; Riazi-Esfahani, H.; Khodabande, A.; Pour, E.K.; Mirshahi, A.; Ghassemi, F.; Mirshahi, R.; Khojasteh, H.; Bazvand, F.; Hashemi, A.; et al. Effect of panretinal photocoagulation on macular vasculature using optical coherence tomography angiography. *Eur. J. Ophthalmol.* **2021**, *31*, 1877–1884. [CrossRef] [PubMed]
21. Li, Y.-F.; Ren, Q.; Sun, C.-H.; Li, L.; Lian, H.-D.; Sun, R.-X.; Su, X.; Yu, H. Efficacy and mechanism of anti-vascular endothelial growth factor drugs for diabetic macular edema patients. *World J. Diabetes* **2022**, *13*, 532–542. [CrossRef] [PubMed]
22. Boyer, D.S.; Hopkins, J.J.; Sorof, J.; Ehrlich, J.S. Anti-vascular endothelial growth factor therapy for diabetic macular edema. *Ther. Adv. Endocrinol. Metab.* **2013**, *4*, 151–169. [CrossRef] [PubMed]

biomedicines MDPI

Article

Postprandial Plasma Glucose between 4 and 7.9 h May Be a Potential Diagnostic Marker for Diabetes

Yutang Wang [1,*], Yan Fang [1], Christopher L. Aberson [2], Fadi J. Charchar [1] and Antonio Ceriello [3]

1 Discipline of Life Science, Institute of Innovation, Science and Sustainability, Federation University Australia, Ballarat, VIC 3350, Australia
2 Department of Psychology, Humboldt State University, Arcata, CA 91711, USA
3 RCCS MultiMedica, Via Gaudenzio Fantoli, 16/15, 20138 Milan, Italy; antonio.ceriello@hotmail.it
* Correspondence: yutang.wang@federation.edu.au

Abstract: Postprandial glucose levels between 4 and 7.9 h ($PPG_{4–7.9h}$) correlate with mortality from various diseases, including hypertension, diabetes, cardiovascular disease, and cancer. This study aimed to assess if predicted $PPG_{4–7.9h}$ could diagnose diabetes. Two groups of participants were involved: Group 1 (4420 participants) had actual $PPG_{4–7.9h}$, while Group 2 (8422 participants) lacked this measure but had all the diabetes diagnostic measures. Group 1 underwent multiple linear regression to predict $PPG_{4–7.9h}$ using 30 predictors, achieving accuracy within 11.1 mg/dL in 80% of the participants. Group 2 had $PPG_{4–7.9h}$ predicted using this model. A receiver operating characteristic curve analysis showed that predicted $PPG_{4–7.9h}$ could diagnose diabetes with an accuracy of 87.3% in Group 2, with a sensitivity of 75.1% and specificity of 84.1% at the optimal cutoff of 102.5 mg/dL. A simulation on 10,000 random samples from Group 2 revealed that 175 participants may be needed to investigate $PPG_{4–7.9h}$ as a diabetes diagnostic marker with a power of at least 80%. In conclusion, predicted $PPG_{4–7.9h}$ appears to be a promising diagnostic indicator for diabetes. Future studies seeking to ascertain its definitive diagnostic value might require a minimum sample size of 175 participants.

Keywords: postprandial; non-fasting; fasting; glucose; diabetes

Citation: Wang, Y.; Fang, Y.; Aberson, C.L.; Charchar, F.J.; Ceriello, A. Postprandial Plasma Glucose between 4 and 7.9 h May Be a Potential Diagnostic Marker for Diabetes. *Biomedicines* **2024**, *12*, 1313. https://doi.org/10.3390/biomedicines12061313

Academic Editors: Ana Dascalu and Dragos Serban

Received: 29 May 2024
Accepted: 11 June 2024
Published: 13 June 2024

1. Introduction

According to the World Health Organization, an estimated 422 million individuals globally grapple with diabetes [1], a condition linked to severe complications such as heart disease, chronic kidney disease, and blindness [2], culminating in 1.5 million annual deaths [1]. The direct health expenditure for diabetes worldwide reached approximately USD 760 billion in 2019, with projections soaring to around USD 825 billion by 2030 [3]. Surprisingly, in 2021, nearly half of diabetic adults remained undiagnosed, constituting approximately 239.7 million individuals [4]. Consequently, prioritizing research endeavors aimed at enhancing diabetes detection, pinpointing risk factors, and developing therapies is paramount in clinical practice [5].

Postprandial plasma glucose has long been recognized to play a vital role in diabetes-associated complications [6–8] and glycemic control [9–11]. Its positive correlation with cardiovascular disease incidence [12–15] and mortality [16], cancer mortality [17,18], and all-cause mortality [12–14,17,19] underscores its potential as a therapeutic target to mitigate diabetes-associated morbidity and mortality [7]. However, conventional assessments typically focus on early postprandial glucose levels, such as those measured at 1 h [12] or 2 h after a meal [13–15,17–19]. Yet, the susceptibility of these measurements to dietary variations and timing discrepancies poses challenges, potentially skewing results [15–17,20].

Recent findings highlight the stability and significance of postprandial glucose levels measured between 4 and 7.9 h after a meal ($PPG_{4–7.9h}$) [21,22]. Hourly $PPG_{4–7.9h}$ levels were similar across the duration from 4 to 7.9 h [21,22]. Plasma glucose returned to baseline four hours after a meal regardless of the type of the meal (normal or high carbohydrate) or the time of the meal (breakfast, lunch, and dinner) in healthy individuals [20]. These results suggest that the interval of 4 to 7.9 h after a meal may reflect glucose homeostasis irrespective of meal composition or timing, offering a promising window for assessment. Moreover, $PPG_{4–7.9h}$ exhibits positive associations with mortality from prevalent conditions like hypertension, cardiovascular disease, and cancer [21,22], further emphasizing its clinical relevance.

Diabetes diagnosis relies on fasting plasma glucose, 2 h plasma glucose during OGTT, or HbA_{1c} [23]. However, there are some limitations to using these tests. Fasting plasma glucose and OGTT require fasting and, thus, pose practical challenges. Fasting can be inconvenient and even risky, particularly for vulnerable individuals who may experience hypoglycemia while waiting for blood collection [24]. The HbA_{1c} test does not require fasting and minimizing daily fluctuations due to lifestyle changes; however, this test's diagnostic accuracy is compromised. For instance, factors such as hemodialysis, HIV treatment, age, ethnicity, pregnancy, and hemoglobinopathies can influence HbA_{1c} readings, leading to potential misdiagnoses [23]. The American Diabetes Association recommends prioritizing fasting plasma glucose and 2 h plasma glucose if discrepancies arise between HbA_{1c} and glucose values [23]. In addition, HbA_{1c} has poor sensitivity in diabetes diagnosis and misses a large proportion of diabetes that is detected by OGTT [25], the gold standard method for diabetes diagnosis [26,27]. For example, the proportion of OGTT-diagnosed diabetes that HbA_{1c} can detect was reported to be 43% in Denmark, 25% in the UK, 17% in Australia, 30% in Greenland, 20% in Kenya, 78% in India [28], and 30% in China [29]. Moreover, the HbA_{1c} test is more expensive than the glucose test [30,31]. Therefore, there remains a need to explore additional diabetic diagnostic tools that provide diagnostic accuracy while upholding convenience and safety, alongside the capacity to forecast clinical outcomes.

Given the potential diagnostic utility of $PPG_{4–7.9h}$, exploring its feasibility in diagnosing diabetes warrants attention. However, existing datasets lack concurrent measurements of $PPG_{4–7.9h}$ alongside traditional diagnostic indicators for diabetes. This gap impedes the accurate estimation of the sample size necessary for investigating the diagnostic efficacy of $PPG_{4–7.9h}$.

To address this, the present study leveraged the comprehensive National Health and Nutrition Examination Survey (NHANES) dataset in which a large number of predictors of $PPG_{4–7.9h}$ are available. This study aimed to construct a model predicting $PPG_{4–7.9h}$ in one group (Group 1, n = 4420) of participants who had actual $PPG_{4–7.9h}$ values; subsequently, $PPG_{4–7.9h}$ values were estimated using this predictive model in another group (Group 2, n = 8422) of participants who lacked $PPG_{4–7.9h}$ but had complete diabetes diagnostic profiles, i.e., fasting plasma glucose, 2 h plasma glucose during oral glucose tolerance test (OGTT), and hemoglobin A_{1c} (HbA_{1c}) [23,26]. The diagnostic suitability of predicted $PPG_{4–7.9h}$ for diabetes was then investigated in Group 2 participants, and the sample size that would be required by future studies aiming to investigate the true diagnostic value of $PPG_{4–7.9h}$ for diabetes was estimated. Therefore, this study aimed to investigate the diagnostic potential of predicted $PPG_{4–7.9h}$ for diabetes, which may lay the groundwork for future investigations and clinical applications.

Antidiabetic medications have confounding effects on blood glucose levels [32–34]. Therefore, this study excluded those who were taking antidiabetic medications or with unknown medication status.

2. Materials and Methods

2.1. Study Participants

This study included adult participants (aged ≥20 years) from NHANES III (1988–1994) and the subsequent eight cycles of NHANES from 1999 to 2014 [35]. Two groups of participants were selected from the NHANES participants: Group 1 (the postprandial group) and Group 2 (the fasting group).

Group 1 included all participants who had postprandial plasma glucose measured from blood taken between 4 and 7.9 h ($PPG_{4–7.9h}$, n = 5115). Participants using antidiabetic drugs (n = 277) or with unknown status on the use of antidiabetic drugs (n = 31) were subsequently excluded. Individuals who had missing data from the following variables were also excluded: HbA_{1c} (n = 27), insulin (n = 50), body mass index (n = 15), education (n = 30), smoking (n = 1), systolic blood pressure (n = 25), total cholesterol (n = 25), high-density lipoprotein (HDL) cholesterol (n = 30), cancer (n = 1), dietary intake data (carbohydrate, protein, fat, and total energy, n = 130), laboratory profile (n = 53 including n = 50 for potassium, n = 1 for total protein, and n = 2 for bilirubin). Therefore, the remaining 4420 participants were included in the final analysis for Group 1 (Figure 1).

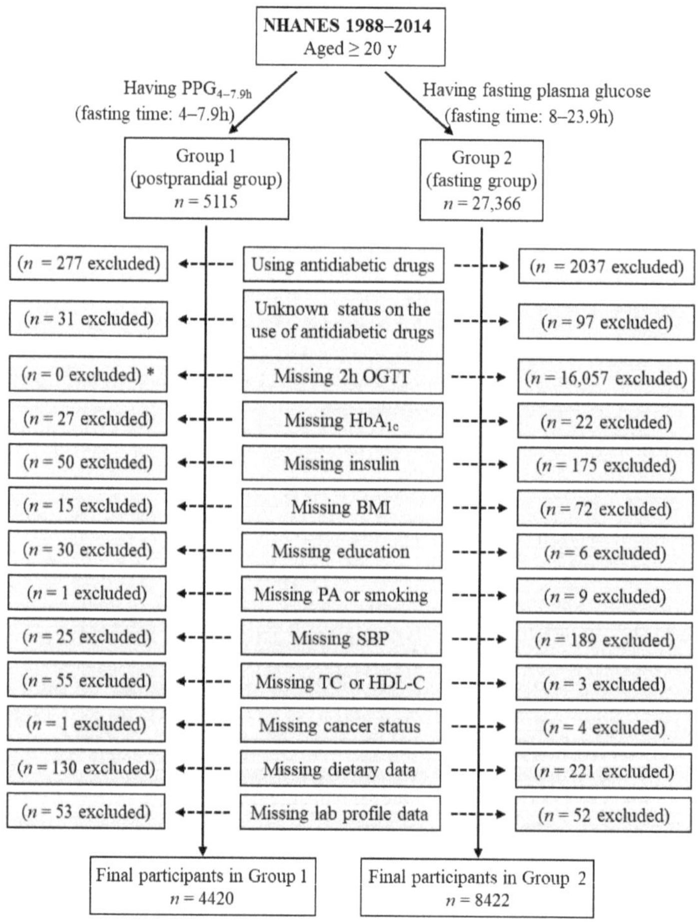

Figure 1. Flow diagram of the study participants. Group 1 participants had postprandial plasma glucose measured from blood taken between 4 and 7.9 h ($PPG_{4–7.9h}$). Group 2 participants had fasting plasma glucose with a fasting time between 8 and 23.9 h. *: participants in Group 1 did not have OGTT data, and OGTT was not an exclusion criterion for Group 1 participants. BMI, body mass index; HbA_{1c}, hemoglobin A_{1c}; HDL-C, high-density lipoprotein cholesterol; NHANES, National Health and Nutrition Examination Survey; OGTT, oral glucose tolerance test; PA, physical activity; SBP, systolic blood pressure; TC, total cholesterol.

Group 2 included those who had fasting plasma glucose (fasting time of 8–23.9 h), n = 27,366. Participants using antidiabetic drugs (n = 2037) or with unknown status on the use of antidiabetic drugs (n = 97) were subsequently excluded. Individuals who had missing data from the following variables were also excluded: 2 h plasma glucose during OGTT (n = 16,057), HbA_{1c} (n = 22), insulin (n = 175), body mass index (n = 72), education (n = 6), physical activity (n = 2), smoking (n = 7), systolic blood pressure (n = 189), total cholesterol (n = 3), cancer (n = 4), dietary intake data (carbohydrate, protein, fat, and total energy, n = 221), and laboratory profile (n = 52 including n = 12 for potassium, n = 12 for calcium, n = 1 for phosphorus, n = 14 for bicarbonate, n = 7 for total protein, and n = 6 for bilirubin). Therefore, the remaining 8422 participants were included in the final analysis for Group 2 (Figure 1). Group 2 participants had all three diabetes diagnostic measures, namely, fasting plasma glucose, 2 h plasma glucose during OGTT, and HbA_{1c}.

2.2. Diabetes Definition

Diabetes was diagnosed based on criteria established by the American Diabetes Association [23,36], which included a fasting plasma glucose level equal to or exceeding 126 mg/dL, a 2 h plasma glucose during OGTT equal to or exceeding 200 mg/dL, or an HbA_{1c} level in whole blood equal to or exceeding 6.5%.

2.3. $PPG_{4–7.9h}$

Blood was drawn from participants. The time of blood collection and last caloric intake were recorded, and the fasting time was calculated. Blood was taken between 4 and 7.9 h after the last caloric intake was used to measure $PPG_{4–7.9h}$ by the hexokinase-mediated reaction method as previously described [37].

2.4. Potential $PPG_{4–7.9h}$ Predictors

The following variables were retrieved from the NHANES data set and treated as potential factors for $PPG_{4–7.9h}$ as they may affect plasma glucose levels: age, sex, ethnicity, body mass index, education, income, physical activity, smoking, alcohol intake, dietary intake (carbohydrate, protein, fat, and total calorie), systolic blood pressure, total cholesterol, HDL cholesterol, family history of diabetes, cancer diagnosis, use of antihypertensive medication, use of cholesterol-lowering medication, circulating ionic profile (potassium, calcium, sodium, phosphorus, bicarbonate, and chloride), circulating enzymatic and metabolic profile (alanine aminotransferase, aspartate aminotransferase, bilirubin, blood urea nitrogen, creatinine, and uric acid), serum protein, serum albumin, serum insulin, HbA_{1c}, and fasting time.

2.5. Statistical Analyses

The participants' baseline characteristics were described using numbers with percentages for categorical variables, median with interquartile range for non-normally distributed continuous variables [38], and mean with standard deviation (SD) for normally distributed continuous variables in the presented data [39].

The associations of $PPG_{4–7.9h}$ with potential predictors were analyzed using simple linear regression [40]. The significant predictors, determined by the simple linear regression, were then added to the multiple linear regression model to predict $PPG_{4–7.9h}$ [41].

The following variables were natural log transformed to improve data distribution prior to linear regression [42]: $PPG_{4–7.9h}$, fasting plasma glucose, 2 h plasma glucose during OGTT, body mass index, systolic blood pressure, total cholesterol, HDL cholesterol, dietary carbohydrate intake, dietary protein intake, dietary fat intake, dietary caloric intake, alanine aminotransferase, aspartate aminotransferase, bilirubin, blood urea nitrogen, serum creatinine, serum insulin, and blood HbA_{1c}.

The performance of predicted $PPG_{4–7.9h}$ for classifying diabetes was assessed by receiver operating characteristic (ROC) curve analysis [43,44]. The optimal cutoff of predicted PPG4–7.9h was determined by the Youden Index [45].

Power estimation was carried out through simulations involving 10,000 randomly generated samples with various sample sizes derived from the pool of 8422 participants in Group 2 [46,47]. Within each sample, the diagnostic accuracy, sensitivity, and specificity of predicted $PPG_{4–7.9h}$ for diabetes diagnosis were computed using the following formulas [48–50]:

Diagnosis accuracy = (number of participants correctly diagnosed with diabetes + number of participants correctly diagnosed without diabetes)/total number of participants in the sample.

Sensitivity = number of participants correctly diagnosed with diabetes/total number of participants with actual diabetes.

Specificity = number of participants correctly diagnosed without diabetes/total number of participants without actual diabetes.

Among the 10,000 random samples, the percentage exhibiting a diagnostic accuracy of 80%, which is deemed a minimum threshold for an excellent diagnostic marker [51], was computed to determine the diagnostic power of $PPG_{4–7.9h}$ in identifying diabetes. Mean sensitivity and specificity values were calculated from the 10,000 samples, and their 95% confidence intervals were derived from the 2.5th and 97.5th percentiles of the sensitivity and specificity readings [52]. Furthermore, an investigation into a diagnostic accuracy of 81% was conducted to assess power and sample size requirements.

The null hypothesis was rejected for two-sided values of $p < 0.05$. Power and sample size were estimated using the R program, and all other analyses were performed using SPSS version 27.0 (IBM SPSS Statistics for Windows, Armonk, NY, USA, IBM Corporation).

3. Results

3.1. Baseline Characteristics

Group 1 (the postprandial group) included 4420 participants with a mean (SD) age of 49 (19) years, and Group 2 (the fasting group) had 8842 participants with a mean (SD) age of 48 (17) years (Table 1). All other characteristics of the participants are described in Table 1.

Table 1. Baseline characteristics of participants.

Variables	Group 1 (Postprandial Group)	Group 2 (Fasting Group)
Sample size	4420	8422
$PPG_{4–7.9h}$, mg/dL, median (IQR)	92 (87–98)	N/A
FPG, mg/dL, median (IQR)	N/A	99 (92–106)
2 h PG during OGTT, mg/dL, median (IQR)	N/A	109 (88–138)
Age, year, mean (SD)	49 (19)	48 (17)
Sex (male), *n* (%)	2042 (46)	4240 (50)
BMI, kg/m^2, median (IQR)	26 (23–30)	28 (24–32)
Ethnicity, *n* (%)		
Non-Hispanic white	2104 (48)	4061 (48)
Non-Hispanic black	1033 (23)	1498 (18)
Hispanic	1220 (28)	2148 (26)
Other	63 (1)	715 (9)
Education, *n* (%)		
<High School	1679 (38)	1985 (24)
High School	1375 (31)	1928 (23)
>High School	1366 (31)	4509 (54)
Poverty/income ratio, *n* (%)		
<130%	1176 (27)	2367 (28)

Table 1. *Cont.*

Variables	Group 1 (Postprandial Group)	Group 2 (Fasting Group)
130–349%	1822 (41)	2888 (34)
≥350%	1081 (25)	2591 (31)
Unknown	341 (8)	576 (7)
Physical activity, *n* (%)		
Active	1594 (36)	2108 (25)
Insufficiently active	1890 (43)	2464 (29)
Inactive	936 (21)	3850 (46)
Alcohol consumption, *n* (%)		
0 drinks/week	755 (17)	1379 (16)
<1 drink/week	532 (12)	2459 (29)
1–6 drinks/week	870 (20)	2032 (24)
≥7 drinks/week	578 (13)	1241 (15)
Unknown	1685 (38)	1311 (16)
Smoking status, *n* (%)		
Current smoker	1094 (25)	1773 (21)
Past smoker	1127 (26)	2018 (24)
Non-smoker	2199 (50)	4631 (55)
Dietary carbohydrate intake, g/day, median (IQR)	235 (171–313)	239 (182–311)
Dietary protein intake, g/day, median (IQR)	71 (51–97)	76 (57–99)
Dietary fat intake, g/day, median (IQR)	68 (45–101)	71 (51–97)
Dietary caloric intake, kcal/day, median (IQR)	1899 (1392–2525)	1969 (1507–2513)
SBP, mm Hg, median (IQR)	123 (112–138)	119 (110–131)
TC, mg/dL, median (IQR)	204 (177–235)	193 (167–220)
HDL-C, mg/dL, median (IQR)	50 (41–60)	52 (43–63)
Use of antihypertensive medication		
No	3384 (77)	6053 (72)
Yes	693 (16)	1916 (23)
Unknown	343 (8)	453 (5)
Use of cholesterol-lowering medication		
No	1655 (37)	4617 (55)
Yes	134 (3)	1202 (14)
Unknown	2631 (60)	2603 (31)
Cancer diagnosis, *n* (%)		
No	4029 (91)	7694 (91)
Yes	391 (9)	728 (9)
Family history of diabetes, *n* (%)		
No	2424 (55)	5178 (62)
Yes	1918 (43)	3082 (37)
Unknown	78 (2)	162 (2)
Serum potassium, mM, mean (SD)	4.0 (0.3)	4.0 (0.3)

Table 1. *Cont.*

Variables	Group 1 (Postprandial Group)	Group 2 (Fasting Group)
Serum calcium, mM, mean (SD)	2.3 (0.1)	2.3 (0.1)
Serum sodium, mM, mean (SD)	141.3 (2.5)	139.4 (2.2)
Serum phosphorus, mM, mean (SD)	1.2 (0.2)	1.2 (0.2)
Serum bicarbonate, mM, mean (SD)	27.8 (3.9)	25.2 (2.2)
Serum chloride, mM, mean (SD)	104 (3)	104 (3)
ALT, U/L, median (IQR)	14 (10–20)	21 (16–29)
AST, U/L, median (IQR)	20 (17–24)	23 (20–28)
Bilirubin, μM, median (IQR)	8.6 (6.8–10.3)	12.0 (10.3–15.4)
Blood urea nitrogen, mM, median (IQR)	5.0 (3.9–6.1)	4.3 (3.6–5.4)
Creatinine, μM, median (IQR)	85 (76–102)	75 (64–88)
Uric acid, μM, mean (SD)	311 (88)	326 (82)
Serum protein, g/L, mean (SD)	74 (5)	72 (5)
Serum albumin, g/L, mean (SD)	41.8 (3.7)	42.4 (3.1)
Serum insulin, μU/mL, median (IQR)	8.5 (5.9–12.7)	9.5 (6.1–15.4)
HbA_{1c}, %, median (IQR)	5.3 (5.0–5.7)	5.4 (5.2–5.7)
Fasting time, h, mean (SD)	6.6 (0.8)	12.1 (1.9)

ALT, alanine aminotransferase; AST, aspartate aminotransferase; BMI, body mass index; FPG, fasting plasma glucose; HbA_{1c}, hemoglobin A_{1c}; HDL-C, high-density lipoprotein cholesterol; IQR, interquartile range; N/A, not available; OGTT, oral glucose tolerance test; PG, plasma glucose; $PPG_{4–7.9h}$, postprandial plasma glucose measured from blood taken between 4 and 7.9 h; SBP, systolic blood pressure; SD, standard deviation; TC, total cholesterol.

3.2. *Factors Associated with $PPG_{4–7.9h}$ in Group 1 of 4420 Participants, Assessed by Simple Linear Regression*

Simple linear regression analysis identified 30 factors associated with $PPG_{4–7.9h}$ (Table 2). These factors included age, sex, ethnicity, body mass index, education, income, physical activity, smoking, alcohol intake, dietary carbohydrate intake, dietary fat intake, dietary caloric intake, systolic blood pressure, total cholesterol, HDL cholesterol, cancer diagnosis, use of antihypertensive medications, and certain circulating biomarkers. These biomarkers included potassium, calcium, phosphorus, bicarbonate, chloride, alanine aminotransferase, aspartate aminotransferase, bilirubin, blood urea nitrogen, creatinine, uric acid, insulin, and HbA_{1c}.

Table 2. Association of potential predictors with $PPG_{4–7.9h}$, analyzed by simple linear regression in Group 1 of 4420 participants.

Variables	B (Coefficient)	*p* Value	Variables	B (Coefficient)	*p* Value
Age	0.002	<0.001	HDL cholesterol	−0.050	<0.001
Sex			Family diabetes history		
Male	0 (ref.)		No	0 (ref.)	
Female	−0.029	<0.001	Yes	0.003	0.50
Ethnicity			Cancer		
Non-Hispanic white	0 (ref.)		No	0 (ref.)	
Non-Hispanic black	−0.028	<0.001	Yes	0.021	0.01
Hispanic	0.022	<0.001	Antihypertension medicative		

Table 2. *Cont.*

Variables	B (Coefficient)	*p* Value	Variables	B (Coefficient)	*p* Value
Other	0.012	0.53	No	0 (ref.)	
Body mass index	0.098	<0.001	Yes	0.052	<0.001
Education			Cholesterol-lowering medication		
<12 years	0 (ref.)		No	0 (ref.)	
12 years	−0.039	<0.001	Yes	0.021	0.12
>12 years	−0.047	<0.001	Dietary carbohydrate intake	−0.019	<0.001
Income			Dietary protein intake	−0.007	0.07
<130%	0 (ref.)		Dietary fat intake	−0.013	<0.001
130%–349%	−0.003	0.59	Dietary caloric intake	−0.018	<0.001
≥350%	−0.015	0.02	Serum potassium	0.026	<0.001
Unknown	0.025	0.01	Serum calcium	0.157	<0.001
Physical activity			Serum sodium	−0.001	0.16
Inactive	0 (ref.)		Serum phosphorus	−0.05	<0.001
Active	−0.023	<0.001	Serum bicarbonate	0.002	0.002
Insufficiently active	−0.017	0.005	Serum chloride	−0.004	<0.001
Alcohol consumption			Alanine aminotransferase	0.028	<0.001
0 drinks/week	0 (ref.)		Aspartate aminotransferase	0.02	<0.001
<1 drink/week	−0.023	0.006	Bilirubin	0.019	<0.001
1–6 drinks/week	−0.028	<0.001	Blood urea nitrogen	0.052	<0.001
≥7 drinks/week	−0.012	0.15	Creatinine	0.026	0.01
Unknown	−0.016	0.01	Uric acid	0.0002	<0.001
Smoking status			Serum protein	0.001	0.14
Nonsmoker	0 (ref.)		Serum albumin	−0.0002	0.75
Current smoker	−0.005	0.36	Serum insulin	0.070	<0.001
Past smoker	0.024	<0.001	Hemoglobin A_{1c}	0.705	<0.001
SBP	0.235	<0.001	Fasting time	−0.002	0.70
Total cholesterol	0.075	<0.001			

The following variables were natural log transformed to improve data distribution prior to simple linear regression: $PPG_{4–7.9h}$, body mass index, systolic blood pressure, total cholesterol, HDL cholesterol, dietary carbohydrate intake, dietary protein intake, dietary fat intake, dietary caloric intake, alanine aminotransferase, aspartate aminotransferase, bilirubin, blood urea nitrogen, creatinine, serum insulin, and blood hemoglobin A_{1c}. HDL, high-density lipoprotein; $PPG_{4–7.9h}$, postprandial plasma glucose measured from blood taken between 4 and 7.9 h; Ref., reference; SBP, systolic blood pressure.

Simple linear regression showed that the following seven factors were not associated with $PPG_{4–7.9h}$: family history of diabetes, use of cholesterol-lowering medications, dietary protein intake, serum sodium, serum protein, serum albumin, and fasting time (Table 2).

3.3. Predictive Model for $PPG_{4–7.9h}$ Using Multiple Linear Regression in Group 1 of 4420 Participants

The predictive model was constructed using multiple linear regression (Table 3). The predictors were the 30 factors that were identified as significantly associated with $PPG_{4–7.9h}$ in simple linear regression (Table 2). These 30 predictors accounted for 42.9% of the variation in $PPG_{4–7.9h}$ (R square, Model 7, Table 3). The individual coefficients for each predictor in the final model (Model 7, Table 3) are listed in Table 4.

Table 3. Multiple linear regression model in predicting $PPG_{4-7.9h}$ in Group 1 of 4420 participants.

Models	R Square	R Square Change	Significance of R Square Change
1	0.095	0.095	<0.001
2	0.118	0.023	<0.001
3	0.123	0.005	<0.001
4	0.14	0.017	<0.001
5	0.16	0.02	<0.001
6	0.203	0.042	<0.001
7	0.429	0.226	<0.001

Model 1: adjusted for age, sex, and ethnicity; Model 2: adjusted for all the factors in Model 1 plus body mass index, education, income, physical activity, smoking, alcohol intake, dietary carbohydrate intake, dietary fat intake, dietary caloric intake; Model 3: adjusted for all the factors in Model 2 plus systolic blood pressure, total cholesterol, HDL cholesterol, cancer diagnosis, and use of antihypertensive medication; Model 4: adjusted for all the factors in Model 3 plus circulating ionic profile (potassium, calcium, phosphorus, bicarbonate, and chloride); Model 5: adjusted for all the factors in Model 4 plus circulating enzymatic and metabolic profile (alanine aminotransferase, aspartate aminotransferase, bilirubin, blood urea nitrogen, creatinine, and uric acid); Model 6: adjusted for all the factors in Model 5 plus serum insulin; Model 7: adjusted for all the factors in Model 6 plus blood hemoglobin A_{1c}.

Table 4. Coefficients of predictors in the $PPG_{4-7.9h}$ predictive model in Group 1 of 4420 participants analyzed by multiple linear regression.

Variables	B (Coefficient)	Variables	B (Coefficient)
Age	0.001	Smoking status	
Sex		Nonsmoker	0 (reference)
Male	0 (reference)	Current smoker	−0.006
Female	−0.021	Past smoker	−0.003
Ethnicity		Systolic blood pressure	0.048
Non-Hispanic white	0 (reference)	Total cholesterol	−0.042
Non-Hispanic black	−0.05	HDL cholesterol	0.015
Hispanic	0.001	Cancer	
Other	−0.008	No	0 (reference)
Body mass index	−0.042	Yes	−0.012
Education		Antihypertensive medication	
<12 years	0 (reference)	No	0 (reference)
12 years	−0.003	Yes	−0.002
>12 years	−0.0001	Dietary carbohydrate intake	−0.02
Income		Dietary fat intake	0.031
<130%	0 (reference)	Dietary caloric intake	−0.017
130%–349%	0.001	Serum potassium	−0.004
≥350%	−0.009	Serum calcium	0.081
Unknown	0.005	Serum phosphorus	−0.03
Physical activity		Serum bicarbonate	0.001
Inactive	0 (reference)	Serum chloride	−0.001
Active	−0.002	Alanine aminotransferase	0.014
Insufficiently active	0.001	Aspartate aminotransferase	−0.024

Table 4. *Cont.*

Variables	B (Coefficient)	Variables	B (Coefficient)
Alcohol consumption		Bilirubin	0.032
0 drinks/week	0 (reference)	Blood urea nitrogen	0.004
<1 drink/week	0.001	Creatinine	−0.038
1–6 drinks/week	−0.002	Uric acid	−0.0001
≥7 drinks/week	0.015	Serum insulin	0.052
Unknown	−0.013	Hemoglobin A_{1c}	0.661

The following variables were natural log transformed to improve data distribution prior to multiple linear regression: $PPG_{4–7.9h}$, body mass index, systolic blood pressure, total cholesterol, HDL cholesterol, dietary carbohydrate intake, dietary fat intake, dietary caloric intake, alanine aminotransferase, aspartate aminotransferase, bilirubin, blood urea nitrogen, creatinine, serum insulin, and blood hemoglobin A_{1c}. HDL, high-density lipoprotein; $PPG_{4–7.9h}$, postprandial plasma glucose measured from blood taken between 4 and 7.9 h.

In Group 1, the predicted $PPG_{4–7.9h}$ values were generated utilizing the predictive model comprising 30 predictors, along with their respective coefficients listed in Table 4. To assess the model's performance, the difference between the predicted and actual $PPG_{4–7.9h}$ values was calculated. Analysis revealed that approximately 80% of participants exhibited predicted $PPG_{4–7.9h}$ values within a margin of 11.1 mg/dL from the actual values (Table 5). These findings indicated that the predictive model demonstrated a commendable level of accuracy.

Table 5. Distribution of delta $PPG_{4–7.9h}$ in Group 1 of 4420 participants.

Percentiles	Delta $PPG_{4–7.9h}$ (mg/dL)
10	−11.1
20	−6.5
30	−3.8
40	−1.6
50	0.6
60	2.7
70	4.7
80	7.2
90	10.9

$PPG_{4–7.9h}$, postprandial plasma glucose measured from blood taken between 4 and 7.9 h; delta $PPG_{4–7.9h}$ = predicted value–actual value.

3.4. Predicted $PPG_{4–7.9h}$ for Diabetes Diagnosis in Group 2 of 8422 Participants

Predicted $PPG_{4–7.9h}$ values were computed for Group 2 of 8422 participants utilizing the predictive model incorporating 30 predictors along with their corresponding coefficients (Table 4). Diabetes diagnosis followed the diagnostic criteria outlined by the American Diabetes Association. The utility of predicted $PPG_{4–7.9h}$ in diagnosing diabetes was analyzed through ROC curve analysis. Results revealed that predicted $PPG_{4–7.9h}$ could discern diabetes with an accuracy of 87.3% (95% confidence interval: 86.0%–88.7%), as indicated by the area under the curve (AUC, Figure 2). Further analysis via the Youden index indicated that the optimal cutoff point of predicted $PPG_{4–7.9h}$ for diabetes diagnosis was 102.5 mg/dL. This threshold was associated with a diagnostic sensitivity of 75.1% and specificity of 84.1% (Figure 2).

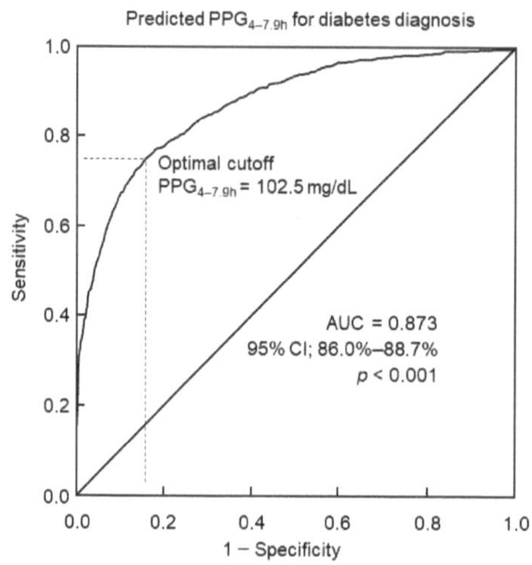

Figure 2. ROC curve analysis of predicted $PPG_{4\text{–}7.9h}$ for diabetes diagnosis. The optimal cutoff was 102.5 mg/dL, with a sensitivity of 75.1% and specificity of 84.1%. The area under the curve (AUC) = 0.873. CI, confidence interval; $PPG_{4\text{–}7.9h}$, postprandial plasma glucose measured from blood taken between 4 and 7.9 h; ROC, receiver operating characteristic.

3.5. Power and Sample Size Estimation for Predicted $PPG_{4\text{–}7.9h}$ to Diagnose Diabetes in Group 2 of 8422 Participants

Power analysis for diagnosing diabetes using predicted $PPG_{4\text{–}7.9h}$ was conducted in Group 2 through the simulation of 10,000 random samples, each with varying sample sizes ranging from 50 to 300 participants. Diabetes prediction was defined as a predicted $PPG_{4\text{–}7.9h}$ equal to or above the optimal cutoff of ≥102.5 mg/dL (Figure 2), and actual diabetes status was determined based on the criteria outlined by the American Diabetes Association. The accuracy of predicted diagnoses for each of the 10,000 random samples was assessed by comparing them with the actual diabetes status.

In evaluating the accuracy, it is notable that an accuracy falling within the range of 0.8 to 0.9 is considered excellent, while an accuracy between 0.9 and 1.0 is deemed outstanding [51]. This study employed an accuracy threshold of 80% to conduct power and sample size estimations. Additionally, a slightly improved accuracy of 81% was also explored for these estimations (Table 6).

Table 6. Power estimation for predicted $PPG_{4\text{–}7.9h}$ to diagnose diabetes.

Sample Size	$n = 50$	$n = 100$	$n = 170$	$n = 175$	$n = 200$	$n = 300$
Power for 80% accuracy	79.9%	84.2%	89.3%	89.9	91.2%	94.5%
Power for 81% accuracy	68.7%	77.8%	79.7%	81.0%	82.9%	87.6%
Sensitivity (95% CI)	75.2% (25.0%–100%)	75.4% (42.7%–100%)	75.0% (50.0%–94.4%)	75.4% (52.9%–94.4%)	75.1% (53.9%–93.8%)	75.1% (57.7%–90.5)
Specificity (95% CI)	84.2% (72.9%–93.6%)	84.1% (76.3%–91.2%)	84.1% (78.3%–89.6%)	84.1% (78.3%–89.4%)	84.1% (78.8%–89.1%)	84.1% (79.8%–88.2%)

Power was estimated using simulations on 10,000 random samples for each sample size. CI, confidence interval.

Analysis revealed that as the sample size increased, there was a corresponding rise in power and a reduction in the confidence interval range for sensitivity and specificity (Table 6). The findings suggest that a sample size of 175 participants may be necessary to achieve over 80% power in detecting a diagnostic accuracy of 81% (Table 6).

4. Discussion

This study revealed that predicted $PPG_{4–7.9h}$ demonstrated a commendable diagnostic accuracy of 87.3% for identifying diabetes. At the optimal cutoff of 102.5 mg/dL, predicted $PPG_{4–7.9h}$ exhibited a sensitivity of 75.1% and specificity of 84.1%. Utilizing simulation on 10,000 random samples, power and sample size estimations indicated that future investigations into $PPG_{4–7.9h}$ as a diagnostic marker for diabetes may require a minimum of 175 participants.

This study demonstrated an accuracy of 87.3% (indicated by the area under the ROC curve) for predicted $PPG_{4–7.9h}$ in diagnosing diabetes with a sensitivity of 75.1% and specificity of 84.1% at the optimal cut-off. This indicates that the capacity of $PPG_{4–7.9h}$ for diabetes is within the excellent accuracy range of 80% to 90% [51]. This accuracy is higher than HbA_{1c}. For example, it has been reported that in 2332 Chinese individuals, the diagnostic accuracy of HbA_{1c} for diabetes was 67%, with a sensitivity of about 63% and a specificity of about 62% [29]. In the Finnish Diabetes Prevention Study, HbA_{1c} of $\geq$6.5% diagnosed diabetes with a sensitivity of 35% in women and 47% in men [53]. Another report showed that the average sensitivity of HbA_{1c} of $\geq$6.5% in diagnosing diabetes among studies from six countries (Denmark, UK, Australia, Greenland, Kenya and India) was 44% [28].

In addition, the sensitivity of fasting plasma glucose of $\geq$126 mg/dL to detect OGTT-diagnosed diabetes was 44.7% in Japanese individuals [54]. The corresponding figure was 70.1% in UK individuals [55] and 41% in US individuals [56].

Therefore, predicted $PPG_{4–7.9h}$ may have a better sensitivity and accuracy than HbA_{1c} and fasting plasma glucose in diabetes diagnosis. However, whether this is the case for actual $PPG_{4–7.9h}$ needs to be investigated in the future.

$PPG_{4–7.9h}$ displays positive correlations with mortality across various diseases, including hypertension, diabetes, cardiovascular disease, and cancer [21,22]. Notably, $PPG_{4–7.9h}$ appears to exhibit stronger associations with certain disease outcomes compared to HbA_{1c}. Specifically, the relationship between $PPG_{4–7.9h}$ and mortality from hypertension, cardiovascular disease [22], and cancer [21] are independent of HbA_{1c}. However, HbA_{1c} is not associated with cancer mortality [21] or all-cause mortality [17]. In addition, fasting plasma glucose and 2 h plasma glucose during OGTT were not associated with cancer mortality [21]. These results suggest that $PPG_{4–7.9h}$ may be superior to the current diabetes diagnostic markers in predicting clinical outcomes.

In addition, unlike fasting plasma glucose and 2 h plasma glucose during OGTT, $PPG_{4–7.9h}$ offers the convenience of measurement without requiring fasting, further underscoring its potential clinical utility. Moreover, the glucose test is cheaper than the HbA_{1c} test [30,31]. Consequently, validating $PPG_{4–7.9h}$ as an additional diagnostic marker for diabetes may hold significant promise for future clinical practice.

This study found that $PPG_{4–7.9h}$ was stable over the duration of 4 to 7.9 h, which was evidenced by the observation that fasting time did not influence its levels. This finding aligns with previous research indicating consistent hourly $PPG_{4–7.9h}$ levels within this time frame [16,21,22]. Additionally, it echoes findings from Eichenlau et al. [20], who showed that plasma glucose returned to baseline four hours after a meal regardless of meal type and meal time, suggesting that $PPG_{4–7.9h}$ may reflect an individual's state of glucose homeostasis.

The optimal cutoff of 102.5 mg/dL for predicted $PPG_{4–7.9h}$ falls below the current fasting plasma glucose cutoff for diabetes diagnosis (126 mg/dL) [23,26]. This observation is consistent with prior reports indicating lower $PPG_{4–7.9h}$ values compared to fasting plasma glucose in individuals with diabetes under good control [57,58]. For example, Avignon et al. [57] reported that in patients with type 2 diabetes who had good diabetic control (HbA_{1c} < 7.0%), the $PPG_{4–7.9h}$ level (measured 5 h after lunch) was 104 mg/dL while the fasting plasma glucose level in those patients was 133 mg/dL. Similarly, Peter et al. [58] reported that in patients with type 2 diabetes who had good diabetic control (HbA_{1c} < 7.3%), the $PPG_{4–7.9h}$ level (measured 4 h after breakfast, lunch, or dinner) was 102 mg/dL while the fasting plasma glucose level in those patients was 127 mg/dL. The common observation of higher fasting plasma glucose than $PPG_{4–7.9h}$ in those with type 2 diabetes may result from a transient increase in both glycogenolysis and gluconeogenesis in the early morning [59], a phenomenon termed "dawn phenomenon" [60].

The identified cutoff of 102.5 mg/dL for diabetes diagnosis corresponds closely to $PPG_{4–7.9h}$ levels of 102–104 mg/dL observed in type 2 diabetes patients maintaining relatively good control [57,58]. Furthermore, this cutoff mirrors the $PPG_{4–7.9h}$ threshold associated with cancer mortality (101 mg/dL) [21].

Strengths of the study include its relatively large sample size (n = 4420 for the postprandial group and n = 8422 for the fasting group) and the incorporation of numerous variables to estimate $PPG_{4–7.9h}$ levels. However, a limitation lies in the use of prediction of $PPG_{4–7.9h}$ while investigating its utility for diabetes diagnosis. Nevertheless, the predictive model, consisting of 30 predictors, performed satisfactorily, with 80% of participants having a predicted $PPG_{4–7.9h}$ within 11.1 mg/dL of the true value. By providing insights into sample size estimation, this study enables researchers to properly design future studies aimed at elucidating the true value of $PPG_{4–7.9h}$ in diabetes diagnosis.

5. Conclusions

Predicted $PPG_{4–7.9h}$ appears to serve as a promising diagnostic indicator for diabetes. Subsequent studies seeking to ascertain its definitive diagnostic value might require a minimum sample size of 175 participants.

Author Contributions: Conceptualization, Y.W.; formal analysis, Y.W.; data curation, Y.W. and Y.F.; writing—original draft preparation, Y.W., C.L.A., F.J.C. and A.C.; writing—review and editing, Y.W., Y.F., C.L.A., F.J.C. and A.C.; funding acquisition, Y.W. All authors have read and agreed to the published version of the manuscript.

Funding: Y.W. was supported by a grant from the National Health and Medical Research Council of Australia (1062671).

Institutional Review Board Statement: The study was conducted in accordance with the Declaration of Helsinki and approved by the NHANES Institutional Review Board. Approval Code: NHANES Protocol #98-12, #2005-06, and #2011-17.

Informed Consent Statement: All participants provided written informed consent. The participants' records were anonymized before being accessed by the author.

Data Availability Statement: All data in the current analysis are publicly available on the NHANES website (https://www.cdc.gov/nchs/nhanes/index.htm), accessed on 3 July 2023.

Conflicts of Interest: The authors declare no conflicts of interest.

References

1. World Health Organization. Diabetes Overview. 2024. Available online: https://www.who.int/health-topics/diabetes#tab=tab_1 (accessed on 3 April 2024).
2. Centers for Disease Control and Prevention (CDC). Prevent Diabetes Complications. 2022. Available online: https://www.cdc.gov/diabetes/managing/problems.html (accessed on 4 April 2024).
3. Williams, R.; Karuranga, S.; Malanda, B.; Saeedi, P.; Basit, A.; Besançon, S.; Bommer, C.; Esteghamati, A.; Ogurtsova, K.; Zhang, P.; et al. Global and regional estimates and projections of diabetes-related health expenditure: Results from the International Diabetes Federation Diabetes Atlas, 9th edition. *Diabetes Res. Clin. Pract.* **2020**, *162*, 108072. [CrossRef] [PubMed]
4. Ogurtsova, K.; Guariguata, L.; Barengo, N.C.; Ruiz, P.L.-D.; Sacre, J.W.; Karuranga, S.; Sun, H.; Boyko, E.J.; Magliano, D.J. IDF diabetes Atlas: Global estimates of undiagnosed diabetes in adults for 2021. *Diabetes Res. Clin. Pract.* **2022**, *183*, 109118. [CrossRef] [PubMed]
5. Dagogo-Jack, S. Preventing diabetes-related morbidity and mortality in the primary care setting. *J. Natl. Med. Assoc.* **2002**, *94*, 549–560. [PubMed]
6. Ceriello, A.; Colagiuri, S.; Gerich, J.; Tuomilehto, J. Guideline for management of postmeal glucose. *Nutr. Metab. Cardiovasc. Dis.* **2008**, *18*, S17–S33. [CrossRef] [PubMed]
7. Peter, R.; Okoseime, O.E.; Rees, A.; Owens, D.R. Postprandial glucose—A potential therapeutic target to reduce cardiovascular mortality. *Curr. Vasc. Pharmacol.* **2009**, *7*, 68–74. [CrossRef] [PubMed]

8. American Diabetes Association. Postprandial Blood Glucose. *Diabetes Care* **2001**, *24*, 775–778. [CrossRef]
9. Bell, D.S. Importance of postprandial glucose control. *South. Med. J.* **2001**, *94*, 804–809. [CrossRef] [PubMed]
10. Monnier, L.; Colette, C.; Owens, D. Postprandial and basal glucose in type 2 diabetes: Assessment and respective impacts. *Diabetes Technol. Ther.* **2011**, *13* (Suppl. S1), S25–S32. [CrossRef] [PubMed]
11. Veciana, M.d.; Major, C.A.; Morgan, M.A.; Asrat, T.; Toohey, J.S.; Lien, J.M.; Evans, A.T. Postprandial versus Preprandial Blood Glucose Monitoring in Women with Gestational Diabetes Mellitus Requiring Insulin Therapy. *N. Engl. J. Med.* **1995**, *333*, 1237–1241. [CrossRef]
12. Hanefeld, M.; Fischer, S.; Julius, U.; Schulze, J.; Schwanebeck, U.; Schmechel, H.; Ziegelasch, H.J.; Lindner, J. Risk factors for myocardial infarction and death in newly detected NIDDM: The Diabetes Intervention Study, 11-year follow-up. *Diabetologia* **1996**, *39*, 1577–1583. [CrossRef]
13. Takao, T.; Suka, M.; Yanagisawa, H.; Iwamoto, Y. Impact of postprandial hyperglycemia at clinic visits on the incidence of cardiovascular events and all-cause mortality in patients with type 2 diabetes. *J. Diabetes Investig.* **2017**, *8*, 600–608. [CrossRef]
14. Cavalot, F.; Pagliarino, A.; Valle, M.; Di Martino, L.; Bonomo, K.; Massucco, P.; Anfossi, G.; Trovati, M. Postprandial blood glucose predicts cardiovascular events and all-cause mortality in type 2 diabetes in a 14-year follow-up: Lessons from the San Luigi Gonzaga Diabetes Study. *Diabetes Care* **2011**, *34*, 2237–2243. [CrossRef]
15. Cavalot, F.; Petrelli, A.; Traversa, M.; Bonomo, K.; Fiora, E.; Conti, M.; Anfossi, G.; Costa, G.; Trovati, M. Postprandial blood glucose is a stronger predictor of cardiovascular events than fasting blood glucose in type 2 diabetes mellitus, particularly in women: Lessons from the San Luigi Gonzaga Diabetes Study. *J. Clin. Endocrinol. Metab.* **2006**, *91*, 813–819. [CrossRef] [PubMed]
16. Wang, Y.; Fang, Y. Late non-fasting plasma glucose predicts cardiovascular mortality independent of hemoglobin A1c. *Sci. Rep.* **2022**, *12*, 7778. [CrossRef] [PubMed]
17. Takao, T.; Takahashi, K.; Suka, M.; Suzuki, N.; Yanagisawa, H. Association between postprandial hyperglycemia at clinic visits and all-cause and cancer mortality in patients with type 2 diabetes: A long-term historical cohort study in Japan. *Diabetes Res. Clin. Pract.* **2019**, *148*, 152–159. [CrossRef]
18. Abe, H.; Aida, Y.; Ishiguro, H.; Yoshizawa, K.; Miyazaki, T.; Itagaki, M.; Sutoh, S.; Aizawa, Y. Alcohol, postprandial plasma glucose, and prognosis of hepatocellular carcinoma. *World J. Gastroenterol.* **2013**, *19*, 78–85. [CrossRef] [PubMed]
19. Takao, T.; Suka, M.; Yanagisawa, H.; Kasuga, M. Thresholds for postprandial hyperglycemia and hypertriglyceridemia associated with increased mortality risk in type 2 diabetes patients: A real-world longitudinal study. *J. Diabetes Investig.* **2021**, *12*, 886–893. [CrossRef]
20. Eichenlaub, M.M.; Khovanova, N.A.; Gannon, M.C.; Nuttall, F.Q.; Hattersley, J.G. A Glucose-Only Model to Extract Physiological Information from Postprandial Glucose Profiles in Subjects with Normal Glucose Tolerance. *J. Diabetes Sci. Technol.* **2022**, *16*, 1532–1540. [CrossRef]
21. Wang, Y.; Fang, Y.; Habenicht, A.; Golledge, J.; Giovannucci, E.; Ceriello, A. Postprandial Plasma Glucose and Associated Cancer Mortality. *Preprints* **2024**, 2024011578. [CrossRef]
22. Wang, Y. Postprandial Plasma Glucose Measured from Blood Taken between 4 and 7.9 h Is Positively Associated with Mortality from Hypertension and Cardiovascular Disease. *J. Cardiovasc. Dev. Dis.* **2024**, *11*, 53. [CrossRef]
23. ElSayed, N.A.; Aleppo, G.; Aroda, V.R.; Bannuru, R.R.; Brown, F.M.; Bruemmer, D.; Collins, B.S.; Hilliard, M.E.; Isaacs, D.; Johnson, E.L.; et al. 2. Classification and Diagnosis of Diabetes: Standards of Care in Diabetes—2023. *Diabetes Care* **2023**, *46*, S19–S40. [CrossRef]
24. Darras, P.; Mattman, A.; Francis, G.A. Nonfasting lipid testing: The new standard for cardiovascular risk assessment. *Can. Med. Assoc. J.* **2018**, *190*, E1317–E1318. [CrossRef] [PubMed]
25. Bonora, E.; Tuomilehto, J. The Pros and Cons of Diagnosing Diabetes with A1C. *Diabetes Care* **2011**, *34*, S184–S190. [CrossRef] [PubMed]
26. American Diabetes Association. 2. Classification and Diagnosis of Diabetes: Standards of Medical Care in Diabetes—2021. *Diabetes Care* **2021**, *44*, S15–S33. [CrossRef]
27. Phillips, P.J. Oral glucose tolerance testing. *Aust. Fam. Physician* **2012**, *41*, 391–393.
28. Christensen, D.L.; Witte, D.R.; Kaduka, L.; Jørgensen, M.E.; Borch-Johnsen, K.; Mohan, V.; Shaw, J.E.; Tabák, A.G.; Vistisen, D. Moving to an A1C-based diagnosis of diabetes has a different impact on prevalence in different ethnic groups. *Diabetes Care* **2010**, *33*, 580–582. [CrossRef]
29. Zhou, X.; Pang, Z.; Gao, W.; Wang, S.; Zhang, L.; Ning, F.; Qiao, Q. Performance of an A1C and fasting capillary blood glucose test for screening newly diagnosed diabetes and pre-diabetes defined by an oral glucose tolerance test in Qingdao, China. *Diabetes Care* **2010**, *33*, 545–550. [CrossRef]
30. Tonelli, M.; Pottie, K. Diabetes guidelines. *Can. Med. Assoc. J.* **2013**, *185*, 238. [CrossRef] [PubMed]
31. Robinson, C.A.; Sohal, P. Diabetes guidelines. *Can. Med. Assoc. J.* **2013**, *185*, 237–238. [CrossRef]
32. Luna, B.; Feinglos, M.N. Oral agents in the management of type 2 diabetes mellitus. *Am. Fam. Physician* **2001**, *63*, 1747–1756.
33. von Nicolai, H.; Brickl, R.; Eschey, H.; Greischel, A.; Heinzel, G.; König, E.; Limmer, J.; Rupprecht, E. Duration of action and pharmacokinetics of the oral antidiabetic drug gliquidone in patients with non-insulin-dependent (type 2) diabetes mellitus. *Arzneimittelforschung* **1997**, *47*, 247–252. [PubMed]
34. Padhi, S.; Nayak, A.K.; Behera, A. Type II diabetes mellitus: A review on recent drug based therapeutics. *Biomed. Pharmacother.* **2020**, *131*, 110708. [CrossRef] [PubMed]

35. Wang, Y. Higher fasting triglyceride predicts higher risks of diabetes mortality in US adults. *Lipids Health Dis.* **2021**, *20*, 181. [CrossRef]
36. American Diabetes Association. Classification and Diagnosis of Diabetes: Standards of Medical Care in Diabetes—2019. *Diabetes Care* **2019**, *42*, S13–S28. [CrossRef]
37. NHANES. Hexokinase-Mediated Reaction Roche/Hitachi Cobas C Chemistry Analyzer. Laboratory Procedure Manual 2014. Available online: https://wwwn.cdc.gov/nchs/data/nhanes/2013-2014/labmethods/GLU_H_MET.pdf (accessed on 15 January 2024).
38. Jungo, K.T.; Meier, R.; Valeri, F.; Schwab, N.; Schneider, C.; Reeve, E.; Spruit, M.; Schwenkglenks, M.; Rodondi, N.; Streit, S. Baseline characteristics and comparability of older multimorbid patients with polypharmacy and general practitioners participating in a randomized controlled primary care trial. *BMC Fam. Pract.* **2021**, *22*, 123. [CrossRef] [PubMed]
39. Wang, Y. Stage 1 hypertension and risk of cardiovascular disease mortality in United States adults with or without diabetes. *J. Hypertens.* **2022**, *40*, 794–803. [CrossRef] [PubMed]
40. Qian, T.; Sun, H.; Xu, Q.; Hou, X.; Hu, W.; Zhang, G.; Drummond, G.R.; Sobey, C.G.; Charchar, F.J.; Golledge, J.; et al. Hyperuricemia is independently associated with hypertension in men under 60 years in a general Chinese population. *J. Hum. Hypertens.* **2021**, *35*, 1020–1028. [CrossRef] [PubMed]
41. Wang, Y.; Zhang, W.; Qian, T.; Sun, H.; Xu, Q.; Hou, X.; Hu, W.; Zhang, G.; Drummond, G.R.; Sobey, C.G.; et al. Reduced renal function may explain the higher prevalence of hyperuricemia in older people. *Sci. Rep.* **2021**, *11*, 1302. [CrossRef] [PubMed]
42. Wang, Y. Definition, prevalence, and risk factors of low sex hormone-binding globulin in US adults. *J. Clin. Endocrinol. Metab.* **2021**, *106*, e3946–e3956. [CrossRef]
43. Brancato, D.; Saura, G.; Fleres, M.; Ferrara, L.; Scorsone, A.; Aiello, V.; Di Noto, A.; Spano, L.; Provenzano, V. Prognostic accuracy of continuous glucose monitoring in the prediction of diabetes mellitus in children with incidental hyperglycemia: Receiver operating characteristic analysis. *Diabetes Technol. Ther.* **2013**, *15*, 580–585. [CrossRef]
44. Wang, Y.; Fang, Y. Postabsorptive homeostasis model assessment for insulin resistance is a reliable biomarker for cardiovascular disease mortality and all-cause mortality. *Diabetes Epidemiol. Manag.* **2021**, *6*, 100045. [CrossRef]
45. Perkins, N.J.; Schisterman, E.F. The inconsistency of "optimal" cutpoints obtained using two criteria based on the receiver operating characteristic curve. *Am. J. Epidemiol.* **2006**, *163*, 670–675. [CrossRef] [PubMed]
46. Arnold, B.F.; Hogan, D.R.; Colford, J.M.; Hubbard, A.E. Simulation methods to estimate design power: An overview for applied research. *BMC Med. Res. Methodol.* **2011**, *11*, 94. [CrossRef]
47. Wilson, D.T.; Hooper, R.; Brown, J.; Farrin, A.J.; Walwyn, R.E. Efficient and flexible simulation-based sample size determination for clinical trials with multiple design parameters. *Stat. Methods Med. Res.* **2021**, *30*, 799–815. [CrossRef] [PubMed]
48. Šimundić, A.M. Measures of Diagnostic Accuracy: Basic Definitions. *J. Int. Fed. Clin. Chem. Lab. Med.* **2009**, *19*, 203–211.
49. Shreffler, J.; Huecker, M. Diagnostic Testing Accuracy: Sensitivity, Specificity, Predictive Values and Likelihood Ratios. StatPearls. 2023. Available online: https://www.ncbi.nlm.nih.gov/books/NBK557491/ (accessed on 5 April 2024).
50. Eusebi, P. Diagnostic Accuracy Measures. *Cerebrovasc. Dis.* **2013**, *36*, 267–272. [CrossRef]
51. Mandrekar, J.N. Receiver Operating Characteristic Curve in Diagnostic Test Assessment. *J. Thorac. Oncol.* **2010**, *5*, 1315–1316. [CrossRef]
52. Ialongo, C. Confidence interval for quantiles and percentiles. *Biochem. Med.* **2019**, *29*, 010101. [CrossRef] [PubMed]
53. Pajunen, P.; Peltonen, M.; Eriksson, J.G.; Ilanne-Parikka, P.; Aunola, S.; Keinänen-Kiukaanniemi, S.; Uusitupa, M.; Tuomilehto, J.; Lindström, J. HbA(1c) in diagnosing and predicting Type 2 diabetes in impaired glucose tolerance: The Finnish Diabetes Prevention Study. *Diabet. Med.* **2011**, *28*, 36–42. [CrossRef]
54. Sato, Y.; Ohfusa, H.; Katakura, M.; Komatsu, M.; Yamada, S.; Yamauchi, K.; Ichikawa, K.; Aizawa, T.; Hashizume, K. A problem with the diagnosis of diabetes mellitus based on fasting plasma glucose. *Diabet. Med.* **2002**, *19*, 82–83. [CrossRef]
55. Gatling, W.; Begley, J.P. Diagnosing diabetes mellitus in clinical practice: Is fasting plasma glucose a good initial test?*. *Pract. Diabetes Int.* **2001**, *18*, 89–93. [CrossRef]
56. Huang, J.; Ou, H.Y.; Karnchanasorn, R.; Samoa, R.; Chuang, L.M.; Chiu, K.C.; Feng, W. Clinical implication of fasting and post-challenged plasma glucose in diagnosis of diabetes mellitus. *Endocrine* **2015**, *48*, 511–518. [CrossRef] [PubMed]
57. Avignon, A.; Radauceanu, A.; Monnier, L. Nonfasting plasma glucose is a better marker of diabetic control than fasting plasma glucose in type 2 diabetes. *Diabetes Care* **1997**, *20*, 1822–1826. [CrossRef] [PubMed]
58. Peter, R.; Dunseath, G.; Luzio, S.D.; Chudleigh, R.; Roy Choudhury, S.; Owens, D.R. Daytime variability of postprandial glucose tolerance and pancreatic B-cell function using 12-h profiles in persons with Type 2 diabetes. *Diabet. Med.* **2010**, *27*, 266–273. [CrossRef] [PubMed]
59. O'Neal, T.B.; Luther, E.E. Dawn Phenomenon. StatPearls. 2022. Available online: https://www.ncbi.nlm.nih.gov/books/NBK430893/ (accessed on 5 April 2024).
60. Monnier, L.; Colette, C.; Dejager, S.; Owens, D. The dawn phenomenon in type 2 diabetes: How to assess it in clinical practice? *Diabetes Metab.* **2015**, *41*, 132–137. [CrossRef]

biomedicines

MDPI

Article

Association of Endotoxemia with Low-Grade Inflammation, Metabolic Syndrome and Distinct Response to Lipopolysaccharide in Type 1 Diabetes

Aleksejs Fedulovs [1], Leonora Pahirko [2], Kaspars Jekabsons [1], Liga Kunrade [1], Jānis Valeinis [2], Una Riekstina [1], Valdis Pīrāgs [1,3] and Jelizaveta Sokolovska [1,*]

[1] Faculty of Medicine, University of Latvia, LV-1004 Riga, Latvia; aleksejs.fedulovs@lu.lv (A.F.); kaspars.jekabsons@lu.lv (K.J.); liga.kunrade@lu.lv (L.K.); una.riekstina@lu.lv (U.R.); valdis.pirags@lu.lv (V.P.)

[2] Faculty of Physics, Mathematics and Optometry, University of Latvia, LV-1004 Riga, Latvia; leonora.pahirko@lu.lv (L.P.); janis.valeinis@lu.lv (J.V.)

[3] Pauls Stradins Clinical University Hospital, LV-1002 Riga, Latvia

* Correspondence: jelizaveta.sokolovska@lu.lv

Abstract: The association of endotoxemia with metabolic syndrome (MS) and low-grade inflammation in type 1 diabetes (T1D) is little-studied. We investigated the levels of lipopolysaccharide (LPS), lipopolysaccharide-binding protein (LBP), endogenous anti-endotoxin core antibodies (EndoCAb IgG and IgM) and high-sensitivity C-reactive protein (hsCRP) in 74 T1D patients with different MS statuses and 33 control subjects. Within the T1D group, 31 patients had MS. These subjects had higher levels of LPS compared to patients without MS (MS 0.42 (0.35–0.56) or no MS 0.34 (0.3–0.4), $p = 0.009$). MS was associated with LPS/HDL (OR = 6.5 (2.1; 20.0), $p = 0.036$) and EndoCAb IgM (OR = 0.32 (0.11; 0.93), $p = 0.036$) in patients with T1D. LBP ($\beta = 0.30$ (0.09; 0.51), $p = 0.005$), EndoCAb IgG ($\beta = 0.29$ (0.07; 0.51), $p = 0.008$) and the LPS/HDL ratio ($\beta = 0.19$ (0.03; 0.41, $p = 0.084$) were significantly associated with log-transformed hsCRP in T1D. Higher levels of hsCRP and EndoCAb IgG were observed in T1D compared to the control ($p = 0.002$ and $p = 0.091$, respectively). In contrast to the situation in the control group, LPS did not correlate with LBP, EndoCAb, leukocytes or HDL in T1D. To conclude, endotoxemia is associated with low-grade inflammation, MS and a distinct response to LPS in T1D.

Keywords: type 1 diabetes; metabolic syndrome; endotoxin; low-grade inflammation; lipopolysaccharide-binding protein; endogenous anti-endotoxin core antibodies

Citation: Fedulovs, A.; Pahirko, L.; Jekabsons, K.; Kunrade, L.; Valeinis, J.; Riekstina, U.; Pīrāgs, V.; Sokolovska, J. Association of Endotoxemia with Low-Grade Inflammation, Metabolic Syndrome and Distinct Response to Lipopolysaccharide in Type 1 Diabetes. *Biomedicines* **2023**, *11*, 3269. https://doi.org/10.3390/biomedicines11123269

Academic Editors: Ana Dascalu and Dragos Serban

Received: 26 September 2023
Revised: 10 November 2023
Accepted: 11 November 2023
Published: 10 December 2023

1. Introduction

The prevalence of autoimmune conditions, including type 1 diabetes (T1D), has markedly increased in the recent decades [1]. T1D is characterized by absolute insulin deficiency caused by the loss of insulin-producing pancreatic beta cells and increased morbidity and mortality mainly due to long-term diabetes complications [2]. It has been shown that metabolic syndrome (MS)—a combination of cardiovascular risk factors [3] and a state of low-grade inflammation [4]—is associated with the progression of both micro- and macrovascular complications of diabetes [5–7] in T1D. Increased bacterial lipopolysaccharide (LPS, endotoxin) leakage into the circulation, or endotoxemia, is considered to be a possible source of the chronic inflammatory reaction in T1D [8] and is associated with the progression of diabetic kidney disease (DKD) [9,10], cardiovascular disease (CVD), atherosclerosis [11] and MS components [12]. In addition to possible factors predisposing T1D patients to increased endotoxemia and low-grade inflammation (such as leaky gut syndrome, intestinal microbiota dysbiosis, hyperglycemia [8,13–16], etc.), impaired response to infection is of particular interest [17].

In addition to LPS activity itself, the serum concentration of its binding proteins such as LPS-binding protein (LBP) and the endogenous anti-endotoxin core antibodies (EndoCAb)

IgM and IgG can serve as markers of infection severity and the host's anti-inflammatory response [18,19]. Unfortunately, only a few studies report concentrations of some of these markers in T1D compared to healthy subjects [20–22] and in relation to MS components in T1D [23]. Also, no study to date reports correlations of the levels of fecal calprotectin as a marker of intestinal inflammation [15] with endotoxemia markers in T1D.

Finally, the potential link of non-alcoholic fatty liver disease (NAFLD) or its surrogate markers with endotoxemia in T1D has not yet been studied. This gap in knowledge should also be filled, as there is evidence of links between MS, NAFLD [24] and endotoxemia in the general population [25]. The latter data would be of high importance due to the increasing prevalence of NAFLD in T1D and its association with the complications of T1D [26,27].

To summarize, a better understanding of associations between low-grade inflammation, endotoxins and their binding proteins and MS components is important for the development of future treatment and prevention strategies for vascular T1D complications.

In this work, we utilized several novel approaches of statistical analysis in addition to conventionally used methods; for example, for data normalization between the plates, a transformation based on the two-sample location-scale model was used [28,29]. For two-distribution comparisons, we used the smoothed empirical likelihood method for the probability–probability (PP) plot [30–32]. In addition, the empirical likelihood method for the equality test of all deciles of two distributions, which allows for obtaining information about the differences or similarities of two samples, was applied for data analysis [33].

To conclude, the aim of this study was to investigate the levels of endotoxins and their binding proteins in T1D patients with different MS statuses and to study associations between endotoxemia markers, low-grade inflammation and MS using both conventional and novel approaches of statistical analysis.

2. Materials and Methods

2.1. Study Groups and Subjects

All participants in this investigation took part in the longitudinal LatDiane study, which is a part of the international InterDiane consortium. Adult patients with T1D diagnosed before the age of 40 years, with insulin treatment initiated within one year of diagnosis and C-peptide levels below 0.3 nmol/L, were recruited for LatDiane. Patients with a history of chronic kidney disease apart from DKD were excluded [34]. Follow-up visits and re-assessment of the status of complications of diabetes were performed every three years or more frequently.

We performed recruitment, biobanking and sample storage in agreement with the procedures of the Genome Database of the Latvian Population [35,36].

Recruitment took place between 15th January 2021 and 31st August 2021 in Latvia, Riga, Pilsoņu 13 str. building 10 (the rooms of the Laboratory for Personalized Medicine of the University of Latvia). The group of subjects with normal glucose metabolism (control group) included non-smoking spouses and friends of patients participating in this study, university personnel and students willing to participate in this study. Information about the study was disseminated via a webpage and social media of the University of Latvia.

The inclusion criterion for the group of patients with T1D was a diabetes duration of at least 8 years.

The inclusion criteria for the control group were normal glucose metabolism, defined as fasting glucose $\leq$ 5.6 mmol and HbA1c $\leq$ 5.7% [2], and no known or documented autoimmune, cardiovascular or other chronic diseases.

Exclusion criteria for both study groups were pregnancy, history of inflammatory bowel disease (Crohn's disease or ulcerative colitis) and coeliac disease, acute intestinal infection within 2 months of the planned fecal collection, asymptomatic coeliac disease (detected via screening of serum transglutaminase IgA antibodies), clinical signs of acute inflammation, and fever.

On the study day, patients were investigated for the collection of anthropometric measures, blood samples and information on disease history. They also received instructions and vials for the collection of the fecal sample.

2.2. Clinical Definitions

Patients were investigated to document weight, height, waist circumference and blood pressure. The body mass index (BMI, weight (kg)/height $(m)^2$) and waist/height ratio were calculated. Patients with systolic blood pressure $\geq$ 140 mmHg (18.7 kPa) or diastolic blood pressure $\geq$ 90 mmHg (12.0 kPa), or a history of antihypertensive drug usage, were defined to have arterial hypertension.

Smoking was self-reported, and the "smokers" group referred to patients currently smoking at least one cigarette per day.

Medical files were investigated for assessment of CVD and complications of diabetes such as retinopathy, neuropathy and DKD. We defined CVD as a history of acute myocardial infarction, coronary bypass/percutaneous transluminal coronary angioplasty stroke, amputation or peripheral vascular disease.

The albumin-to-creatinine ratio in two out of the three morning spot urine samples was used for the definition of albuminuria status. The estimated glomerular filtration rate (eGFR) was calculated according to the Chronic Kidney Disease Epidemiology Collaboration (CKD-EPI).

MS was assessed according to Alberti et al. [3]. The waist criterion was fulfilled for men if $\geq$102 cm and for women if $\geq$88 cm. Patients with serum triglycerides $\geq$ 1.7 mmol/L, serum high-density lipoproteins < 1.0 mmol/L (for men) and <1.3 mmol/L (for women) or medication to manage these dyslipidemia disorders fulfilled these respective criteria. Furthermore, subjects with blood pressure exceeding 130/85 mmHg or antihypertensive treatment fulfilled the blood pressure criterion. The elevated fasting blood glucose criterion was positive for all subjects with T1D. In the case of $\geq$3 positive criteria, MS was confirmed.

2.3. Sampling of Blood for Serum Preparation and Fecal Collection

Blood samples and morning spot urine samples were sent to a certified clinical lab for assessment of clinical markers (blood count, CRP, clinical chemistry, albuminuria).

For serum preparation for further analysis of endotoxemia markers and hsCRP, peripheral venous blood was collected. The blood samples were incubated undisturbed for 30 min at room temperature and then centrifuged. The serum was removed from the pellet and transferred into fresh 2 mL tubes, frozen and stored at -20 °C until analysis.

Participants collected their fecal samples at home within two weeks after blood collection, using sterile collection tubes without buffer (collection date and time were marked). Within 24 h, samples were delivered to the laboratory for calprotectin measurement in unfrozen samples.

2.4. Determination of Inflammatory and Endotoxemia Markers

In serum, the LPS activity was measured using a Hycult LAL chromogenic endpoint assay (HIT302, Hycult Biotech, Uden, The Netherlands). EndoCAb IgG and EndoCAb IgM were measured using Hycult EndoCAb IgG and IgM Elisa kits (HK504-IGG; HK504-IGM; Hycult Biotech, Uden, The Netherlands). LPB was measured using a Hycult LPB Human Elisa kit (HK315-02, Hycult Biotech, Uden, The Netherlands). hsCRP was measured using a Hycult Human hsCRP kit (HK369 Hycult Biotech, Uden, The Netherlands). Measurements were performed according to the manufacturer's instructions. Fecal calprotectin was measured using an Alegria® Calprotectin Elisa kit (REF ORG280, Organotech Diagnostika GmbH, Budapest, Hungary) in a certified clinical lab.

2.5. Indices of Insulin Resistance and NAFLD

Estimated glucose disposal rate (eGDR) allows for the estimation of insulin resistance in patients with T1D. It was calculated according to the following formula:

$$\text{eGDR} = 24.4 - 12.97 \times \text{Waist/Hips} - 3.39 \times \text{Hypertension} - 0.6 \times \text{HbA1c\%},$$

where hypertension is 1 if blood pressure is $\geq$140/90 mmHg and/or the patient takes antihypertensive drugs regularly; otherwise, it is 0 [37,38]. The lower the eGDR, the higher the insulin resistance.

The NAFLD hepatic steatosis index (HSI) and fatty liver index (FLI) were calculated according to the following formulas:

$$\text{HSI} = 8 \times \text{ALT/AST} + \text{BMI} + 2\ (\text{if DM}) + 2(\text{if female}),$$

with values < 30 ruling out and values $\geq$ 36 ruling in steatosis [39], and

$$\text{FLI} = logistic(0.953 \times ln(\text{TG}) + 0.139 \times \text{BMI} + 0.718 + ln(\text{GGT}) \\ +0.053 \times \text{Waist} - 15.745) \times 100,$$

where $logistic(x) = 1/(1 + e^{-x})$ denotes the logistic function and $ln(x)$ the natural logarithm. Values < 30 rule out and values $\geq$ 60 rule in steatosis [40].

2.6. Statistical Analysis

All statistical analysis was performed using statistical open-source software R version 4.3.1 (http://www.r-project.org accessed on 20 September 2023) [41]. The Shapiro–Wilk test was used to test the normality assumption. We report all p-values below 0.1 as statistically significant throughout the paper.

2.6.1. Data Normalization across the Plates

Data on five serum inflammatory markers were accessed on 1 February 2023. They were examined for homogeneity across three plates. The distributions were highly skewed to the left (the skewness coefficient ranged from 0.66 to 1.93); the normality assumption was violated for all five markers ($p < 0.001$). The equality of locations across three plates was tested using the Kruskal–Wallis test and was not rejected for hsCRP ($p = 0.46$), LPS ($p = 0.24$), EndoCAb IgG ($p = 0.82$) and EndoCAb IgM ($p = 0.22$). However, variability was detected for LBP measurements across the plates ($p < 0.001$) (Supplemental Figure S1). Post hoc comparisons demonstrated that measurements of plate 3 differed significantly from both plate 1 and plate 2. It was decided to normalize observations of LBP via simultaneous location and scale transformation for plate 3 based on a two-sample location-scale model. We verified the simultaneous location-scale change between two distributions using the test developed by Hall et al. [29]. The hypothesis of location-scale change between plates 1 and 2 and plate 3 was not rejected ($p = 0.76$). To obtain normalized data, we transformed observations on plate 3 by multiplying with the scale estimate 6.475 and by adding the location estimate −24.83. The equality of locations across all the plates of normalized LBP was not rejected (Kruskal–Wallis $p = 0.93$).

2.6.2. Descriptive Statistics and Comparisons between Two Samples

Categorical variables were summarized as counts and percentages. The equality of proportions between the two study groups was tested using the chi-square test for proportions. Most of the continuous variables analyzed violated the normality assumption; therefore, data are presented as medians with interquartile range (IQR). For descriptive statistics, we used several plots to compare the two samples: probability–probability plots, smoothed kernel density estimates, one-dimensional scatter plots and quantile difference plots. The pointwise confidence intervals were added for PP plots and quantile difference plots.

Comparisons between the two samples were performed by using the Kolmogorov–Smirnov, the Wilcoxon and the Lepage tests. The Wilcoxon test compares two medians, the Lepage test determines the statistically significant change in location and scale and the Kolmogorov–Smirnov test compares whole distributions. The Wilcoxon test was followed by ANCOVA on ranks to perform the covariate adjustment. Finally, the empirical likelihood method was used to test for equality of all deciles of two distributions. All tests were implemented in the statistical software R. Package EL was used to produce the empirical likelihood-based statistical inference.

2.6.3. Correlations and Regression Analysis

Correlation analysis was carried out separately in the control and T1D groups. Spearman's rank correlation coefficient ρ with a 95% confidence interval and the p-value for the significance of the association were calculated. The results are illustrated in a separate forest plot for each variable of interest.

A linear multivariate regression model was fitted on log-transformed hsCRP in the T1D group with LBP, EndoCAb IgG, EndoCAb IgM, LPS/HDL, sex, BMI and diabetes duration as the predictors (Model 1). Logistic multivariate regression was performed to predict the MS in T1D using LBP, EndoCAb IgG, EndoCAb IgM, LPS, hsCRP, sex, BMI and diabetes duration as the predictors (Model 2). All predictors were standardized before the model fitting in both cases. Multicollinearity was examined using variance inflation factor calculations. The R-squared and Nagelkerke R-squared were calculated to assess the fit for Model 1 and Model 2, respectively.

3. Results

3.1. Characteristics of Subjects

In this study, 74 patients with T1DM were included. In the T1D group, the median diabetes duration was 21 (13–32) years and 31 subjects had MS. Compared to patients without MS, subjects with MS had a higher BMI, waist/height ratio and prevalence of diabetic complications and used antihypertensive and lipid-lowering medications more frequently. In addition, patients with MS had lower insulin sensitivity as assessed using the eGDR formula, higher HbA1c and had higher scores of NAFLD surrogate markers FLI and HSI (Table 1).

We also included 33 generally healthy adults (control) in this study. A detailed comparison of the control vs. T1D group can be found in Supplemental Table S1. In brief, patients with diabetes and controls did not differ in anthropometric measures and gender distribution. Compared to subjects in the control group, patients with diabetes were older (control: 35 (30–44) years; T1D: 43 (34–51), $p = 0.013$); had higher HbA1c levels (control: 5.2 (5.0–5.5), T1D: 7.7 (6.9–9.3), $p < 0.001$); had a higher prevalence of MS (control: 4 (12.1%), T1D: 31 (41.9%), $p = 0.005$), hypertension (control: 7 (21.2%), T1D: 35 (27.0%), $p = 0.019$) CVD (control: 0 (0%), T1D: 9 (12.2%), $p = 0.086$) and autoimmune thyroid disease (control: 1 (3%), T1D: 19 (25.7%), $p = 0.012$); and had a higher level of albuminuria (control: 0.32 (0.11–0.76), T1D: 0.32 (0.11–0.76), $p = 0.013$) and lower level of insulin sensitivity as assessed using eGDR (control: 7.1 (5.5–7.7), T1D: 3.3 (1.6–5.7), $p < 0.001$).

3.2. Levels of Serum Inflammatory Markers in Patients with T1D Stratified According to MS Presence and Controls

Patients with T1D and MS had higher levels of LPS compared to patients without MS (MS: 0.42 (0.35–0.56) EU/mL, no MS: 0.34 (0.30–0.40) EU/mL, $p = 0.009$) (Figure 1, Table 2, Supplemental Table S3). LPS/HDL was also higher in MS vs. no MS patients (MS: 0.28 (0.24–0.38), no MS: 0.22 (0.17–0.26), $p = 0.01$) (Figure 2, Table 2, Supplemental Table S3). Most of the LPS/HDL and LPS deciles differed significantly between the groups in line with the median comparisons (Supplemental Table S2, Figures 1 and 2). The levels of LPS-binding proteins and hsCRP did not differ between MS groups.

Table 1. Characteristics of subjects.

Phenotype	No Metabolic Syndrome (n = 43)	Metabolic Syndrome (n = 31)	p
Male/Female, n (%)	19/24 (44/56)	9/22 (29/71)	0.28
Age, years	38 (32–50)	45 (36–54)	0.065
BMI, kg/m^2	23.5 (21.45–25.65)	28 (25.4–32.65)	<0.001
Waist/height ratio	0.457 (0.431–0.497)	0.56 (0.509–0.621)	<0.001
Smoker, n (%)	11 (26)	9 (29)	0.95
Hypertension, n (%)	15 (35)	20 (65)	0.022
Length of diabetes, years	20 (11–32)	24 (17–34)	0.062
Retinopathy, n (%)	13 (30)	19 (61)	0.015
CVD, n (%)	3 (7)	6 (19)	0.21
Macroalbuminuria, n (%)	2 (5)	5 (16)	0.21
On ACEI/ARB, n (%)	4 (9)	11 (35)	0.013
On lipid-lowering medication, n (%)	5 (12)	12 (39)	0.014
Autoimmune thyroid disease, n (%)	9 (21)	10 (32)	0.41
Hemoglobin A1C, %	7.3 (6.8–9.0)	8.4 (7.4–9.3)	0.087
Hemoglobin A1C, mmol/mol	53 (47–67)	64 (54–64)	0.087
Estimated glomerular filtration rate, mL/min/1.73m^2	114 (104–119)	98 (74–108)	<0.001
Albumin/creatinine ratio in urine, mg/mmol	0.38 (0.19–1.05)	0.97 (0.37–2.34)	0.072
Total cholesterol, mmol/L	4.72 (4.29–5.18)	5.51 (4.89–6.04)	0.006
Low-density lipoproteins, mmol/L	2.68 (2.08–3.24)	3.22 (2.47–3.75)	0.02
High-density lipoproteins, mmol/L	1.71 (1.38–2.01)	1.44 (1.21–1.73)	0.012
Triglycerides, mmol/L	1.02 (0.79–1.18)	1.72 (1.30–2.11)	<0.001
Alanine aminotransaminase, U/L	19 (134–26)	21 (17–28)	0.087
Aspartate aminotransferase, U/L	22 (19–28)	25 (17–31)	0.879
Gamma-glutamyltransferase, U/L	15 (13–22)	18 (15–26)	0.212
Bilirubin, µmol/L	10.3 (8.0–13.1)	8.4 (6.3–10.6)	0.015
Hemoglobin, g/L	141 (134–149)	139 (126–150)	0.321
Erythrocytes, 10 × 12/L	4.7 (4.5–4.9)	4.7 (4.4–5.1)	0.583
Leukocytes, 10 × 9/L	6.2 (5.2–7.5)	6.1 (5.0–7.1)	0.576
Thrombocytes, 10 × 9/L	262 (238–280)	254 (226–295)	0.518
eGDR	4.9 (2.8–6.7)	2.1 (1.0–3.3)	0.001
FLI	13.2 (6.0–22.1)	54.0 (30.1–78.9)	<0.001
HSI	31.2 (28.8–33.7)	39.6 (34.5–42.0)	<0.001

For categorical variables, data are presented as counts (%), and the equality of proportions between study groups was tested using the chi-square test for the equality of proportions. For continuous variables, data are presented as medians (IQR), and the comparisons between the control and T1D group were performed using the Wilcoxon test. BMI—body mass index. CVD—cardiovascular disease. ACEI—angiotensin-converting enzyme inhibitors. ARB—angiotensin receptor blockers. eGDR—estimated glucose disposal rate. FLI—fatty liver index. HSI—hepatic steatosis index.

Table 2. Inflammatory marker analysis in patients with T1D stratified by metabolic syndrome.

Phenotype	No Metabolic Syndrome (n = 43)	Metabolic Syndrome (n = 31)	p
hsCRP, mg/L	0.80 (0.49–1.53)	1.23 (0.38–2.17)	0.41
LPS, EU/mL	0.34 (0.30–0.40)	0.42 (0.35–0.56)	0.009
LPS/HDL ratio	0.22 (0.17–0.26)	0.28 (0.24–0.38)	0.001
EndoCAb IgG, GMU/mL	89.5 (67.1–150.2)	96.0 (59.1–141.6)	0.84
EndoCAb IgM, MMU/mL	46.6 (34.2–85.2)	43.7 (31.3–60.4)	0.33
LBP, µg/mL	11.1 (7.9–16.3)	11.1 (7.9–13.3)	0.89
Calprotectin, µg/g	3.8 (2.7–8.6)	12.5 (3.2–18.4)	0.18

Data are presented as medians (IQR). p-values of the Wilcoxon test between the study groups are presented. The results did not differ significantly after adjusting for sex, diabetes duration and BMI (ANCOVA on ranks).

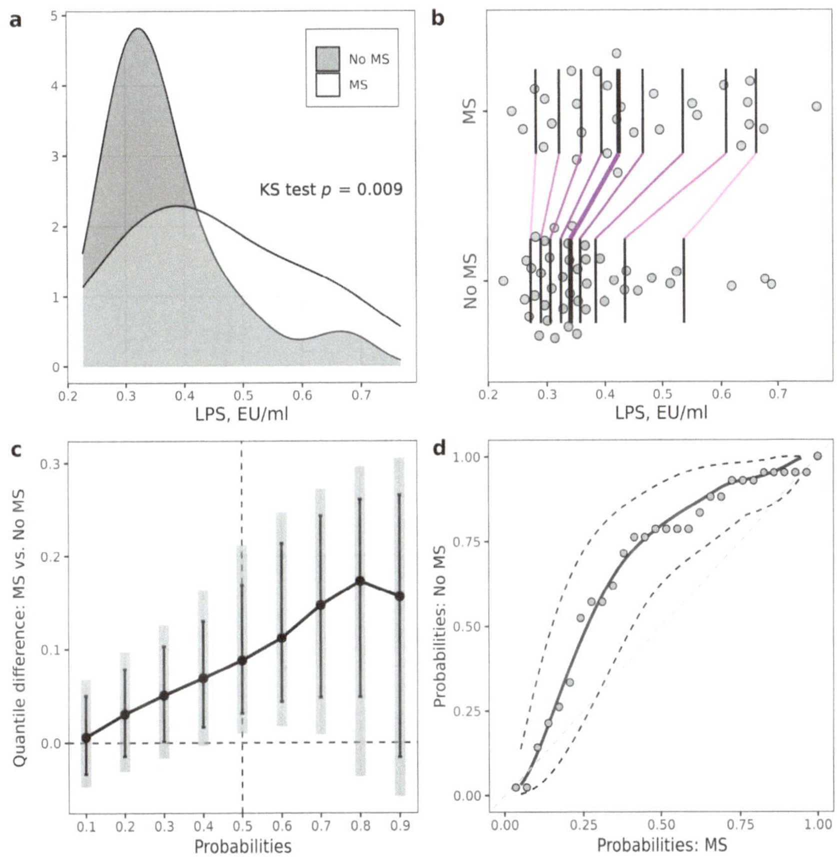

Figure 1. Illustrations for comparison of LPS levels in patients with T1D stratified according to the presence of metabolic syndrome (MS). (**a**) Plots of densities in patients with and MS, with p-value of Kolmogorov–Smirnov (KS) test for the equality of distributions; (**b**) one-dimensional scatterplots for patients with and without MS, with vertical bars representing deciles in each group (R library rogme); (**c**) quantile difference (MS—no MS) plot constructed at deciles, with estimates obtained using two-sample smoothed empirical likelihood method (R library EL), error bars representing 95% pointwise CIs and shaded bars representing 95% simultaneous confidence bands for 9 deciles; (**d**) probabilities of patients with MS versus probabilities of patients without MS (PP-plot) with estimates and 95% pointwise CIs obtained using the two-sample smoothed empirical likelihood method (R library EL). Bandwidths for kernel estimation in (**c**,**d**) were selected using the method of Sheather and Jones (function bw.SJ in R).

In comparison to controls, the T1D group had statistically significantly higher levels of hsCRP (control: 0.47 (0.26–0.75) mg/L, T1D: 0.84 (0.49–1.78) mg/L, p = 0.002) and EndoCAb IgG (control 71.1 (55.7–101.5) GMU/mL, T1D 92.6 (65.8–148.6) GMU/mL, p = 0.091) (Supplemental Table S4, Supplemental Figures S2 and S3).

In the T1D group, only six patients had calprotectin levels above 50 µg/g, and five of them had calprotectin levels above 200 µg/g. Therefore, to reduce data variability, we considered calprotectin measurements above 200 µg/g as outliers for further statistical analysis.

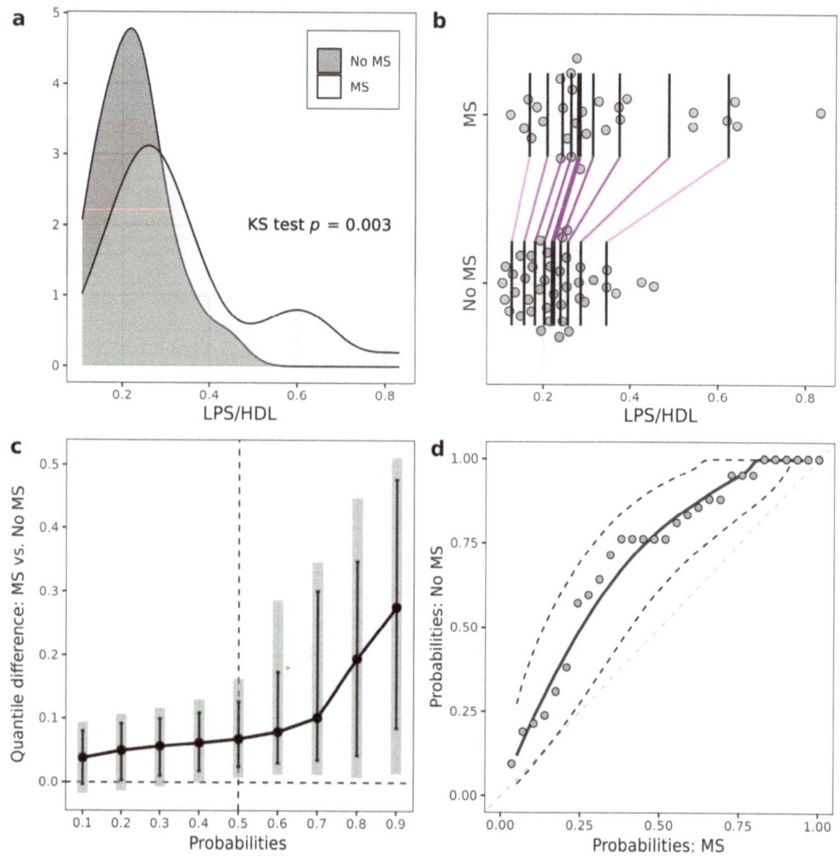

Figure 2. Illustrations for comparison of LPS/HDL in patients with T1D stratified according to the presence of metabolic syndrome (MS). (**a**) Plots of densities in patients with and without MS, with *p*-value of Kolmogorov–Smirnov (KS) test for the equality of distributions; (**b**) one-dimensional scatterplots for patients with and without MS, with vertical bars representing deciles in each group (R library rogme); (**c**) quantile difference (MS — No MS) plot constructed at deciles, with estimates obtained using two-sample smoothed empirical likelihood method (R library EL), error bars representing 95% pointwise CIs and shaded bars representing 95% simultaneous confidence bands for 9 deciles; (**d**) probabilities of patients with MS versus probabilities of patients without MS (PP-plot) with estimates and 95% pointwise CIs obtained using the two-sample smoothed empirical likelihood method (R library EL). Bandwidths for kernel estimation in (**c**,**d**) were selected using the method of Sheather and Jones (function bw.SJ in R).

Within the T1D group, patients with MS had statistically significantly higher calprotectin levels at middle deciles (quantile difference 40%: 4.3 (0.23, 9.98), $p = 0.036$; 50%: 6.9 (1.24, 11.81), $p = 0.014$; 60%: 8.72 (2.47,13.13), $p < 0.011$) (Figure 3), although the medians of calprotectin did not differ statistically significantly between the groups.

The median level of fecal calprotectin also did not differ between T1D and controls. However, patients with T1D had statistically significantly lower calprotectin levels in the first three deciles (quantile difference 10%: −2.14 (−3.34, −0.44), $p = 0.017$; 20%: −1.86 (−3.26, −0.40), $p = 0.016$; 30%: −1.54 (−3.21, 0.14), $p = 0.071$) (Supplemental Table S4, Supplemental Figure S4).

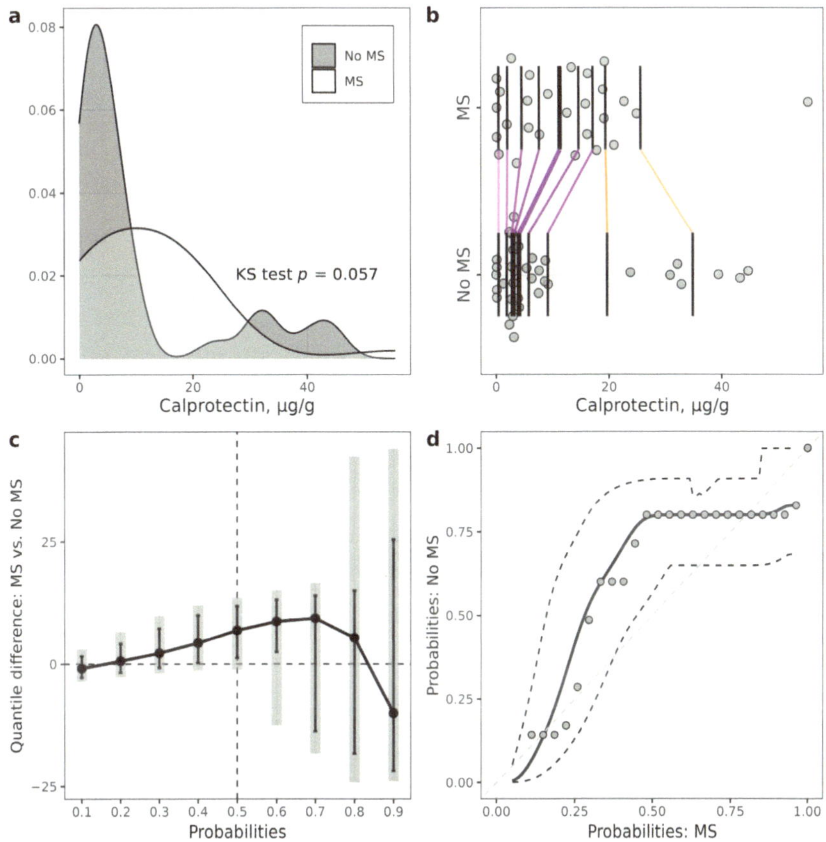

Figure 3. Illustrations for comparison of calprotectin in patients with T1D stratified according to the presence of metabolic syndrome (MS). (**a**) Plots of densities in patients with and without MS, with *p*-value of Kolmogorov–Smirnov (KS) test for the equality of distributions; (**b**) one-dimensional scatterplots for patients with and without MS, with vertical bars representing deciles in each group (R library rogme); (**c**) quantile difference (MS—No MS) plot constructed at deciles, with estimates obtained using two-sample smoothed empirical likelihood method (R library EL), error bars representing 95% pointwise CIs and shaded bars representing 95% simultaneous confidence bands for 9 deciles; (**d**) probabilities of patients with MS versus probabilities of patients without MS (PP-plot), with estimates and 95% pointwise CIs obtained using two-sample smoothed empirical likelihood method (R library EL). Bandwidths for kernel estimation in (**c**,**d**) were selected using the method of Sheather and Jones (function bw.SJ in R).

3.3. Correlation Analysis in Study Groups

Patterns of correlations were compared between the T1D group and control group (data presented in Figure 4, Supplemental Tables S5 and S6). In the T1D group, hsCRP correlated positively with EndoCAb IgG (r = 0.22, p = 0.07) and LBP (r = 0.36, p = 0.002). hsCRP did not correlate with any of the endotoxemia markers in the control group. Both in the control and diabetes groups, hsCRP correlated positively with waist/height ratio and NAFLD index HSI. hsCRP correlated positively with HbA1c (r = 0.5, p = 0.004) and BMI (r = 0.38, p = 0.034) only in the control group. The LPS and LPS/HDL ratio did not correlate with other inflammatory markers in the T1D group. In the control group, LPS correlated positively with LBP (r = 0.36, p = 0.045) and negatively with EndoCAb IgM (r = −0.41, p = 0.021), and the LPS/HDL ratio correlated positively with LBP (r = 0.39, p = 0.03). LPS correlated positively with FLI, as well as with serum triglycerides in both study groups. In the control group, it also correlated positively with leukocytes ($r = 0.32$, $p = 0.075$) and negatively with HDL ($r = -0.032$, $p = 0.073$). In the T1D group, LPS/HDL correlated positively with FLI ($r = 0.23$, $p = 0.053$) and triglycerides ($r = 0.52$, $p < 0.001$), and negatively with diabetes duration ($r = -0.22$, $p = 0.069$). In the control group, LPS/HDL ratio correlated positively with weight ($r = 0.35$, $p = 0.05$), ALT ($r = 0.31$, $p = 0.087$), GGT ($r = 0.3$, $p = 0.097$), leukocytes ($r = 0.33$, $p = 0.062$) and triglycerides ($r = 0.67$, $p = 0 < 0.001$), and negatively with eGDR ($r = -0.32$, $p = 0.077$). Similar to endotoxin, LBP correlated positively with leukocytes ($r = 0.34$, $p = 0.057$) in the control group, but not in the T1D group. Fecal calprotectin did not correlate with any of the inflammatory or endotoxemia markers in any of the groups.

3.4. Regression Analysis in T1D Group

The normality assumption was not rejected for log-transformed hsCRP in the T1D group ($p = 0.134$). The linear regression model results demonstrated that the significant predictors of log(hsCRP)) were LBP ($\beta = 0.30$ (0.09; 0.51), $p = 0.005$) and EndoCAb IgG ($\beta = 0.29$ (0.07; 0.50), $p = 0.008$); however, LPS/HDL ($\beta = 0.19$ (−0.03; 0.41), $p = 0.084$) and diabetes duration ($\beta = 0.24$ (−0.01; 0.49), $p = 0.059$) were significant at the 10% significance level. Overall, Model 1 was significant with $F(7, 60) = 3.97$ ($p = 0.001$), and 32% of the variability in data was explained by the fitted model (Table 3).

Table 3. Regression analysis in type 1 diabetic patients.

Predictor	Model 1 β	p	Predictor	Model 2 OR (95% CI)	p
LBP	0.30 (0.09; 0.51)	0.005	LBP	0.92 (0.45; 1.90)	0.82
EndoCAb IgG	0.29 (0.07; 0.50)	0.008	EndoCAb IgG	1.67 (0.81; 3.45)	0.16
EndoCAb IgM	−0.06 (−0.27; 0.16)	0.61	EndoCAb IgM	0.32 (0.11; 0.93)	0.036
LPS/HDL	0.19 (−0.03; 0.41)	0.084	LPS/HDL	6.5 (2.1; 20.0)	0.001
hsCRP	-	-	hsCRP	0.57 (0.26; 1.25)	0.16
Sex (female)	0.23 (−0.22; 0.68)	0.31	Sex (female)	2.7 (0.6; 12.7)	0.21
Diabetes duration	0.24 (−0.01; 0.49)	0.059	Diabetes duration	3.4 (1.2; 9.8)	0.021
BMI	0.10 (−0.13; 0.32)	0.39	BMI	2.3 (1.1; 4.9)	0.025
$R^2 = 0.32$, $F(7, 60) = 3.97$, $p = 0.001$				AIC = 72.8, Nagelkerke $R^2 = 0.56$	

Model 1—multivariate linear regression for log-transformed hsCRP as the response variable in T1D group, where data are presented as coefficient estimates with 95% CI; Model 2—multivariate logistic regression for the presence of metabolic syndrome as the response variable in T1D group, where data are presented as odds ratio (OR) with 95% CI. All predictors were standardized before model fitting, and multicollinearity was examined using variance inflation factor calculations.

Logistic regression model results demonstrated that the significant predictors of MS were EndoCAb IgM (OR = 0.32 (0.11; 0.93), $p = 0.036$), LPS/HDL (OR = 6.5 (2.1; 20.0), $p = 0.001$), diabetes duration (OR = 3.43 (1.21; 9.8), $p = 0.021$) and body mass index (OR = 2.3 (1.1; 4.9), $p = 0.025$). The Nagelkerke R2 of Model 2 was 0.56 and AIC = 72.8 (Table 3).

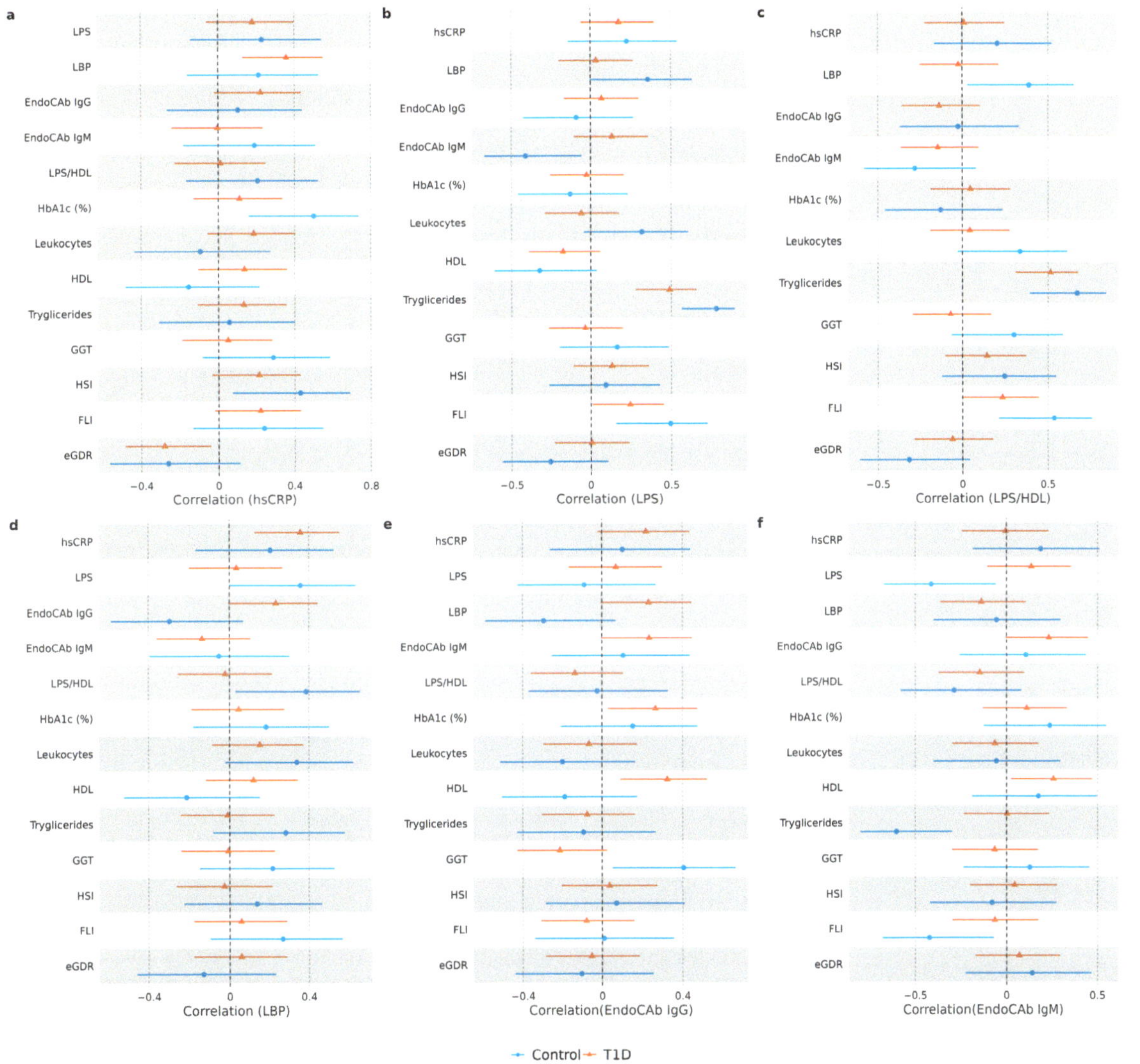

Figure 4. Forest plots demonstrating correlations between hsCRP and endotoxemia markers in each study group. Spearman correlation coefficients (95% CI) are illustrated with forest plots for hsCRP (**a**), LPS (**b**), LPS/HDL (**c**), LBP (**d**), EndoCAb IgG (**e**) and EndoCAb IgM (**f**).

4. Discussion

In this work, we reported higher levels of endotoxemia in patients with T1D and MS, as compared to T1D subjects with no MS. In addition, we observed statistically significant associations between endotoxemia markers, hsCRP and MS in T1D. Finally, we described distinct relationships between LPS, its binding proteins, HDL and leukocytes in T1D, as compared to the control group, which might indicate an impaired response to endotoxins in T1D.

We demonstrated here that patients with T1D and MS had higher LPS and LPS/HDL levels, as compared to subjects with T1D and no MS. This is in line with the findings of the previous studies [9]. However, the levels of hsCRP and LPS-binding proteins did not differ between MS groups in our study. This can be explained by the course of events in the inflammatory reaction. Indeed, the rise in CRP starts at 4–6 h and peaks at 36 h after infection [42]. Therefore, CRP, LPS and LPS-binding protein levels measured in one sample might not correlate. Nevertheless, the results of the correlation and regression analysis demonstrated the link between increased CRP and endotoxemia in the T1D group in our study, which is in agreement with the published data [8,9,14,15,43].

We are the first to report a positive correlation of NAFLD index FLI with hsCRP, LPS and LPS/HDL ratio in T1D. These findings might indicate that similarly to the situation in NAFLD in other population groups [25], this condition is linked to endotoxemia in T1D. It seems that the observed correlation of LPS with triglycerides as a component of FLI is largely "responsible" for LPS correlation with FLI in our sample. Associations between triglycerides and fat consumption were found to be associated with LPS in the general population [44] and T1D previously [12].

To gain more insights into mechanisms of response to endotoxins in T1D, we compared correlation patterns of measured biomarkers in T1D and control groups. As a result, we identified a distinct response to endotoxemia in T1D compared to controls. For example, endotoxins in the control group correlated positively with leukocytes and negatively with HDL, possibly illustrating a "healthy" response to LPS. None of these correlations were observed in the T1D group. One of the functions of HDL is to scavenge and neutralize LPS [45]; therefore, individuals with higher HDL levels might have lower LPS concentrations. A rise in leukocyte number is a well-known body response to bacterial invasion [42]. Lack of the above correlations in the T1D group might be linked to differences in HDL size and functionality previously reported in T1D [21].

LBP, one of the acute-phase response proteins, is involved in LPS binding, and its levels are usually increased in endotoxemia along with changes in EndoCAb IgM and IgG levels [18,46]. Interestingly, we did not observe a correlation of LPS with any of these endotoxemia-response proteins in the T1D group. In contrast, in the control group, LPS and the LPS/HDL ratio correlated positively with LBP and negatively with EndoCAb IgM, as reported by others [19].

To summarize, the observed differences in response to endotoxins between T1D and control might be related to changes in the immune response in T1D [17].

In this study, we reported several novel approaches to statistical analysis. For example, in two-group comparisons, utilization of PP plots and quantile (or decile) difference plots helped to detect the group effect when the Wilcoxon test failed to do so. For example, in the T1D group stratified according to MS presence, median calprotectin levels did not differ significantly between groups according to the Wilcoxon test ($p = 0.18$). However, there were significant differences at the three middle deciles (40th, 50th and 60th percentiles) according to the EL test. The explanation of the insufficient power of the Wilcoxon test to detect these effects can be based on the skewness of the data distributions and observations relatively far from the median. For example, in T1D patients, we can see from the one-dimensional scatterplot (Figure 3b) that seven observations of calprotectin were relatively far from the central location of distribution in patients without MS. Also, from the kernel density plots, we can see that robust statistical methods should be used for some further investigations. As another example, according to the Wilcoxon test, calprotectin medians did not differ significantly between the control and T1D. However, the variance for the T1D group was much higher (it was confirmed by the Lepage test), leading to big differences at lower and higher quantiles. The variance difference can be seen from the quantile difference plot and PP plot. Thus, more detailed statistical analysis using different data visualization tools and additional tests to compare lower and higher deciles or quantiles should give us a better understanding of the differences between two samples [33].

Our study's limitations include a relatively low number of subjects and the cross-sectional nature of this study, which did not allow us to evaluate the causal relationship between MS, markers of endotoxemia and hsCRP in T1D.

The major strength of the study is the analysis of the LPS, LBP, EndoCAb IgM, EndoCAb IgG and fecal calprotectin in patients with T1D, in comparison to healthy subjects and when stratified according to MS status, which has not been reported previously and allowed us to identify signs of a distinct response to endotoxins in T1D. Finally, the non-traditional approaches of statistical analysis presented in this paper might be useful for other researchers in biomedicine for gaining a deeper understanding during two-sample comparisons.

5. Conclusions

To conclude, we report higher levels of endotoxemia in subjects with T1D and MS and statistically significant associations between endotoxemia markers, hsCRP and MS in T1D. We also observed distinct patterns of response to LPS in T1D, as compared to the control group. These findings are important for clinical practice as they underline the added value of screening and treatment of MS as a state predisposing to endotoxemia. On the other hand, our data warrant more research on inflammatory response mechanisms in T1D, which might lead to the development of novel treatment options for low-grade inflammation and MS to delay the progression of T1D complications.

Supplementary Materials: The following supporting information can be downloaded at: https://www.mdpi.com/article/10.3390/biomedicines11123269/s1. Supplemental Figure S1. Illustrations of LBP before and after the normalization. Supplemental Table S1. Characteristics of T1D subjects and controls. Supplemental Table S2. Comparisons of deciles of inflammatory markers in T1D patients stratified according to the presence of metabolic syndrome. Supplemental Figure S2. Illustrations for comparison of hsCRP levels between control and T1D groups. Supplemental Figure S3. Illustrations for comparison of EndoCAb IgG levels between control and T1D groups. Supplemental Table S3. Inflammatory marker analysis in patients with T1D stratified by metabolic syndrome. Supplemental Table S4. Markers of endotoxemia and inflammation in the study groups. Supplemental Figure S4. Illustrations for comparison of calprotectin levels between control and T1D groups. Supplemental Table S5. Correlations in the type 1 diabetes group. Supplemental Table S6. Correlations in the control group.

Author Contributions: A.F. investigation, data curation, writing—original draft preparation; L.P. data curation, formal analysis, visualization, writing—original draft preparation; L.K. and K.J. investigation, data curation, writing—review and editing; J.V.—methodology, visualization, writing—original draft; U.R. methodology, supervision, writing—review and editing; V.P. writing—review and editing; and J.S. conceptualization, methodology, supervision, project administration, funding acquisition, writing—original draft, review and editing. All authors have read and agreed to the published version of the manuscript.

Funding: This work was funded by the European Regional fund project No. 1.1.1.2/VIAA/3/19/525 "Intestinal inflammation as a potentially modifiable risk factor for complications in type 1 diabetes" and by the project "Strengthening of the Capacity of Doctoral Studies at the University of Latvia within the Framework of the New Doctoral Model" (identification No. 8.2.2.0/20/I/006), along with receiving a MikroTik donation administrated by the University of Latvia Foundation.

Institutional Review Board Statement: This study was carried out in line with the 1964 Declaration of Helsinki and its later amendments. We obtained written informed consent from all study participants prior to their inclusion in the study. The protocols of the general LatDiane and sub-study on inflammatory markers described in this work were approved by the Latvian Central Ethics Committee and received permissions No 01-29.1/3 (dated 10 June 2013), Nr.A-17/19-10-17 (dated 17 October 2019) and Nr. 01-29.1/2226 (dated 30 April 2020).

Informed Consent Statement: All study participants provided written informed consent for their participation.

Data Availability Statement: The data are not available publicly due to restrictions of privacy, but they are available from the corresponding author on request.

Acknowledgments: We thank Irēna Puzirevska for coordination of patient recruitment. We acknowledge the Latvian Biomedical Research and Study Center and Latvian National Biobank—Genome Database of Latvian population—for providing materials for recruitment and sample processing.

Conflicts of Interest: The authors declare no conflict of interest.

References

1. Miller, F.W. The increasing prevalence of autoimmunity and autoimmune diseases: An urgent call to action for improved understanding, diagnosis, treatment, and prevention. *Curr. Opin. Immunol.* **2023**, *80*, 102266. [CrossRef]
2. ElSayed, N.A.; Aleppo, G.; Aroda, V.R.; Bannuru, R.R.; Brown, F.M.; Bruemmer, D.; Collins, B.S.; Cusi, K.; Das, S.R.; Gibbons, C.H.; et al. Introduction and Methodology: Standards of Care in Diabetes-2023. *Diabetes Care* **2023**, *46*, S1–S4. [CrossRef]
3. Alberti, K.G.; Eckel, R.H.; Grundy, S.M.; Zimmet, P.Z.; Cleeman, J.I.; Donato, K.A.; Fruchart, J.C.; James, W.P.; Loria, C.M.; Smith, S.C., Jr. Harmonizing the metabolic syndrome: A joint interim statement of the International Diabetes Federation Task Force on Epidemiology and Prevention; National Heart, Lung, and Blood Institute; American Heart Association; World Heart Federation; International Atherosclerosis Society; and International Association for the Study of Obesity. *Circulation* **2009**, *120*, 1640–1645. [CrossRef]
4. Fahed, G.; Aoun, L.; Bou Zerdan, M.; Allam, S.; Bou Zerdan, M.; Bouferraa, Y.; Assi, H.I. Metabolic Syndrome: Updates on Pathophysiology and Management in 2021. *Int. J. Mol. Sci.* **2022**, *23*, 786. [CrossRef] [PubMed]
5. Lee, A.S.; Twigg, S.M.; Flack, J.R. Metabolic syndrome in type 1 diabetes and its association with diabetes complications. *Diabet Med.* **2021**, *38*, e14376. [CrossRef] [PubMed]
6. Huang, Q.; Yang, D.; Deng, H.; Liang, H.; Zheng, X.; Yan, J.; Xu, W.; Liu, X.; Yao, B.; Luo, S.; et al. Association between Metabolic Syndrome and Microvascular Complications in Chinese Adults with Type 1 Diabetes Mellitus. *Diabetes Metab. J.* **2022**, *46*, 93–103. [CrossRef] [PubMed]
7. Chillarón, J.J.; Flores Le-Roux, J.A.; Benaiges, D.; Pedro-Botet, J. Type 1 diabetes, metabolic syndrome and cardiovascular risk. *Metabolism* **2014**, *63*, 181–187. [CrossRef]
8. Lehto, M.; Groop, P.H. The Gut-Kidney Axis: Putative Interconnections Between Gastrointestinal and Renal Disorders. *Front. Endocrinol.* **2018**, *9*, 553. [CrossRef]
9. Nymark, M.; Pussinen, P.J.; Tuomainen, A.M.; Forsblom, C.; Groop, P.H.; Lehto, M. Serum lipopolysaccharide activity is associated with the progression of kidney disease in finnish patients with type 1 diabetes. *Diabetes Care* **2009**, *32*, 1689–1693. [CrossRef]
10. Riddle, M.; Pencek, R.; Charenkavanich, S.; Lutz, K.; Wilhelm, K.; Porter, L. Randomized comparison of pramlintide or mealtime insulin added to basal insulin treatment for patients with type 2 diabetes. *Diabetes Care* **2009**, *32*, 1577–1582. [CrossRef]
11. Gorabi, A.M.; Kiaie, N.; Khosrojerdi, A.; Jamialahmadi, T.; Al-Rasadi, K.; Johnston, T.P.; Sahebkar, A. Implications for the role of lipopolysaccharide in the development of atherosclerosis. *Trends Cardiovasc. Med.* **2022**, *32*, 525–533. [CrossRef]
12. Lassenius, M.I.; Ahola, A.J.; Harjutsalo, V.; Forsblom, C.; Groop, P.H.; Lehto, M. Endotoxins are associated with visceral fat mass in type 1 diabetes. *Sci. Rep.* **2016**, *6*, 38887. [CrossRef] [PubMed]
13. Winther, S.A.; Henriksen, P.; Vogt, J.K.; Hansen, T.H.; Ahonen, L.; Suvitaival, T.; Hein Zobel, E.; Frimodt-Møller, M.; Hansen, T.W.; Hansen, T.; et al. Gut microbiota profile and selected plasma metabolites in type 1 diabetes without and with stratification by albuminuria. *Diabetologia* **2020**, *63*, 2713–2724. [CrossRef] [PubMed]
14. Lassenius, M.I.; Fogarty, C.L.; Blaut, M.; Haimila, K.; Riittinen, L.; Paju, A.; Kirveskari, J.; Järvelä, J.; Ahola, A.J.; Gordin, D.; et al. Intestinal alkaline phosphatase at the crossroad of intestinal health and disease—A putative role in type 1 diabetes. *J. Intern. Med.* **2017**, *281*, 586–600. [CrossRef] [PubMed]
15. Winther, S.A.; Mannerla, M.M.; Frimodt-Møller, M.; Persson, F.; Hansen, T.W.; Lehto, M.; Hörkkö, S.; Blaut, M.; Forsblom, C.; Groop, P.H.; et al. Faecal biomarkers in type 1 diabetes with and without diabetic nephropathy. *Sci. Rep.* **2021**, *11*, 15208. [CrossRef] [PubMed]
16. Rhee, S.H. Lipopolysaccharide: Basic biochemistry, intracellular signaling, and physiological impacts in the gut. *Intestig. Res.* **2014**, *12*, 90–95. [CrossRef] [PubMed]
17. Mora, V.P.; Loaiza, R.A.; Soto, J.A.; Bohmwald, K.; Kalergis, A.M. Involvement of trained immunity during autoimmune responses. *J. Autoimmun.* **2023**, *137*, 102956. [CrossRef]
18. Schumann, R.R.; Zweigner, J. A novel acute-phase marker: Lipopolysaccharide binding protein (LBP). *Clin. Chem. Lab. Med.* **1999**, *37*, 271–274. [CrossRef] [PubMed]
19. Barclay, G.R. Endogenous endotoxin-core antibody (EndoCAb) as a marker of endotoxin exposure and a prognostic indicator: A review. *Prog. Clin. Biol. Res.* **1995**, *392*, 263–272.
20. de Groot, P.F.; Belzer, C.; Aydin, Ö.; Levin, E.; Levels, J.H.; Aalvink, S.; Boot, F.; Holleman, F.; van Raalte, D.H.; Scheithauer, T.P.; et al. Distinct fecal and oral microbiota composition in human type 1 diabetes, an observational study. *PLoS ONE* **2017**, *12*, e0188475. [CrossRef] [PubMed]
21. Lassenius, M.I.; Mäkinen, V.P.; Fogarty, C.L.; Peräneva, L.; Jauhiainen, M.; Pussinen, P.J.; Taskinen, M.R.; Kirveskari, J.; Vaarala, O.; Nieminen, J.K.; et al. Patients with type 1 diabetes show signs of vascular dysfunction in response to multiple high-fat meals. *Nutr. Metab.* **2014**, *11*, 28. [CrossRef] [PubMed]
22. Aravindhan, V.; Mohan, V.; Arunkumar, N.; Sandhya, S.; Babu, S. Chronic Endotoxemia in Subjects with Type-1 Diabetes Is Seen Much before the Onset of Microvascular Complications. *PLoS ONE* **2015**, *10*, e0137618. [CrossRef] [PubMed]
23. Watanabe, H.; Katsura, T.; Takahara, M.; Miyashita, K.; Katakami, N.; Matsuoka, T.A.; Kawamori, D.; Shimomura, I. Plasma lipopolysaccharide binding protein level statistically mediates between body mass index and chronic microinflammation in Japanese patients with type 1 diabetes. *Diabetol. Int.* **2020**, *11*, 293–297. [CrossRef] [PubMed]

24. Portincasa, P. NAFLD, MAFLD, and beyond: One or several acronyms for better comprehension and patient care. *Intern. Emerg. Med.* **2023**, *18*, 993–1006. [CrossRef]
25. Nier, A.; Huber, Y.; Labenz, C.; Michel, M.; Bergheim, I.; Schattenberg, J.M. Adipokines and Endotoxemia Correlate with Hepatic Steatosis in Non-Alcoholic Fatty Liver Disease (NAFLD). *Nutrients* **2020**, *12*, 699. [CrossRef]
26. Bulum, T.; Kolarić, B.; Duvnjak, M.; Duvnjak, L. Alkaline phosphatase is independently associated with renal function in normoalbuminuric type 1 diabetic patients. *Ren. Fail.* **2014**, *36*, 372–377. [CrossRef] [PubMed]
27. Targher, G.; Mantovani, A.; Pichiri, I.; Mingolla, L.; Cavalieri, V.; Mantovani, W.; Pancheri, S.; Trombetta, M.; Zoppini, G.; Chonchol, M.; et al. Nonalcoholic fatty liver disease is independently associated with an increased incidence of chronic kidney disease in patients with type 1 diabetes. *Diabetes Care* **2014**, *37*, 1729–1736. [CrossRef]
28. Freitag, G.; Munk, A. On Hadamard differentiability in k-sample semiparametric models—With applications to the assessment of structural relationships. *J. Multivar. Anal.* **2005**, *94*, 123–158. [CrossRef]
29. Hall, P.; Lombard, F.; Potgieter, C.J. A New Approach to Function-Based Hypothesis Testing in Location-Scale Families. *Technometrics* **2013**, *55*, 215–223. [CrossRef]
30. Claeskens, G.; Jing, B.-Y.; Peng, L.; Zhou, W. Empirical Likelihood Confidence Regions for Comparison Distributions and ROC Curves. *Can. J. Stat./La Rev. Can. De Stat.* **2003**, *31*, 173–190. [CrossRef]
31. Doksum, K.A.; Sievers, G.L. Plotting with Confidence: Graphical Comparisons of Two Populations. *Biometrika* **1976**, *63*, 421–434. [CrossRef]
32. Valeinis, J.; Cers, E.; Cielens, J. Two-sample problems in statistical data modelling. *Math. Model. Anal.-Math. Model Anal.* **2010**, *15*, 137–151. [CrossRef]
33. Rousselet, G.A.; Pernet, C.R.; Wilcox, R.R. Beyond differences in means: Robust graphical methods to compare two groups in neuroscience. *Eur. J. Neurosci.* **2017**, *46*, 1738–1748. [CrossRef] [PubMed]
34. Ahola, A.J.; Radzeviciene, L.; Zaharenko, L.; Bulum, T.; Skrebinska, S.; Prakapiene, E.; Blaslov, K.; Roso, V.; Rovite, V.; Pirags, V.; et al. Association between symptoms of depression, diabetes complications and vascular risk factors in four European cohorts of individuals with type 1 diabetes—InterDiane Consortium. *Diabetes Res. Clin. Pr.* **2020**, *170*, 108495. [CrossRef] [PubMed]
35. Rovite, V.; Wolff-Sagi, Y.; Zaharenko, L.; Nikitina-Zake, L.; Grens, E.; Klovins, J. Genome Database of the Latvian Population (LGDB): Design, Goals, and Primary Results. *J. Epidemiol.* **2018**, *28*, 353–360. [CrossRef]
36. Sokolovska, J.; Dekante, A.; Baumane, L.; Pahirko, L.; Valeinis, J.; Dislere, K.; Rovite, V.; Pirags, V.; Sjakste, N. Nitric oxide metabolism is impaired by type 1 diabetes and diabetic nephropathy. *Biomed. Rep.* **2020**, *12*, 251–258. [CrossRef]
37. Wadén, J.; Tikkanen, H.; Forsblom, C.; Fagerudd, J.; Pettersson-Fernholm, K.; Lakka, T.; Riska, M.; Groop, P.H. Leisure time physical activity is associated with poor glycemic control in type 1 diabetic women: The FinnDiane study. *Diabetes Care* **2005**, *28*, 777–782. [CrossRef] [PubMed]
38. Williams, K.V.; Erbey, J.R.; Becker, D.; Arslanian, S.; Orchard, T.J. Can clinical factors estimate insulin resistance in type 1 diabetes? *Diabetes* **2000**, *49*, 626–632. [CrossRef] [PubMed]
39. Sviklāne, L.; Olmane, E.; Dzērve, Z.; Kupčs, K.; Pīrāgs, V.; Sokolovska, J. Fatty liver index and hepatic steatosis index for prediction of non-alcoholic fatty liver disease in type 1 diabetes. *J. Gastroenterol. Hepatol.* **2018**, *33*, 270–276. [CrossRef] [PubMed]
40. Bedogni, G.; Bellentani, S.; Miglioli, L.; Masutti, F.; Passalacqua, M.; Castiglione, A.; Tiribelli, C. The Fatty Liver Index: A simple and accurate predictor of hepatic steatosis in the general population. *BMC Gastroenterol.* **2006**, *6*, 33. [CrossRef] [PubMed]
41. Cers, E.; Valeinis, J. *EL: Two-Sample Empirical Likelihood*, R package version, 1.2; R Core Team: Vienna, Austria, 2022.
42. Niehues, T. C-reactive protein and other biomarkers—The sense and non-sense of using inflammation biomarkers for the diagnosis of severe bacterial infection. *LymphoSign J.* **2018**, *5*, 35–47. [CrossRef]
43. Pearson, T.A.; Mensah, G.A.; Alexander, R.W.; Anderson, J.L.; Cannon, R.O., 3rd; Criqui, M.; Fadl, Y.Y.; Fortmann, S.P.; Hong, Y.; Myers, G.L.; et al. Markers of inflammation and cardiovascular disease: Application to clinical and public health practice: A statement for healthcare professionals from the Centers for Disease Control and Prevention and the American Heart Association. *Circulation* **2003**, *107*, 499–511. [CrossRef] [PubMed]
44. Erlanson-Albertsson, C.; Stenkula, K.G. The Importance of Food for Endotoxemia and an Inflammatory Response. *Int. J. Mol. Sci.* **2021**, *22*, 9562. [CrossRef] [PubMed]
45. Meilhac, O.; Tanaka, S.; Couret, D. High-Density Lipoproteins Are Bug Scavengers. *Biomolecules* **2020**, *10*, 598. [CrossRef] [PubMed]
46. Ayyappan, P.; Harms, R.Z.; Buckner, J.H.; Sarvetnick, N.E. Coordinated Induction of Antimicrobial Response Factors in Systemic Lupus Erythematosus. *Front. Immunol.* **2019**, *10*, 658. [CrossRef]

biomedicines

Review

A Closer Look at the Perivascular Unit in the Development of Enlarged Perivascular Spaces in Obesity, Metabolic Syndrome, and Type 2 Diabetes Mellitus

Melvin R. Hayden

Department of Internal Medicine, Endocrinology Diabetes and Metabolism, Diabetes and Cardiovascular Disease Center, University of Missouri School of Medicine, One Hospital Drive, Columbia, MO 65211, USA; mrh29pete@gmail.com; Tel.: +1-573-346-3019

Abstract: The recently described perivascular unit (PVU) resides immediately adjacent to the true capillary neurovascular unit (NVU) in the postcapillary venule and contains the normal-benign perivascular spaces (PVS) and pathological enlarged perivascular spaces (EPVS). The PVS are important in that they have recently been identified to be the construct and the conduit responsible for the delivery of metabolic waste from the interstitial fluid to the ventricular cerebrospinal fluid for disposal into the systemic circulation, termed the glymphatic system. Importantly, the outermost boundary of the PVS is lined by protoplasmic perivascular astrocyte endfeet (pvACef) that communicate with regional neurons. As compared to the well-recognized and described neurovascular unit (NVU) and NVU coupling, the PVU is less well understood and remains an emerging concept. The primary focus of this narrative review is to compare the similarities and differences between these two units and discuss each of their structural and functional relationships and how they relate not only to brain homeostasis but also how they may relate to the development of multiple clinical neurological disease states and specifically how they may relate to obesity, metabolic syndrome, and type 2 diabetes mellitus. Additionally, the concept and importance of a perisynaptic astrocyte coupling to the neuronal synapses with pre- and postsynaptic neurons will also be considered as a perisynaptic unit to provide for the creation of the information transfer in the brain via synaptic transmission and brain homeostasis. Multiple electron microscopic images and illustrations will be utilized in order to help explain these complex units.

Keywords: enlarged perivascular spaces; glymphatic system; metabolic syndrome; obesity; neurovascular unit; perivascular unit; perivascular space; postcapillary venules; precapillary arterioles; type 2 diabetes mellitus

Citation: Hayden, M.R. A Closer Look at the Perivascular Unit in the Development of Enlarged Perivascular Spaces in Obesity, Metabolic Syndrome, and Type 2 Diabetes Mellitus. *Biomedicines* **2024**, *12*, 96. https://doi.org/10.3390/biomedicines12010096

Academic Editors: Ana Dascalu and Dragos Serban

Received: 10 December 2023
Revised: 28 December 2023
Accepted: 28 December 2023
Published: 2 January 2024

1. Introduction

The perivascular unit (PVU) has been recently defined and described by Troili et al. (2020) [1]. They described the PVU as "a key anatomical and functional substrate for the interaction between neuronal, immune, and vascular mechanisms of brain injury, which are shared across different neurological disease" [1]. They defined the PVU in order to emphasize the contributions that are made by both the cellular (structural) and molecular (functional) activities that surround the perforating vessels (pial arteries, arterioles, and precapillary arterioles) and the effluxing vessels (pial postcapillary venules, venules, and veins). This is in addition to their interactions, which determine the function of the normal or benign perivascular spaces (PVS) in health and the pathologic remodeling of dilated or enlarged perivascular spaces (EPVS) that are associated with many neurologic diseases [1]. Much of the focus regarding the PVU, PVS, and dilated EPVS has been on pial arteries, arterioles, and precapillary arterioles. However, this review intends to focus more on the pial postcapillary venules, venules, and veins because the PVU and their PVS serve as the conduit for the recently described glymphatic system (GS) of waste removal that is essential for proper brain homeostasis (Figure 1) [1,2].

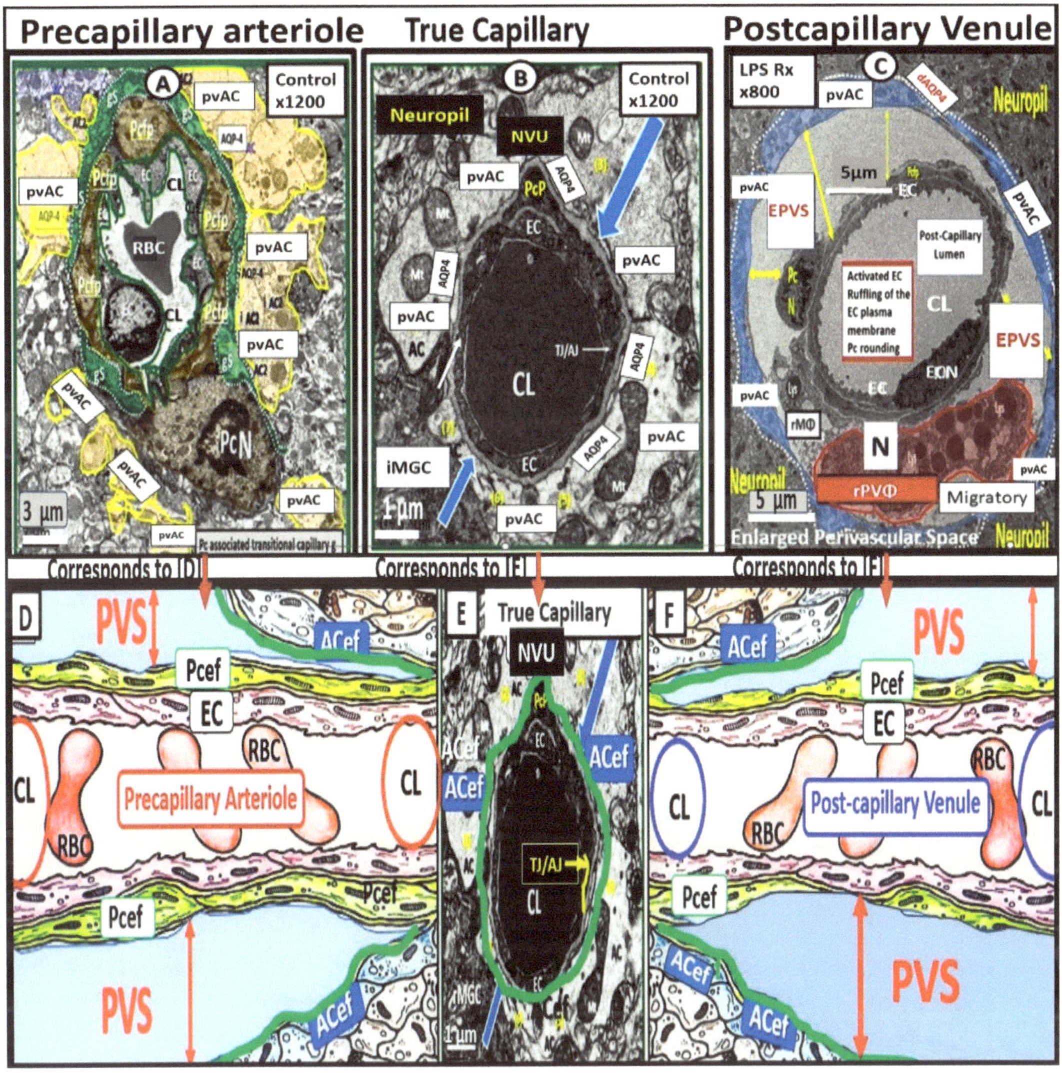

Figure 1. Comparing the ultrastructure transmission electron microscopy (TEM) appearance and illustrations of the true capillary neurovascular unit to the precapillary arteriole and the postcapillary venule with its perivascular spaces (PVS) and enlarged perivascular spaces (EPVS) that exists in the post capillary perivascular unit (PVU). (**A**) A precapillary arteriole, which initially has a larger perivascular space (PVS) at its origin from the subarachnoid space that narrows to a nanometer sized PVS (pseudo-colored green) that ends at the true capillary in the cortical grey and subcortical white matter. (**B**) The primary or true capillary, which has lost the pia mater layer and does not have a PVS. In the true capillary, note how the perivascular astrocyte endfeet (pvACef) tightly abut and are directly adherent to the NVU mural cells, brain endothelial cells (ECs), and pericytes foot processes (PcP–Pcfp) basement membrane(s) (BMs) via the pvACef dystroglycans. The true capillary is the substrate for both the neurovascular unit (NVU) and its blood–brain barrier. Note the interrogating microglia cell (iMGC, white closed arrow). (**C**) A postcapillary venule which is identified via the presence of PVS

and in this image depicts a resident perivascular macrophage (rPVMΦ pseudo-colored red). Importantly, the perivascular spaces serve as the construct and structure responsible for carrying the metabolic waste from the interstitial spaces to the cerebrospinal fluid and are known as the glymphatic system pathway that also forms the perivascular unit (PVU). Scale bars = 3 μm; 0.5 μm; 5 μm, respectively. Panels (**A–C**) are in cross section and downward red arrows indicate corresponding illustrations in (**D–F**). Panels (**D–F**) illustrate longitudinal views of the precapillary arterioles, true capillary, and postcapillary venules, respectively, while the cyan green lines represent the glia limitans of the pvACef. Importantly, note the presence of contractile pericytes and their processes in (**C**,**F**) that allow for neurovascular coupling in postcapillary venules. Note that TEM (**A–C**) correspond to illustrated (**D–F**) respectively and the compressed (**E**) of (**B**) was inserted to illustrate the natural progression of the precapillary arteriole, to the true capillary, and to the postcapillary venule. The modified TEM figures are provided with permission by utilizing the graphic abstract by CC 4.0 [2]. AC = perivascular astrocytes; ACef = astrocyte endfeet—perivascular astrocyte endfeet; AQP4 = aquaporin 4; CL = capillary lumen; dAQP4 = dysfunction aquaporin 4 red lettering; EC = brain endothelial cell; EPVS = enlarged perivascular spaces; gS = glymphatic space—perivascular space; iMGC = interrogating microglial cell; Lys = lysosome; Mt = mitochondria; N = nucleus; NVU = neurovascular unit; Pc = pericyte; PcP = pericyte foot processes; PcN = pericyte nucleus; RBC = red blood cell; rMΦ = reactive macrophage; TJ/AJ = tight and adherens junctions.

PVS are also referred to as Virchow–Robins spaces and are the fluid-filled spaces that ensheathe the pial penetrating vessels, both those entering (arteries, arterioles, precapillary arterioles) and those leaving the brain (postcapillary venules, venules, and veins), as they exit the brain parenchyma back to the subarachnoid space (SAS) allowing the contents of their spaces (interstitial fluid (ISF) and metabolic waste) to eventually enter the cerebrospinal fluid space (CSF) and exit the brain to the systemic circulation (Figure 2) [2].

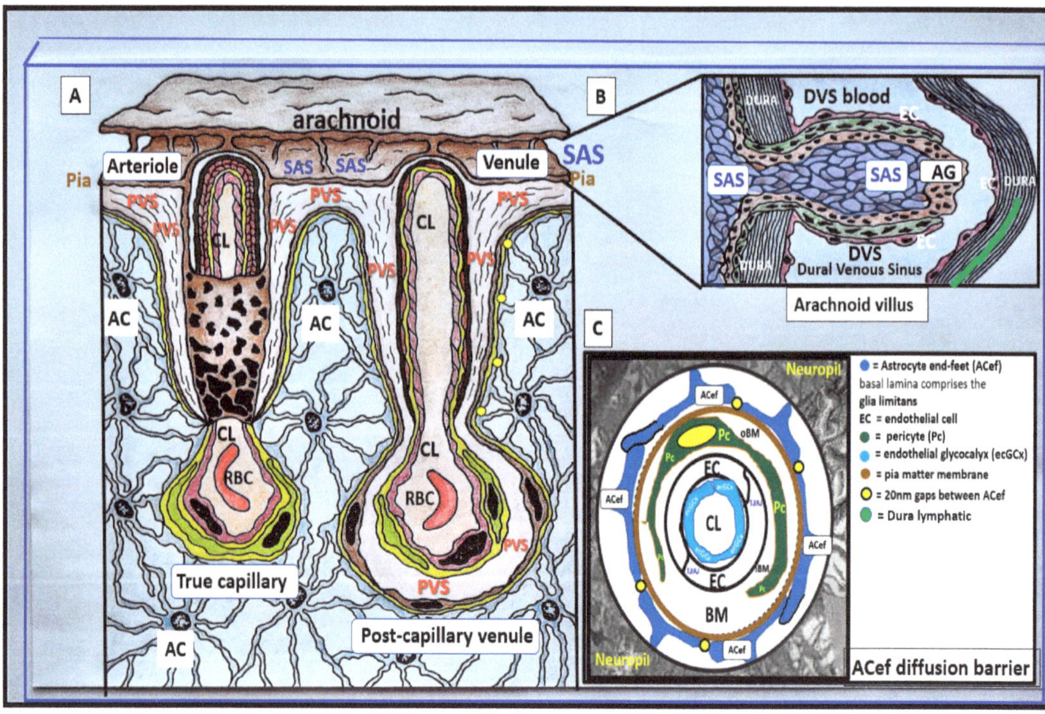

Figure 2. Illustrations of precapillary influx, postcapillary efflux venules, and perivascular spaces (PVS). (**A**) The PVS, which is bounded by the arterial and venous endothelial/pericyte basement membranes

and the pia mater/astrocyte endfeet (ACef) (glia limitans) that is responsible for the influx of cerebrospinal fluid (CSF) to the interstitial fluid (ISF) spaces. Likewise, the PVS of postcapillary venules, venules, and veins are responsible for delivery of the ISF admixed with metabolic waste to the SAS and eventually to the systemic circulation via arachnoid granulations. The outermost pia mater abruptly stops at the true capillary and does not exist in the postcapillary venules and veins. Panel (**B**) illustrates the important role of the arachnoid villus and its granulations for exchange of ISF and metabolic waste with the dural venous sinus blood and dural lymphatics (cyan color) and/or the paranasal sinuses (not shown) to reach the systemic circulation bloodstream. (**C**) NVU and the perivascular astrocyte endfeet (pvACef with blue coloring) barrier with a few 20 nm gaps creating a rate-limiting barrier for water and solute exchange. Notably, the pvACef contain the polarized aquaporin 4 (AQP4) water channels, which are known to be important in fluid and solute exchange in addition to the transfer of metabolic waste to the CSF. Note the key in the image. Image provided by CC 4.0 [2].

Throughout this narrative review, the term "true capillary" is utilized in order to distinguish its ultrastructure characteristics from the precapillary arterioles and postcapillary venules that manifest a PVU, which contains the normal PVS and the pathologic dilated EPVS, which associates with pathologic remodeling and many neurologic diseases as in Figure 1. Importantly, the true capillary of the NVU delivers its peripheral blood and cellular contents into the immediately adjacent postcapillary venule perivascular unit (PVU).

The neurovascular unit (NVU) and NVU coupling are well recognized and accepted structural and functional units in the brain, which are responsible for cerebral autoregulation and are essential for the proper maintenance of regional cerebral blood flow (CBF) and brain homeostasis (Figures 2–4) [2–5].

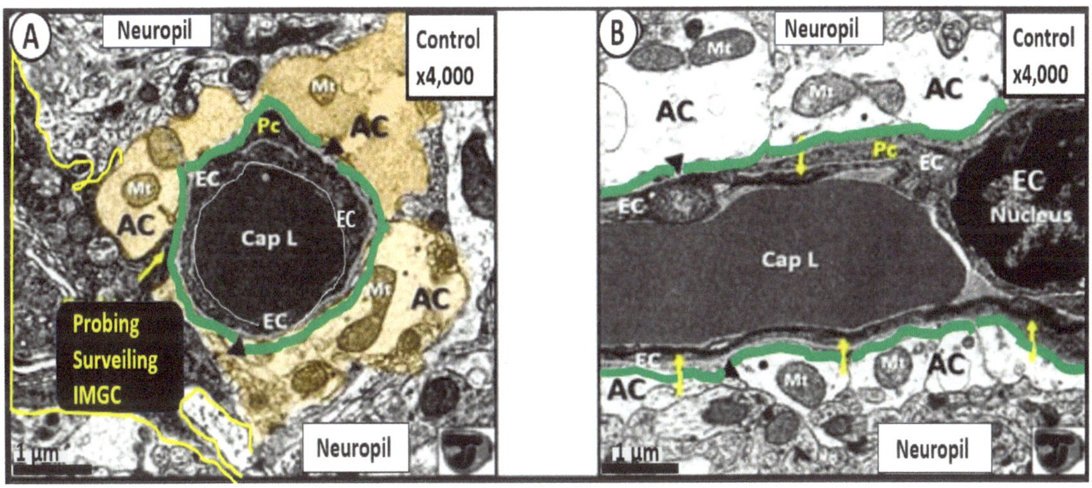

Figure 3. Cross and longitudinal sections of transmission electron microscopic (TEM) images of the true capillary neurovascular unit (NVU). (**A**) TEM cross section of a true capillary NVU. (**B**) True capillary NVU in longitudinal section. The pseudo-colored cyan green line represents and highlights the basement membrane of protoplasmic perivascular endfeet termed the glia limitans perivascularis of the perivascular astrocyte endfeet (**A**,**B**), and the golden pseudo-colored astrocyte endfeet (**A**) represent the critical importance of the perivascular endfeet to the neurovascular unit. Importantly, the pia mater membrane is lost at the level of the true capillary and also the postcapillary venules, venules, and veins of the perivascular unit. Further, note the perivascular astrocyte endfeet (AC) represent the AC clear zone ((**B**), not pseudo-colored as in (**A**)). Modified images provided by CC 4.0 [5]. Magnification x4000; scale bar = 1 µm. AC = perivascular astrocyte endfeet; cap = capillary microvessel; EC = brain endothelial cell; iMGC = interrogating microglia cell; L = lumen; Mt = mitochondria.

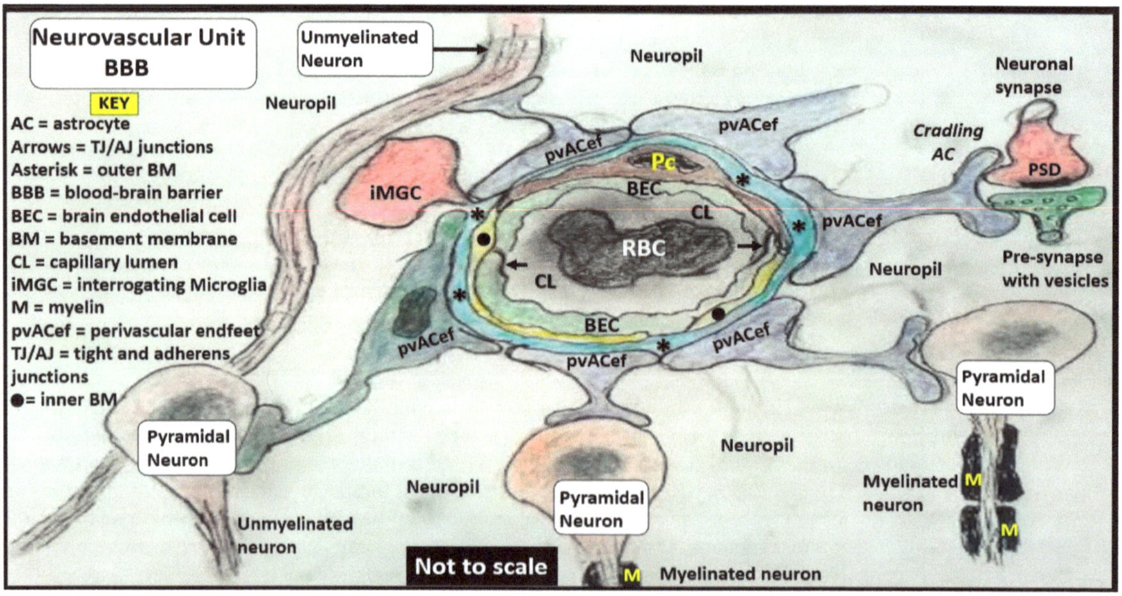

Figure 4. Illustration of the neurovascular unit (NVU) with blood–brain barrier (BBB). Note that not only do the perivascular astrocyte endfeet connect to neurons and dendrites but also to the tripartite synapse (not to scale). Note key for abbreviations within the figure.

Importantly, the PVU that resides immediately adjacent to the true capillary NVU with its blood–brain barrier (BBB) contains the normal PVS that serves as the conduit for the glymphatic system and the pathological EPVS that are more of an emerging concept that is less well understood (Figures 1–6) [1,2,6–10].

Troili et al. only recently introduced this newer term (the PVU) in 2020, and since then, there has been increasing interest in this newly defined unit within the postcapillary venule; however, the PVU still remains an emerging topic of study and deliberation (Figures 3–6) [1]. Importantly, the PVS have been demonstrated to serve as the construct and the structural conduit for the glymphatic system that is responsible for the clearance of metabolic waste from the interstitial spaces to the SAS and CSF [11,12].

In addition to focusing on pial postcapillary venules, venules, and veins, this narrative review also intends to compare the similarities and differences between the NVU and PVU as well as their structural and functional relationships and how they relate not only to brain homeostasis but also how they relate to the development of enlarged perivascular spaces and clinical neurological disease states as they relate to obesity, metabolic syndrome (MetS), and type 2 diabetes mellitus (T2DM) (Figures 7 and 8) [13,14].

While the NVU and PVU have similar and different functions (Figures 7 and 8), they work collaboratively to maintain the proper functioning of the brain's vascular and neural systems to provide proper neurovascular coupling to provide homeostatic CBF to provide nutrients and metabolic waste removal [1]. Additionally, the concept and importance of a perisynaptic astrocyte cradle coupling to the neuronal synapses with pre- and postsynaptic neurons will also be considered as a unit to provide brain homeostasis and the creation of information transfer in the brain via synaptic transmission.

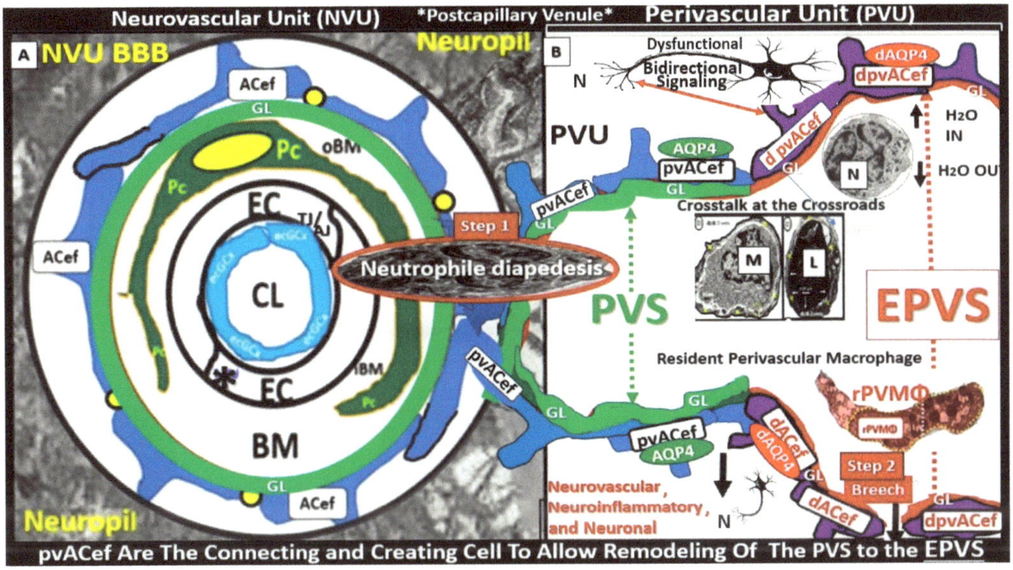

Figure 5. Comparison of the true capillary neurovascular unit (NVU) to the postcapillary venule perivascular unit (PVU). The NVU protoplasmic perivascular astrocyte endfeet (pvACef) (pseudo-colored blue) within the true capillary illustration (**A**) are the connecting and creating cells that allow remodeling of the normal perivascular unit (PVU (**B**)) perivascular spaces (PVS) to transform and remodel into the pathologic enlarged perivascular space (EPVS, which measure 1–3 mm on magnetic resonance imaging). (**A**) Hand-drawn and pseudo-colored control true capillary neurovascular unit (NVU) (representing the transmission electron microscopic (TEM in Figures 1B and 3). Note that when the brain endothelial cells (BECs) become activated and NVU BBB disruption develops, due to BEC activation and dysfunction (BECact/dys) (from multiple causes), there develops an increased permeability of fluids, peripheral cytokines and chemokines, and peripheral immune cells with a neutrophile (N) depicted herein penetrating the tight and adherens junctions (TJ/AJs) paracellular spaces to enter the postcapillary venule along with monocytes (M) and lymphocytes (L) into the postcapillary venule PVS of the PVU (**B**) for step one of the two-step process of neuroinflammation. (**B**) The postcapillary venule that contains the PVU, which includes the normal PVS that has the capability to remodel to the pathological EPVS. Note how the proinflammatory leukocytes enter the PVS along with fluids, solutes, and cytokines/chemokines from an activated, disrupted, and leaky NVU in (**A**). Note how the pvACef (pseudo-colored blue) and its glia limitans (pseudo-colored brown in the control NVU in (**A**) to the cyan color with exaggerated thickness for illustrative purposes in (**B**)) that faces and adheres to the NVU BM extracellular matrix and faces the PVS PVU lumen, since this has detached and separated and allowed the creation of a perivascular space that transforms to an EPVS in (**B**). Also, note how the glia limitans becomes pseudo-colored red, once the EPVS have developed and then become breeched due to activation of matrix metalloproteinases and degradation of the proteins in the glia limitans, which allow neurotoxins and proinflammatory cells to leak into the interstitial spaces of the neuropil and mix with the ISF and result in neuroinflammation (step two) of the two-step process of neuroinflammation [10]. Note that the dysfunctional pvACef AQP4 water channel is associated with the dysfunctional bidirectional signaling between the neuron (N) and the dysfunctional pvACef AQP4 water channel. Image provided by CC 4.0 graphic abstract [9]. AQP4 = aquaporin 4; Asterisk = tight and adherens junction; BBB = blood–brain barrier; BM = both inner (i) and outer (o) basement membrane; dACef and dpvACef = dysfunctional astrocyte endfeet; EC = brain endothelial cell; ecGCx = endothelial glycocalyx; EVPS = enlarged perivascular space; fAQP4 = functional aquaporin 4; GL = glia limitans; H_2O = water; L = lymphocyte; M = monocyte; N = neutrophile and neuron; Pc = pericyte; PVS = perivascular space; PVU = perivascular unit; rPVMΦ = resident perivascular macrophage; TJ/AJ = tight and adherens junctions.

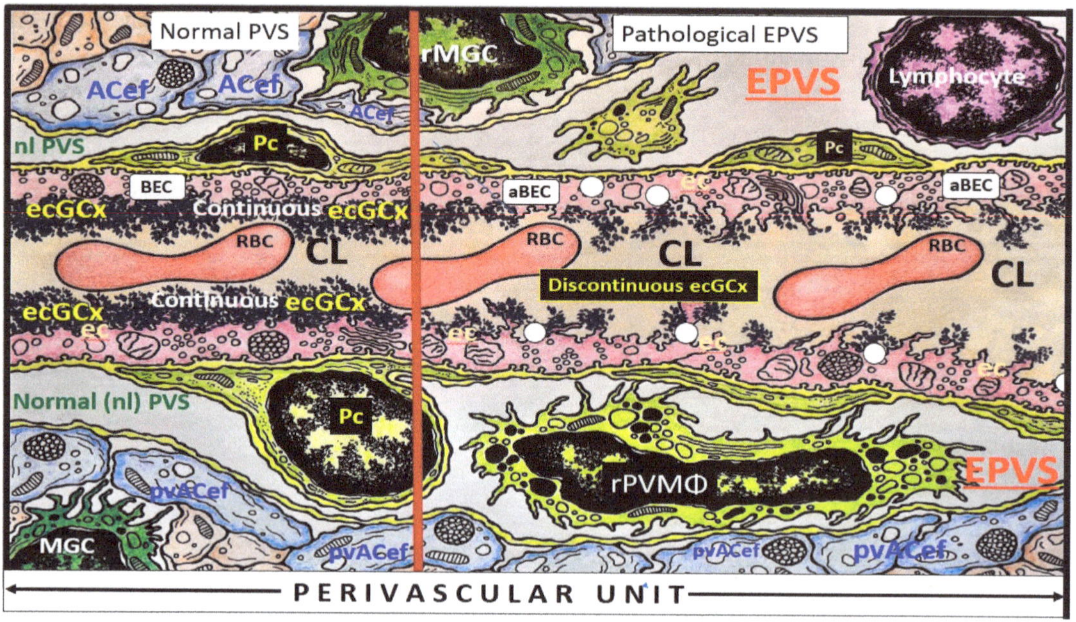

Figure 6. Illustration of the perivascular unit (PVU) with its normal-benign perivascular spaces (PVS) and pathologic enlarged dilated perivascular spaces (EPVS), which lie immediately adjacent to the neurovascular unit true capillary as it transitions from the precapillary arteriole. The vertical red line divides the PVU into the normal PVS on the left-hand side and the pathologic EPVS on the right-hand side of the red line divider. Note the white dots, which represent the attenuation and clumped discontinuous brain endothelial glycocalyx (ecGCx) of the pathologic EPVS in contrast to the continuous ecGCx in the normal PVS. ACef = astrocyte endfeet/pvACef; ecGCx = brain endothelial cell glycocalyx; nl = normal; Pc = pericyte; pvACef = perivascular astrocyte endfeet; RBC = red blood cell; rPVMΦ = resident perivascular macrophage antigen presenting cell; aBEC = activated brain endothelial cell; ACef = astrocyte endfeet/pvACef; ecGCx = brain endothelial cell glycocalyx; nl = normal; Pc = pericyte; pvACef = perivascular astrocyte endfeet; RBC = red blood cell; rMGC = reactive microglial cell; rPVMΦ = resident perivascular macrophage antigen-presenting cell.

Importantly, while the principal focus of this review primarily remains on the pial postcapillary venules and their PVUs with their PVS and EPVS, it is important to note that the PVU and spaces also reside alongside the pia arteries, arterioles, and precapillary arterioles. This arterial system allows for intramural periarterial drainage (IPAD) with removal of metabolic waste to the CSF along the basement membrane(s) (BMs) of the vascular smooth muscle cells and the walls of capillaries and arterioles in a retrograde manner to the SAS [15,16]. Additionally, it is intended to discuss the metabolic waste (MW) clearance from these glymphatic system spaces to the subpial space, SAS, and CSF for delivery into the systemic circulation [1,7,9–11]. Also, newer concepts have been emerging regarding the concept that all brain compartments involved in CSF homeostasis are involved with a functional continuous exchange between them rather than just serving as separate fluid compartment receptacles that are primarily based on hydrostatic pressure [6,17]. Accordingly, aquaporin-4 (AQP4) in the pvACef plays a central role in cerebral fluid homeostasis discussed later in greater detail [6,17].

Neurovascular Unit Neurogliovascular Unit NVU	Perivascular Unit Perigliovascular Unit PVU Adjacent to true capillary NVU
Cellular composition and Function	
Tripartite blood-brain barrier (BBB): **1)** The Brain Endothelial cells (BECs) glycocalyx; **2)** BECs; **3)** extravascular compartment consisting of the BEC, Pc and their joint basement membrane (BM) extracellular matrix and pvACef [Kutuzov et al.]. Interrogating microglial cells (iMGCs) BBB – TJ/AJ restriction of paravascular space movement and a paucity of transcytotic vesicles.	**Single outermost barrier**: a thin BM barrier of the perivascular astrocyte endfeet (pvACef) glia limitans with its luminal boundary by BEC/Pc shared BM Outermost boundary – barrier pvACef thin BM (glia limitans), which serves as the **2nd barrier** for proinflammatory leukocyte penetration into the brain parenchymal interstitial spaces (ISS), which results in neuroinflammation [Horng et al.] Barrier- pvACef BM (glia limitans) **Resident perivascular macrophages (rPVMΦs): APCs**
NVU coupling to neurons via pvACef: **Regional neuronal excitation** promotes NVU vasodilation with increased region cerebral blood flow (CBF) and neuronal quieting leads to NVU normalization.	**PVU coupling** to neurons via pvACef: **Regional neuronal excitation** promotes PVS decreased efflux of the PVS, while neuronal quieting such as sleep or anesthesia leads to increased efflux and metabolic waste removal
NVU BBB microvascular segment-true capillary barrier for solutes	**PVU** postcapillary venule PVS allow for Immune cells to migrate into the parenchymal interstitial spaces and herein lies the importance of the PVS antigen presenting cell the PVMΦ

Figure 7. Similarities and differences between the neurovascular unit (NVU) and the perivascular unit (PVU) cellular composition and function. Note the more vulnerable glia limitans—pvACef basement membrane to being breeched and that the PVU has a unique proinflammatory resident PVMΦ (boxed-in red lettering). APCs = antigen-presenting cells; BM = basement membrane; Pc = pericyte; pvACef = protoplasmic perivascular astrocyte endfeet; TJ/AJ = tight and adherens junctions of the blood–brain barrier [13,14].

Type of "Unit"	Neurovascular Unit (NVU)	Perivascular Unit (PVU)	Perisynaptic Astrocyte Cradle
Function	NVU coupling cerebral blood flow (CBF).	Glymphatic system (GS) metabolic waste (MW) removal.	Synaptic transmission of information
Dysfunction (dys) of pvACef aquaporin 4 water channel (AQP4) or loss of polarization.	Detachment/separation of (pvACef).	Detachment of pvACef with dilation of normal PVS → EPVS and dysfunction.	Impaired synaptic transmission of information
Dys due to loss of pvACef AQP4 and/or dystroglycan.	Detachment/separation of pvACef.	Detachment of pvACef → EPVS and breeching of PVU glia limitans.	Impaired synaptic transmission of information
Impaired function	Impaired regional cerebral blood flow (CBF) with ischemia.	Impaired GS MW removal and neuroinflammation	Impaired synaptic transmission of information
DYSFUNCTIONAL NEUROVASCULAR, NEUROINFLAMMATION, NEURODEGENERATION SYSTEMS, NEUROTOXICITY, AND IMPAIRED COGNITION			

Figure 8. Similarities and differences between function and dysfunction of the neurovascular unit (NVU), perivascular unit (PVU), and the perisynaptic astrocyte endfeet cradle unit (psACef). Note in this table that the psACef and its relation to the tripartite–multipartite synapse with its perisynaptic cradling unit (PSU) may play an important role in the synaptic transmission of information. Further, note that the NVU and PVU function in a collaborative manner to provide homeostatic neurovascular coupling. Kutuzov et al. [13] and Horng et al. [14].

In summary, the NVU consists of the following cellular components: neurons, perivascular astrocytes, microglia, pericytes, blood endothelial cell(s) (BECs), and basement membrane(s) (BMs). The cellular components of the NVU form and share intimate, complex interactions and thus are responsible for the formation of a single functional NVU, while the PVU also contains these same cells with the addition of resident perivascular macrophage(s) (rPVMΦs) as depicted in Figures 1C, 6, 7 and 8 [13,14,18].

2. Obesity, Metabolic Syndrome (MetS), Type 2 Diabetes Mellitus (T2DM), and Global Aging

Obesity, MetS, and T2DM in addition to advanced age are currently global societal problems that are expected to grow over the coming decades [19]. T2DM of this triad and neurodegenerative diseases (including cerebrocardiovascular disease, cerebral small vessel disease (SVD), and thrombotic or hemorrhagic stroke) are also anticipated to develop aging-related EPVS. Currently, the global population is one of the oldest in our history and it is expected to continue to increase over the next 2–3 decades, such that we will observe these four groups to merge and increase in numbers [19].

Obesity with visceral adipose tissue (VAT), MetS, and T2DM predispose to the development of EPVS, impaired synaptic transmission, impaired cognition, and neurodegeneration over time [20,21]; metabolic disorders with MetS are also associated with EPVS [22,23].

T2DM is a heterogeneous, multifactorial, polygenic disease that may be characterized by a defect in insulin secretion (the beta cell secretory defect), insulin action (insulin resistance), and chronic hyperglycemia [24]. T2DM is strongly associated with obesity–visceral adipose tissue (VAT), insulin resistance (IR), and MetS, which is known to have numerous, devastating complications including hypertension, vasculopathy (micro-macrovascular disease) with cerebrocardiovascular disease and stroke, peripheral neuropathy, retinopathy and blindness, neuropathy, non-traumatic amputation, and nephropathy. Importantly, T2DM is also associated with dilated EPVS and impaired glymphatic function of interstitial waste (including multiple neurotoxic substances that include misfolded amyloid beta and tau proteins) [25–29]. Additionally, peripheral and brain IR as well as MetS also play an important role in brain remodeling (Figure 9) [22,23,26,30,31].

Notably, T2DM is known to be associated with significant brain remodeling with cognitive impairment and dysfunction (CID), vascular cognitive impairment and dementia (VCID), and the development of EPVS [25,27–38]. Interestingly, Fulop et al. examined the brain's venous system and its role in the development of enlarged perivascular spaces [39]. They reported that while cerebral microbleeds–microhemorrhages are definitely associated with small arterioles and capillaries, there is increasing evidence that rupture of small veins and venules can also result in microbleeds [39]. Cerebral microbleeds–hemorrhages (CMBs) are associated with the rupture of small intracerebral microvessels and associated with impaired neuronal function and have the potential to contribute to cognitive impairment, older-age psychiatric syndromes, and gait disorders [40]. Interestingly, these multifocal CMBs were readily demonstrated in the 20-week-old female, obese, insulin-resistant, diabetic *db/db* preclinical mouse model of T2DM (Figures 10 and 11) [5].

While we were unable to unravel the possible causes for these cerebral microbleeds ultrastructurally, we were able to demonstrate that the regions associated with microbleeds were definitely also associated with the detachment and separation of pvACef, and there was no ultrastructural evidence of them being associated with congenital vascular malformations [5].

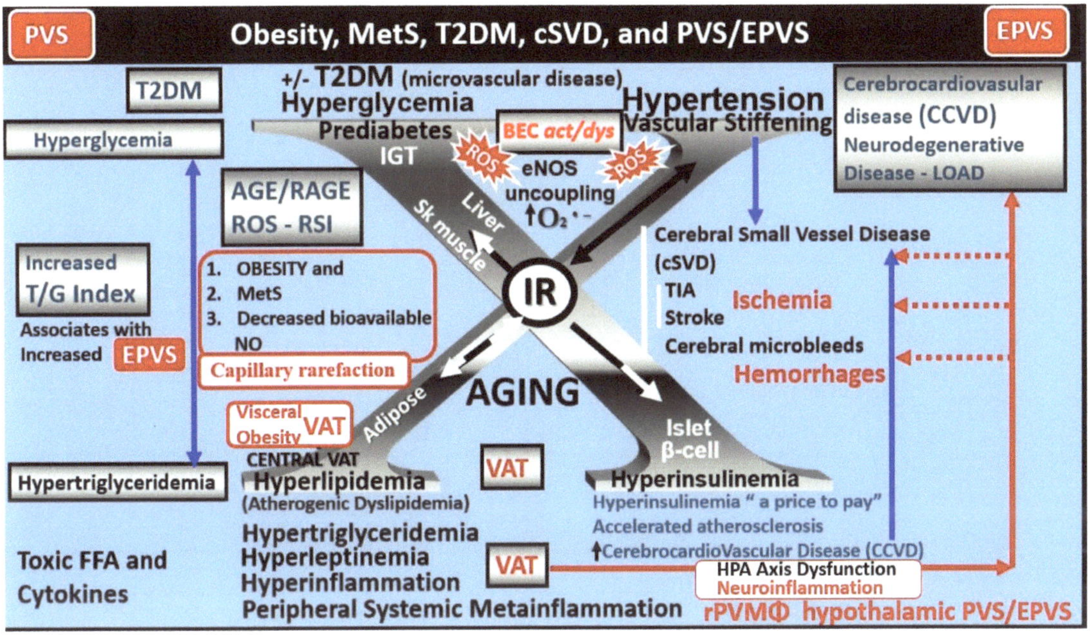

Figure 9. Obesity, metabolic syndrome (MetS), type 2 diabetes mellitus (T2DM), cerebral small vessel disease (SVD), perivascular spaces (PVS), and enlarged perivascular spaces (EPVSs). The visceral adipose tissue (VAT), obesity, and hyperlipidemia (atherogenic dyslipidemia) located in the lower left-hand side of the letter X appears to drive the MetS, peripheral insulin resistance (IR), and brain IR (BIR) that is also located central with the other three arms of the letter X, that includes the associated hyperinsulinemia to compensate for IR (lower right), hypertension and vascular stiffening (upper right), and hyperglycemia (upper left), with impaired glucose tolerance (prediabetes) and with or without manifesting T2DM. Follow the prominent closed red arrows emanating from VAT to cerebrocardiovascular disease (CCVD), SVD, transient ischemic attacks (TIA), stroke, cerebral microbleeds, and hemorrhages. Brain endothelial cell activation and dysfunction (BECact/dys), with its proinflammatory and prooxidative properties, result in endothelial nitric oxide synthesis (eNOS) uncoupling with increased superoxide ($O_2^{\bullet-}$) and decreased nitric oxide (NO) bioavailability in addition to neurovascular unit uncoupling with increased permeability. Importantly, note that obesity, MetS, T2DM, and decreased bioavailable NO interact to result in capillary rarefaction that may allow EPVS to develop, which are biomarkers for cerebral cSVD. While this review does not lend itself to a full discussion of the important role of gut dysbiosis and lipopolysaccharide (LPS) with extracellular vesicle exosomes of LPS producing metainflammation, it was included in this figure. Figure adapted with permission from CC 4.0 [1,8,9,31]. AGE = advanced glycation end-products; RAGE = receptor for AGE; AGE/RAGE = advanced glycation end-products and its receptor interaction; βcell = pancreatic islet insulin-producing beta cell; cSVD = cerebral small vessel disease; FFA = free fatty acids—unsaturated long chain fatty acids; IGT = impaired glucose tolerance; LOAD = late-onset Alzheimer's disease; ROS = reactive oxygen species; RSI = reactive species interactome; Sk = skeletal: TG Index = triglyceride/glucose index; TIA = transient ischemia attack.

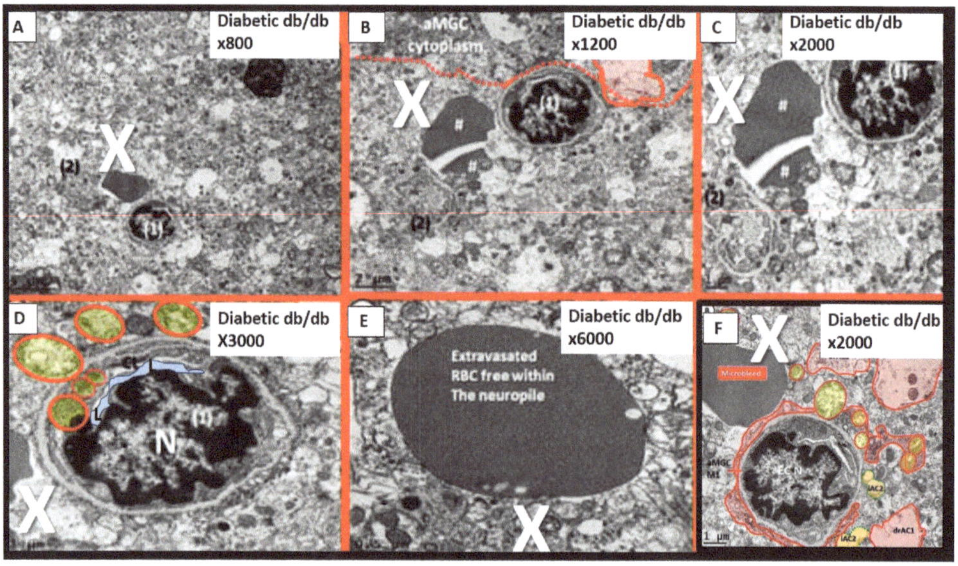

Figure 10. (**A–F**) Cerebral microbleeds–hemorrhages in preclinical female obese metabolic syndrome, and type 2 diabetes mellitus genetic models. Each of these six panels depicts a cerebral microbleed identified by a large white X. Note that in (**E**), the homogeneous material may also represent free plasma within the neuropil. Images provided by CC 4.0 [5]. N = nucleus; RBC = red blood cell; X = microbleed; neuropile = neuropil.

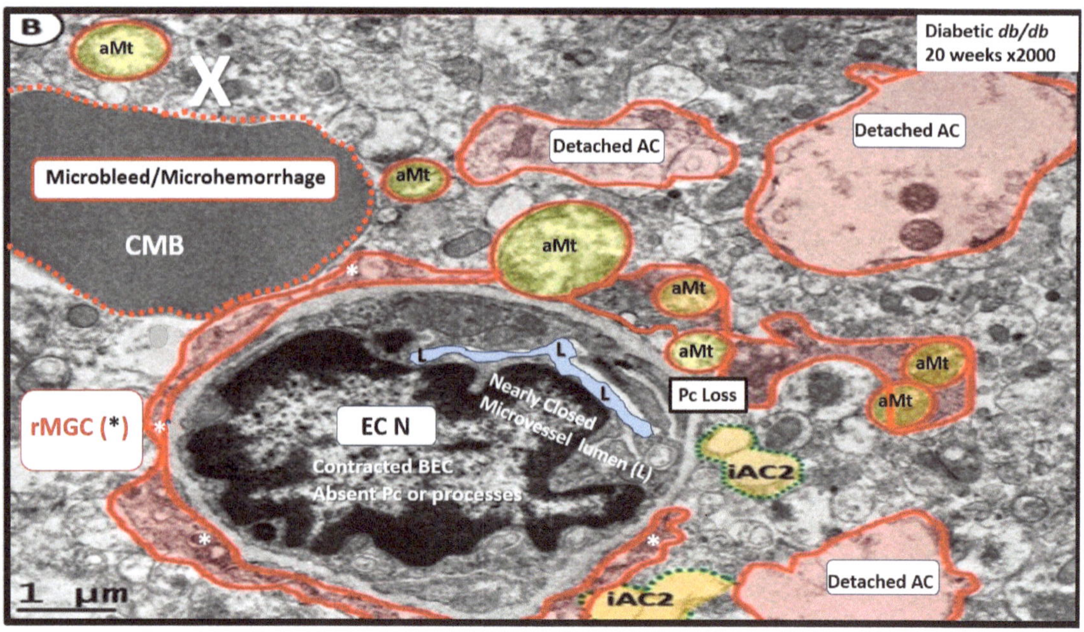

Figure 11. A microbleed (~5 µm) immediately adjacent to a contracted microvessel (~5 µm). Note how the lumen of this microvessel (pseudo-colored light blue) is nearly collapsed and that the brain endothelial cell (BEC) nucleus is contracted with extremely prominent chromatin condensation instead of being heterogeneous, suggesting BEC activation and dysfunction. These similar morphological contracted

BEC remodeling changes and nuclear remodeling changes were observed in the aortic endothelium of activated endothelial cells in Western-diet-fed female mice at 20 weeks of age. Also, note that the reactive microglia (pseudo-colored red) encircle this microvessel which contains multiple aberrant mitochondria (aMt), which provide excessive mitochondria-derived reactive oxygen species that provide BEC injury for the response to injury wound-healing mechanisms at the level of this microvessel to result in BEC activation and dysfunction. Importantly, note reactive astrocyte detachment and separation of reactive perivascular astrocytes. These remodeling changes allow for microvessel disruption and microbleeds. Image provided by CC 4.0 [5]. AC = astrocyte; asterisk = reactive microglia cell; CMB = cerebral microbleed; EC N = brain endothelial nucleus; iAC = intact attached astrocyte; rMGC = reactive microglia cell; Pc = pericyte; X = microbleed-microhemorrhage.

It is important to note that during these studies of the female 20-week-old *db/db* models, we did not examine the venular systems for EPVS or evidence that venular systems may also be involved with both cerebral microbleeds and microinfarcts, since we were not aware of their importance at that time (2018); if you are not looking for a remodeling structure with TEM, you seldom find one [5].

Notably, obesity, MetS, and T2DM have been found to have increased capillary microvascular rarefaction (loss of capillary microvessels) in multiple regions of the brain [2,5,41–44]. Recently, Shulyatnikova and Hayden hypothesized that capillary microvascular rarefaction might be responsible for the development of EPVS [2]. Capillary microvessel loss due to rarefaction would leave an empty space within the confines of the PVUs' PVS that would subsequently fill with ISF, and this could allow for an increase in the percent total fluid volume within the PVS that subsequently results in separation of all surrounding pvACef, leaving an EPVS (Figure 12) [2,8].

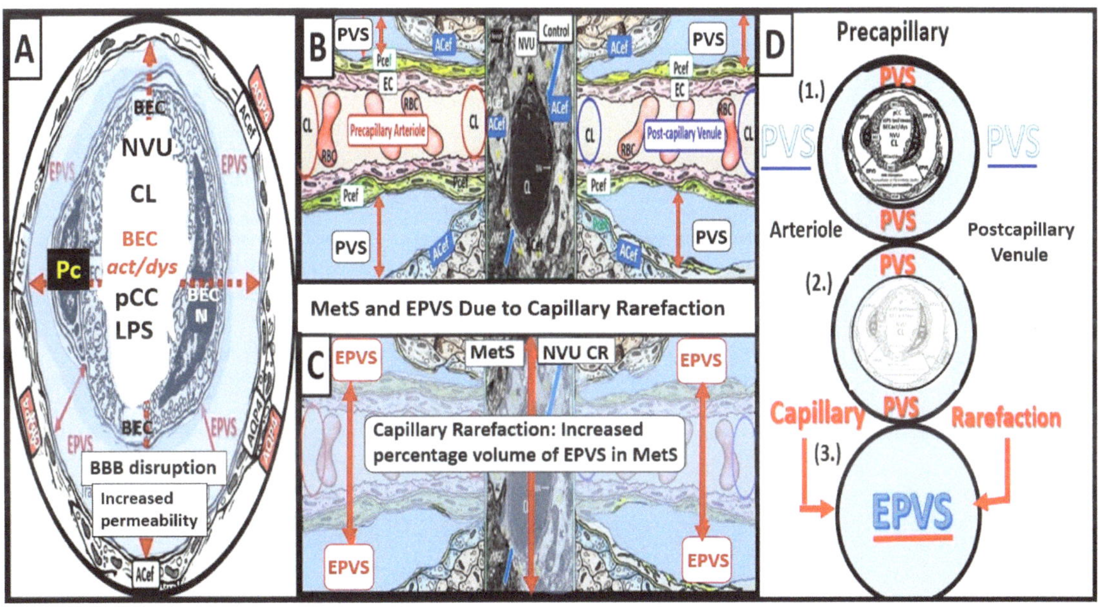

Figure 12. Microvessel rarefaction: Cross and longitudinal sections representative of pre- and postcapillary arterioles and venules with an ensheathing perivascular space (PVS) of the perivascular unit (PVU). (**A**) Cross-section of a capillary microvessel surrounded by PVS (solid double red arrows and light blue color) and its increase in total volume to become an enlarged perivascular space (EPVS) (dashed double red arrows), which represents capillary rarefaction. Note the AQP4 red bars that associate with the perivascular astrocyte endfeet. (**B**) A control longitudinal precapillary arteriole, postcapillary venule, and a neurovascular unit (NVU) capillary that runs through an encompassing PVS

(light blue). (**C**) Capillary microvascular rarefaction (CR) in a longitudinal view; note how the volume of the PVS increases its total percentage volume once the capillary has undergone rarefaction as in obesity, metabolic syndrome, and type 2 diabetes mellitus. (**D**) Progression of a normal precapillary arteriole and postcapillary venule PVS to an EPVS once the capillary has undergone rarefaction, allowing for an increase in its total percentage volume of the PVS (1.–3.). (**B**,**C**) provided with permission by CC 4.0 [9]. ACef = perivascular astrocyte endfeet; AQP4 = aquaporin 4 (red bars); BEC = brain endothelial cells; BECact/dys = brain endothelial cell activation and dysfunction; CL =capillary lumen; EC = endothelial cell; lpsEVexos = lipopolysaccharide extracellular vesicle exosomes; NVU = neurovascular unit; Pcef = pericyte endfeet.

While this capillary microvascular rarefaction still remains a hypothesis that will need to be further tested, it remains an intriguing potential mechanism for the development of EPVS in microvascular disease in the brain.

3. The True Capillary of the NVU Deliverers Its Peripheral Blood and Cellular Contents into the Immediate Adjacent Postcapillary Venule Perivascular Unit

The concept of the NVU and its definition and importance were officially introduced and described in 2001 at the first Stroke Progress Review Group Meeting of the National Institute of Neurological Disorders and Stroke of the National Institutes of Health (NIH) that incorporated neurons and the adjacent vascular cells with the pvACef cells that serve as the connecting cell between the vascular cells and neurons resulting in coupling [4]. The NVU and the NVU coupling are now well accepted and have received great interest in the field of neurobiology. Its cells are comprised of BECs, Pcs, perivascular astrocyte endfeet (pvACef cells), vascular smooth muscle cells in arterioles and arterial microvessels, and interrogating microglia with the pvACef cells being connected to regional neurons to allow for NVU coupling in order to increase oxygen and nutrients to match neuronal excitement demand [4,45–48]. Incidentally, the PVU cells are the same as the NVU except for the presence of the phagocytic and antigen presenting resident perivascular macrophage (rPVMΦ) cells in health and disease [7,49]. While the NVU of the true capillary and PVU of the postcapillary venule appear to be an anatomical continuum with many structural similarities and only a few minor differences, their functions seem to be quite different (Figures 7 and 8).

When one reviews Zlokovic's 2-hit vascular hypothesis for neurodegeneration in Alzheimer's disease [47], it is unquestionable that there is considerable overlap among risk factors for cerebrovascular disorders including cerebral microvascular disease and dysfunction and late onset Alzheimer's disease (LOAD), vascular dementia (VaD), neurodegeneration, and impaired cognition [19]. The two-hit vascular hypothesis for Alzheimer's disease places microvascular disease, and more specifically the NVU BBB, as the first hit of the two-hit hypothesis. Since not only the NVU BBB and the immediately adjacent PVU with its PVS and EPVS are involved, the PVU, PVS, and EPVS could now also be included in this first hit, while the second hit would be the impaired clearance of beta amyloid due to EPVS within the PVU [47].

It has been known for some time that midlife obesity [50], diabetes [51,52], and hypertension [53] are all vascular risk factors that are known to increase the risk for neurodegeneration including LOAD. It is currently well recognized that most cases of LOAD have mixed vascular pathology and SVD [54,55]. Additionally, brain hypoperfusion–hypoxia [56], silent infarcts [57], the presence of one or more infarctions [58], stroke episodes, and transient ischemic or hypoxic attacks all increase the risk of LOAD. Indeed, there may be a continuum of progression in obesity, metabolic syndrome, T2DM to VaD, LOAD, and mixed dementia in addition to the accumulating knowledge that macro-/microvascular disease risk factors might all converge on a common final remodeling disease pathway, involving brain microvascular dysfunction and/or degeneration, as well as amyloid-β and tau pathology [19]. Notably, there has been a trend to soften the once hard-fixed clinical and histopathologic boundary lines drawn between vascular dementia and LOAD. LOAD may be considered to reside under the umbrella of mixed dementias [19].

NVU BBB disruption caused primarily by BECact/dys with activated BECs (Hit-1) allows proinflammatory peripheral cytokines/chemokines (pCC) to enter the PVU and the proinflammatory cells to adhere to the aBEC within the PVU in addition to allowing increased permeability to multiple neurotoxins from the systemic circulation [59]. The neurotoxic molecules are then delivered to the postcapillary venule's PVU with its normal PVS and pathologic remodeled EPVS. These neurotoxic molecules and cells with the ensuing metabolic debris begin to accumulate more and more and may result in PVS obstruction, which results in the PVS becoming dilated, enlarged, and remodeled, which results in EPVS that can be identified by non-invasive magnetic resonance imaging (MRI) studies of the brain that are indicative of impaired waste removal via an impaired glymphatic system (Figure 13).

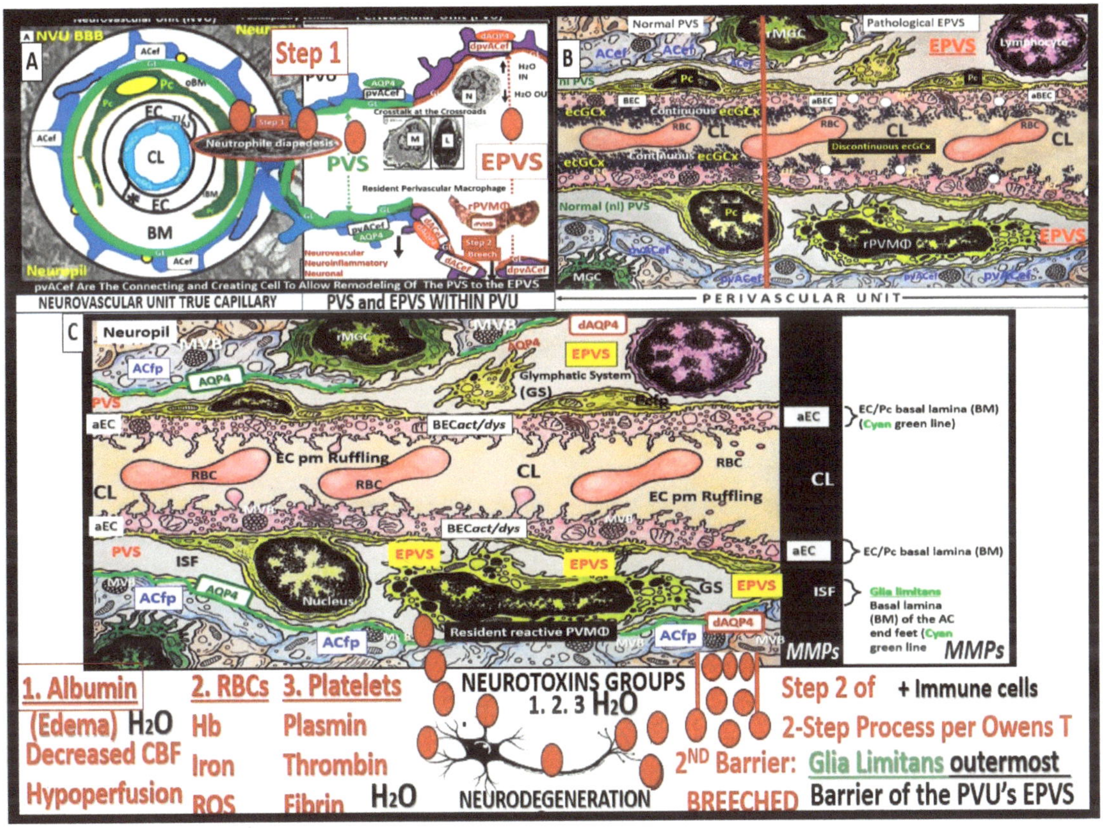

Figure 13. Combining Zlokovic's 2-hit hypothesis with the neurovascular unit (NVU), perivascular unit (PVU), and the development of enlarged perivascular spaces (EPVS). Panel (**A**) illustrates the NVU true capillary and the PVU with its normal perivascular spaces (PVS) and pathologic EPVS. When the true capillary NVU becomes disrupted, it allows neurotoxins including proinflammatory cytokines/chemokines and proinflammatory cells into the perivascular units' PVS. Also, note that (**A**) depicts step-1 of Owens 2-step process of neuroinflammation as well as the first hit of Zlokovic's 2-hit vascular hypothesis [10,47]. Panel (**B**) depicts the PVU with its divisions into the normal PVS and the pathologic EPVS divided by the vertical red line; note the discontinuous endothelial glycocalyx and the presence of the resident perivascular macrophage (rPVMΦ). Panel (**C**) also depicts the PVU; however, its EPVS specifically depicts the breeching of the glia limitans (cyan green) representing

step 2 of neuroinflammation [10] as well as the impaired clearance of amyloid beta and accumulation (hit 2) of Zlokovic's 2-hit hypothesis [47]. Note that the red circles depict various neurotoxin groups that are divided into three groups (1., 2., 3.) that contribute to neuroinflammation, neurodegeneration, and impaired cognition.

These EPVS (1–3 mm by MRI) can now be identified and quantitated via deep learning algorithms that reduce time, effort, and increase specificity in contrast to the earlier manual hand-counting when viewing MRIs [60].

4. The PVU with Its Normal PVS and Pathologic EPVS: Crossroads and Multicellular Crosstalk

Troili et al. were the first to define and describe the perivascular unit (PVU) [1], and this concept is taking better hold now that it has also been learned that the glymphatic system utilizes the postcapillary venule PVS as a conduit for MW clearance [1,11,12,61]. The PVU with its normal PVS and pathologic remodeled EPVS is located in the postcapillary venule and is immediately adjacent to the true capillary (Figures 1, 5, 6 and 11). The PVU allows one to visualize the multicellular crosstalk communication possibilities, which includes BECs, Pcs, resident PVMΦs, the outermost delimiting basement membrane (glia limitans) of the pvACef of this unit, and peripheral cellular leukocytes that are able to penetrate the disrupted NVU BBB (Figures 14 and 15).

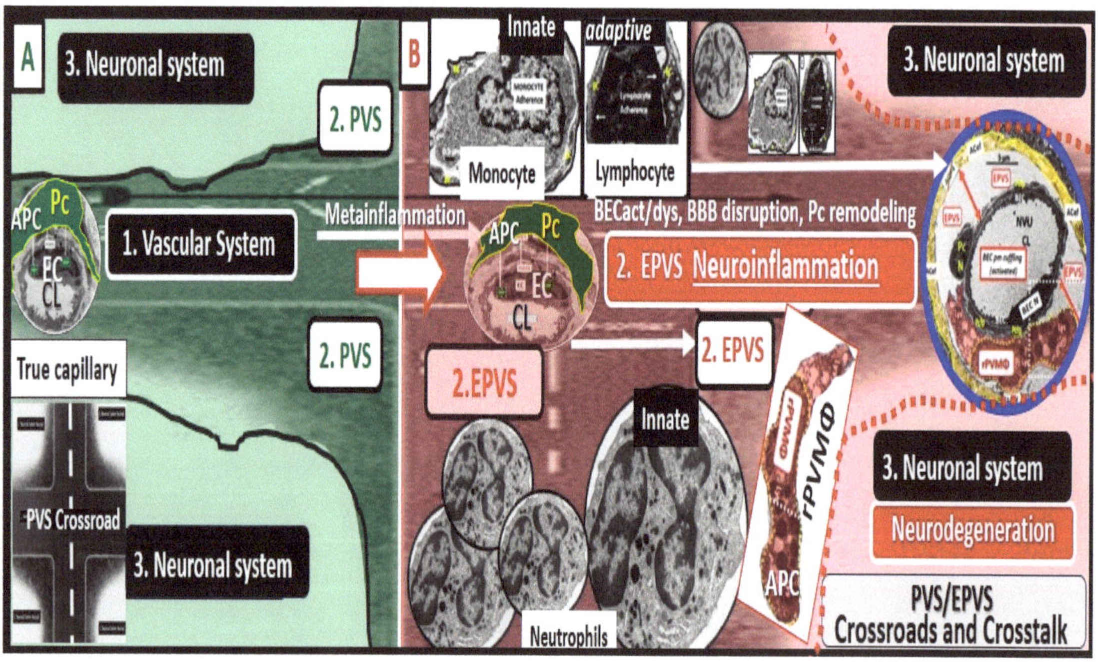

Figure 14. The perivascular unit (PVU) provides a crossroad for multicellular crosstalk communication for vascular, neuroinflammatory, and neuronal systems due to the metainflammation associated with obesity, metabolic syndrome (MetS), and type 2 diabetes mellitus (T2DM). (**A**) Normal-appearing (background pseudo-colored green) perivascular spaces (PVS) indicating their normal function immediately adjacent to the true capillary of the neurovascular unit (NVU). Note the highway intersection icon at the lower left. (**B**) A pathologic enlarged perivascular space (EPVS) (background pseudo-colored red), which suggests pathologic enlargement that resides within the perivascular unit (PVU).

Note that this EPVS contains multiple proinflammatory cells (innate immune neutrophils and monocytes, and adaptive immune lymphocytes) that are induced due to the effects of the visceral obesity-associated peripheral inflammation induced at the NVU with BEC activation and dysfunction with increased permeability to allow the proinflammatory cells and neurotoxic cytokines/chemokines to enter the PVS that results in the pathologic remodeling to create EPVS. PVU and EPVS allow for a crossroad or gathering space to form and create the extensive crosstalk communication between the vascular, neuroinflammatory, and neuronal systems to interact to result in neuroinflammation and neurodegenerative changes with resulting impaired cognition that is associated with obesity, MetS, and T2DM. Note that the red-dashed line represents the glia limitans that is breeched to allow step-two of neuroinflammation and subsequent neuronal remodeling. Image made available by CC 4.0 [7]. BBB = blood–brain barrier; BECact/dys = brain endothelial cell activation and dysfunction; CL = capillary lumen; EC = brain endothelial cells; EPVS = enlarged perivascular space; Pc = pericytes; rPVMΦ = resident reactive perivascular macrophage(s).

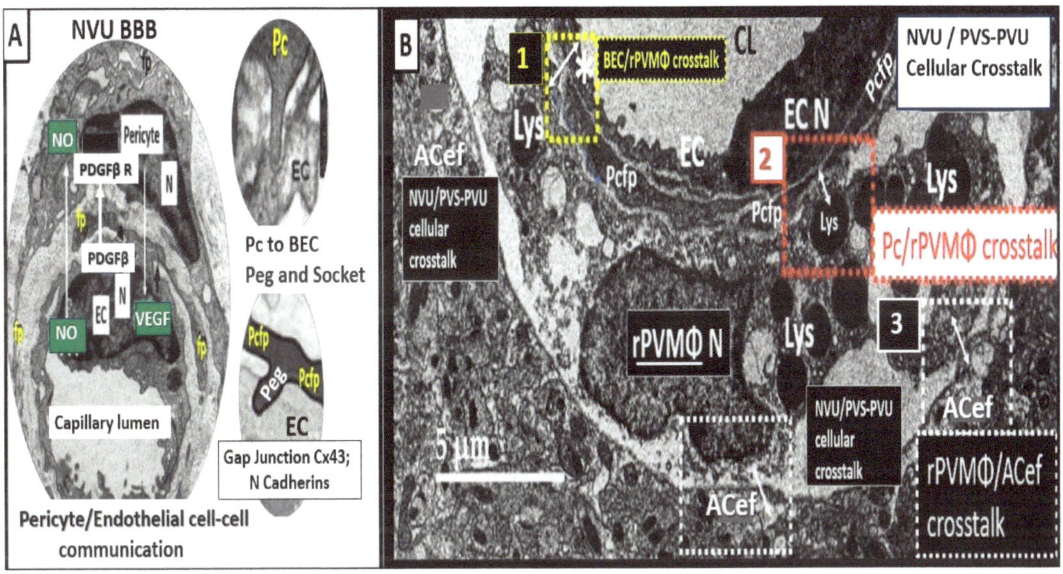

Figure 15. Crosstalk at the crossroads of the perivascular unit (PVU). Multicellular crosstalk between the resident perivascular macrophage (rPVMΦs) and the brain endothelial cells (BECs), pericytes (Pcs), and perivascular astrocyte endfeet (pvACef). (**A**) Normal true capillary of the neurovascular unit (NVU) blood–brain barrier (BBB) interface with peg and socket communicating gap-junctions, connexin 43 (Cx43) and N-cadherin junctions with encircling Pcs, and the cellular signaling utilizing nitric oxide (NO) and platelet-derived growth factor beta (PDGFβ) and vascular endothelial cell growth factor (VEGF). (**B**) Cellular crosstalk between the BECs, Pcs, Pcef, and the pvACef and the rPVMΦs (yellow, red, and white dashed lines, respectively) due to their close proximity within the EPVS. Figure provided by CC 4.0 [7]. ACef = perivascular astrocyte endfeet (pvACef); Asterisk = activated BECs; EC = brain endothelial cell; Lys = lysosomes; N = nucleus; Pcfp = pericyte foot process-endfeet.

The EPVS that result from the multicellular crosstalk and aberrant remodeling are most commonly identified in either the basal ganglia (BG) or the centrum semiovale (CSO) on MRI T2 weighted images. However, EPVS have also been identified in the midbrain (Type III), and they have been identified in the hippocampus and more recently characterized in the subcortical white matter of the anterior superior temporal lobe, cerebellum at the dentate nucleus, and brain stem [62]. With the use of newer more high-intensity 7T MRIs and increased interest in the glymphatic system and EPVS, we will undoubtedly come to learn of even more areas as the venular system is explored and studied more carefully [63].

5. Regional Variation in Enlarged Perivascular Space Locations: Basal Ganglion (BG) and Centrum Semiovale (CSO)

During the remodeling phase in the response to injury wound healing (RTIWH) mechanism [64], multiple remodeling changes occur, including the development of EPVSs, which can be visualized on non-invasive T2-weighted MRI images. These changes are primarily located either in the basal ganglia or the centrum semiovale depending on the clinical disease state of the brain injury and response to injury wound-healing mechanism (Figure 16) [2,64–68].

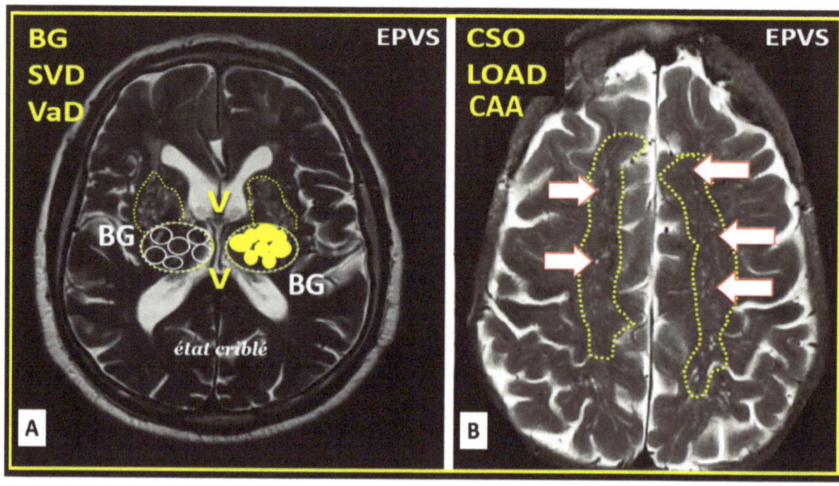

Figure 16. Magnetic resonance imaging (MRI) identification and comparison of basal ganglia (BG) to centrum semiovale (CSO) enlarged perivascular spaces (EPVS). (**A**) Paired EPVSs within the BG that are traced in open circles on the left and solid yellow circles on the right BG. Note the white spaces within the paired dashed lines just above the paired BG structures. MRI image from a 75 year old male post-stroke, recovered with small vessel disease. (**B**) Paired elongated oval structures outlined by yellow dashed lines to enclose multiple white enlarged perivascular spaces. Note the open white arrows outlined in red pointing to prominent EPVSs. MRI image from a 79 year old female with history of transient ischemic attacks. Importantly, note that BG EPVS are strongly associated with cerebral small vessel disease (SVD) in (**A**) and that CSO EPVSs are strongly associated with late-onset Alzheimer's disease and cerebral amyloid angiopathy (CAA) in (**B**). Incidentally, EPVS are more commonly associated with CSO in atherosclerosis, arteriolosclerosis, obesity, metabolic syndrome, and T2DM. Image reproduced with permission from CC 4.0 [64]. Etat cribble = sieve-like state.

EPVS are most commonly viewed in the BG and CSO. This observation may represent different pathologic remodeling, in that BG EPVS are more associated with hypertension-related diseases such as SVD, while CSO EPVS are more commonly associated with misfolded protein diseases such as occurs in LOAD, CAA, and CADSIL [69–71]. Additionally, with the newer use of algorithm-based identification of EPV we may find new regional variation in other clinical diseases as we perform more refined studies in this field of study. Over time, we have found that this aberrant remodeling and development of EPVS may be also be associated with the venular systems [63] within the brain, since we are learning more and more about the glymphatic system at an exponential rate [72].

6. Protoplasmic Perivascular Astrocyte Endfeet and Their Aquaporin 4 (AQP4) Water Channels Play a Crucial Role in the Development of Enlarged Perivascular Spaces

Maintaining the Structural Integrity and Suspension of the Brain

pvACef line the bulk of the brain vasculature at their abluminal surface [9]. Their polarized expression of AQP4 water channels at the plasma membrane of their endfeet are necessary conditions for the functioning glymphatic system pathway of waste removal [11,61]. Actually, Rasmussen, Mestre, and Nedergaard along with Iliff [11,61] were responsible for coining the term glymphatic system that was based on the pvACef and their dependency on the abluminal location of AQP4 at the plasma membrane facing mostly ISF and metabolic waste filled PVS/EPVS at the postcapillary venule without a pial membrane and the CSF filled PVS at the pial-lined precapillary arterioles [11,73]. Loss of AQP4 polarization in the pvACef leads to diminished CSF influx and significantly, a reduced clearance of metabolic waste in the postcapillary venule of the PVU [9,74,75].

The role of AQP4 in the maintenance of CNS homeostasis includes proper CSF circulation and flow, potassium buffering, regulation of extracellular space volume, interstitial fluid resorption, neuroinflammation, osmosensing, calcium signaling, cell migration, and importantly, metabolic waste clearance via the PVS/EPVS-glymphatic system [9,76,77]. When AQP4 is dysfunctional or undergoes loss of polarization as occurs in LOAD [68], SVD, VCI, VCID, and VaD [78–80], there is dysfunction. Also, when AQP4 is dysfunctional and/or lost as in the clinical diseases neuromyelitis optica [81,82] and neuromyelitis optica spectrum disorders [83,84] as a result of autoantibodies against the AQP4 water channel as well as the genetic knockout rodent models, which are associated with EPVS [11,85], there is dysfunction of the brain. Further, Nielsen et al. were able to demonstrate that nanogold particles staining of AQP4 by transmission electron immunochemistry were localized to the plasma membrane of the pvACef where they tightly adhered to the NVUs' pvACef basement membrane [86]. I have only had the opportunity to explore the AQP4 by immunohistochemistry in hepatic cirrhosis individuals and included images from the brains of those individuals with encephalopathy and EPVS (Figure 17) [9,64].

Notably, the glymphatic system is most active during sleep when the clearance of exogenous tracers undergoes a doubling of the clearance rate as compared to wakefulness [87].

EPVS are most commonly viewed in the BG and CSO. This observation may represent different pathologic remodeling, in that BG EPVS are associated with more hypertension-related diseases such as SVD, while CSO EPVS are more commonly associated with misfolded protein diseases such as occurs in LOAD, CAA, and CADSIL [69–71]. Additionally, with the newer use of algorithm-based identification of EPV, we may find new regional variation in other clinical diseases as we perform more refined studies with higher-intensity MRI machines such as 3 and 7T MRIs in this field of study. Over time, we have found that this aberrant remodeling and development of EPVS may be also found to be associated with the venular systems [63] within the brain, since we are learning more and more about the glymphatic system at an exponential rate [72].

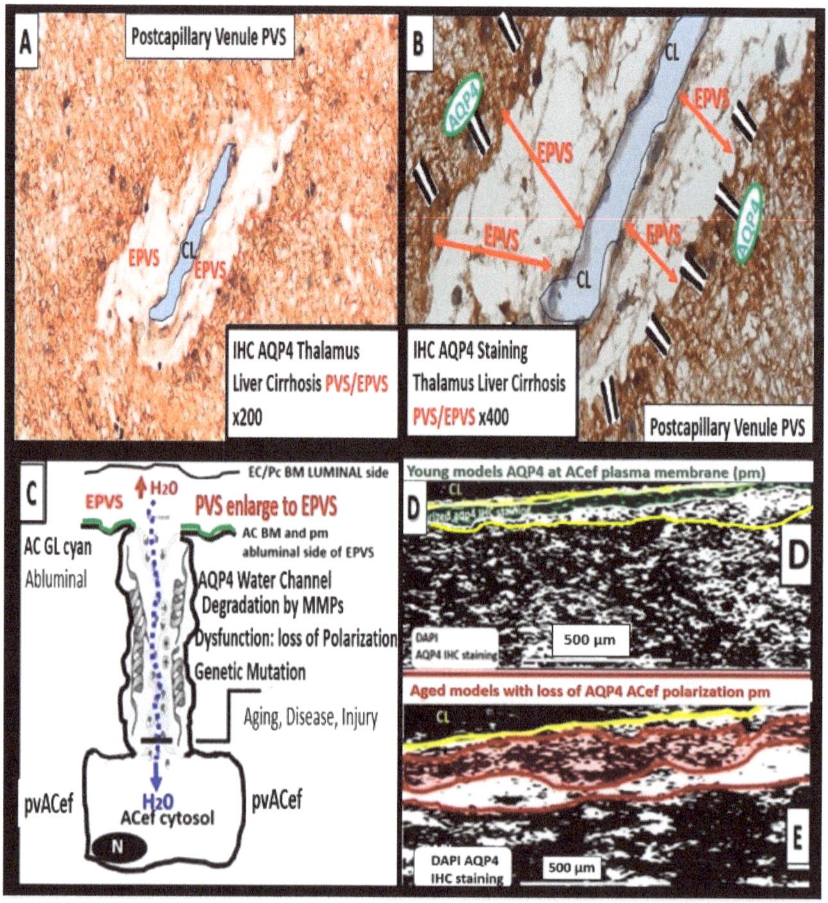

Figure 17. The perivascular astrocyte end feet (pvACef) with its polarized aquaporin 4 (AQP4) water channels delimits the abluminal perivascular unit (PVU) with its perivascular spaces/enlarged perivascular spaces (PVS/EPVS) and perisynaptic astrocyte endfeet (psACef). Panels (**A**,**B**) each demonstrate (via immunohistochemical staining) the presence of AQP4 in the pvACef surrounding a postcapillary venule in an individual with hepatic cirrhosis in the thalamus of the brain. (**C**) Schematic rendering of the AQP4 channel, illustrating water moving into the PVS to contribute the PVS enlargement when AQP4 is dysfunctional and or lost. (**D**) In younger models, AQP4 is tightly polarized to the plasma membrane of the pvACef as compared to (**E**), which depicts a loss of AQP4 polarization in older models. Modified image provided with permission by CC 4.0 [64]. Scale bar = 500 µm. CL = capillary lumen; IHC = immunohistochemistry; N = nucleus.

7. Loss of Polarity of Aquaporin 4 (AQP4) and Dysfunction or Loss of Dystroglycan (DC) Results in Detachment and Separation of pvACef from NVU and the psACef from the Perisynaptic Unit (PSU)

The brain is critically dependent on the homeostatic functions of astrocytes [9]. Astrocytes (AC) are multifunctional and play an essential role in brain development, modeling, and homeostasis [9,88,89]. ACs are among the most abundant cells in the brain and are the master connecting and communicating cells that provide structural and functional support of brain cells at all levels of organization as depicted in Figure 5, in addition to being regarded as the guardians and housekeepers of the brain [9,88]. The large AC cellular presence in the brain and their vast cell–cell communication via gap junctions connexins may be viewed as the brain's functional syncytium [89,90]. Additionally, their role in controlling volume in the brain is of essential importance in maintaining homeostasis, in which their highly polarized AQP4 water channels provide for the maintenance of the CNS CSF, ISF, PVS, glymphatic space for waste removal, and buoyancy [9,89] in addition to controlling its own size and volume due to its highly AQP4 polarized plasma membranes [9,89]. Homeostatic functions of ACs (via pvACef and perisynaptic ACef) include molecular homeostasis, which includes ion homeostasis of (calcium, potassium, chloride, and potassium), regulation of pH, water transport and homeostasis via AQP 4, and neurotransmitter homeostasis (including glutamate, gamma-aminobutyric acid (GABA), adenosine, and monoamines) for further homeostatic functions [9,88,89]. There are known to be three major types of ACs, which include (1) pvACef and occur primarily in cortical grey matter, (2) fibrous ACs, which occur primarily in cortical white matter, and (3) peripheral astroglial processes (PAPs) important for providing cradling the astrocyte leaflets of the perisynaptic unit [88,89,91].

In this review, the AC focus has been on the pvACef that connect the NVU to neurons responsible for neurovascular coupling and maintain regional CBF, pvACef, and the PAPS or perisynaptic astrocyte endfeet (psACef) that cradle synaptic neurons of the perisynaptic cradle unit (PSU) that control synaptic transmission and information transfer between neurons.

7.1. The NVU and PVU pvACef Are Responsible for Neurovascular Coupling and Regional Neuronal Activity-Induced Maintenance of Regional CBF

The pvACef of the NVU and PVUs are known to work in collaborative synergism to maintain the proper functioning of the brain's vascular and neural systems via coupling to provide homeostatic CBF to provide nutrients as well as metabolic waste removal [1]. Indeed, the NVU and PVU consist of cellular networks that control and maintain BBB integrity and tightly regulate CBF, which is known to match energy supply to neuronal demand (neurovascular coupling) [4,92]. Proper polarization and functioning of both the AQP4 water channels (AQP4) and dystroglycan (DG) at the pvACef are absolutely necessary for the normal functioning of both the NVU and PVU. Once pvACef or perisynaptic astrocytes become dysfunctional or lose their polarization of AQP4 or DG the pvACef will undergo detachment and separation from the NVU BEC and pericyte endfeet (Pcef) BMs and separate, which allows for not only NVU but also PVU dysfunction with increased permeability, and become increasingly vulnerable to the response to brain injury wound-healing mechanisms (Figure 15) [7,9,59,64,89]. In a past study of obese, insulin-resistant, 20-week-old female diabetic *db/db* models, I was able to demonstrate pvACef detachment and separation from the NVU BECs and Pcfps outer BM [7]. This detachment and separation are believed to be caused by the dysfunction or degradation of dystroglycan and *a4b6* integrins and are most likely due to increased oxidative stress (ROS) with matrix metalloproteinase(s) (MMPs)-2, 9 activation with partial or complete degradation (Figures 8, 9, 16, 17, 18 and 19A,B) [7,93].

In summary, the NVU-PVU pvACef are responsible for neurovascular coupling, ensuring that regional CBF is tightly coupled to the activity of neurons in specific brain regions (Figure 17A,B). This dynamic regulation is essential for maintaining optimal brain function and responding to the varying metabolic demands of different brain areas.

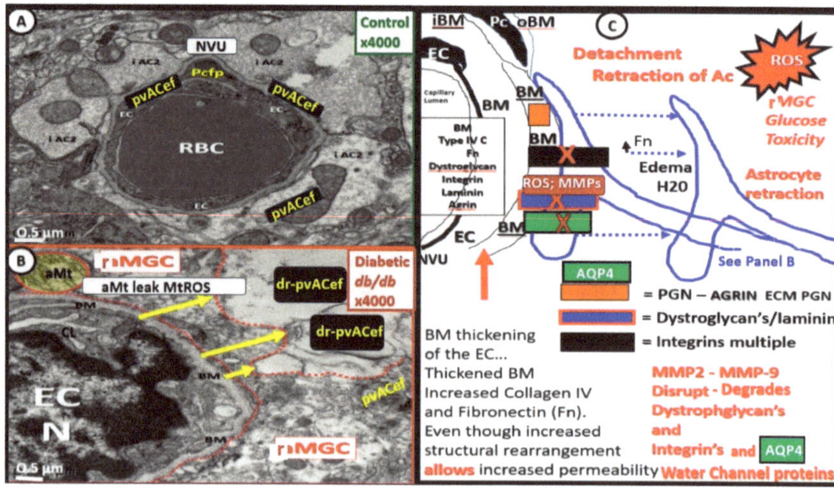

Figure 18. Detachment and retraction of perivascular astrocyte endfeet (pvACef) from the neurovascular unit (NVU) in obese, insulin-resistant, female diabetic *db/db* mice. (**A**) The NVU capillary in control non-diabetic models. Note how the pvACef tightly adhere to the NVU endothelial (EC) and pericyte–pericyte foot processes Pc-Pcfp outer basement membrane (BM). (**B**) Detachment and retraction of reactive pvACef (drpvACef) (yellow arrows) from the NVU. Panel (**C**) illustrates the involved proteins and integrins that are degraded in order for the drpvACef to detach and retract due to increased permeability of the NVU due to NVU disruption. AQP4 = aquaporin 4; AC = astrocyte; BM = basement membrane; EC = brain endothelial cell—endothelium; Fn = fibronectin; MMP2 and MMP 9 = matrix metalloproteinases 2, 9; ROS = reactive oxygen species.

7.2. Cradling Perisynaptic Astrocyte Endfeet (psACef) Are Responsible for Synaptic Transmission of Information

psACef play an essential role in cradling neuronal synapses, synaptic transmission, and plasticity (Figures 19C,D and 20) [89,91,94].

Among the numerous roles of the ACs, they are absolutely essential for controlling the volume of CNS, ISF, and PVS within the PVU as well as the AC itself via its highly polarized plasma membrane AQP4 bidirectional water channel. This is especially true at the pvACef in contact with the vasculature as well as including the PVS and the psACef that are in contact with pre- and postsynapses of the PVU [89,91,95,96]. As the psACef form the cradle of the PSU, they also hide the tripartite synapse from nearby regional injurious stimuli, termed "synaptic isolation", insulation, and shielding (Figure 17C) [91,96]. Further, this pvACef cradling may result in reducing both the "spill-in" of transmitters released during extrasynaptic signaling events and the "spill-out" of transmitters from the synaptic cleft [97]. Importantly, this would contribute to isolating the synapse from the rest of the CNS, i.e., insulating and shielding (Figure 17C).

AQP4 water channel dysfunction, deficiency, and loss of polarization have been demonstrated to show impaired synaptic plasticity and neurotransmission [96,98–100]. Notably, there are at least two clinical neurological diseases that are associated with antibodies against AQP4 water channels: neuromyelitis optica [81–83] and neuromyelitis optica spectrum disorder [83,84] as previously presented in Section 6. psACef may indeed follow the same detachment and separation (Figure 17C,D) as demonstrated for the pvACef detachment and separation (Figures 8, 9 and 17A,B), since both are dependent on the functioning presence of β dystroglycan and polarized AQP4 water channels that will become aberrant under similar clinical diseases and brain-injurious stimuli [9,94].

Up to this point, I have referred to synapses as the tripartite synapses, but it is important to note that there is strong scientific reason to consider the synapses as tetrapartite synapses, since the extracellular matrix plays such an integral role in synaptic transmission; at the very least, one should refer to these synapses as multipartite synapses [89,101].

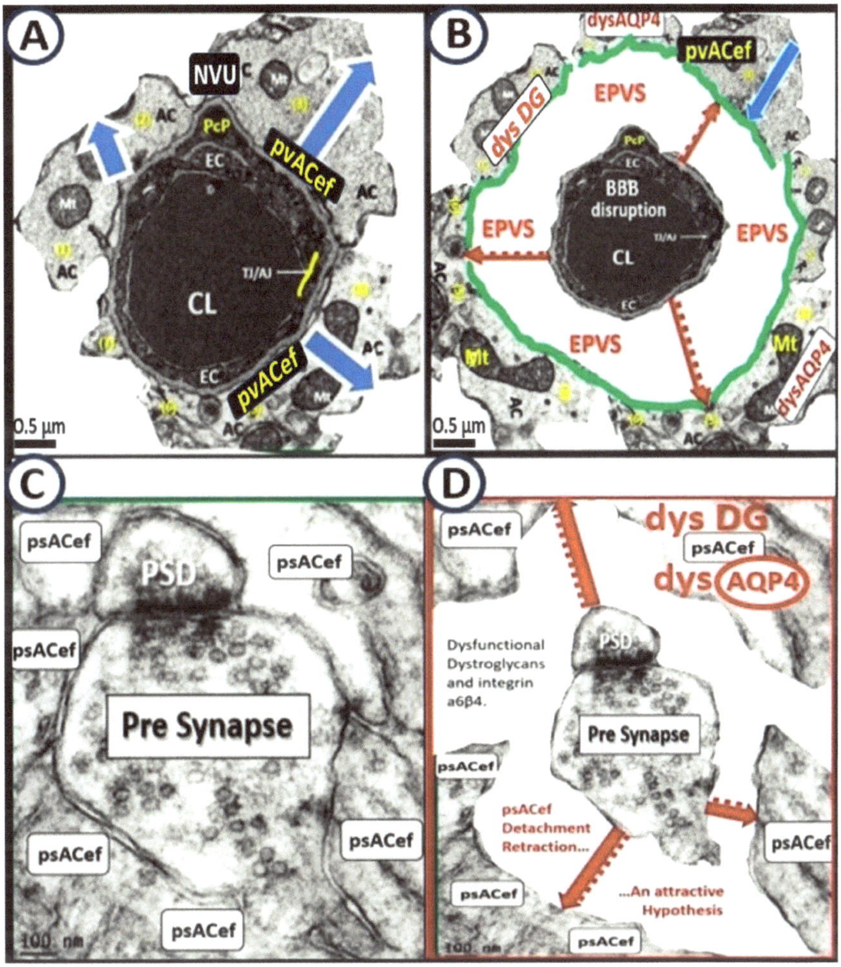

Figure 19. Similarities and comparisons between perivascular astrocyte endfeet (pvACef) and the cradling perisynaptic astrocyte endfeet (psACef) detachment and separation. These similarities implicate damaged or dysfunctional aquaporin 4 (AQP4) either due to activated proteases such as matrix metalloproteinases (MMP-2, 9) or to loss of polarization of AQP4 from the plasma membranes (psACef) resulting in impaired synaptic transmission and impaired cognition or the timing of arrival of incoming information to disturb multiple networks of informational transfer. Panels (**A**,**C**) are from 20-week-old female controls and (**B**,**D**) are from 20-week-old female diabetic *db/db* models with tissues obtained from the frontal cortex, cortical layer III, and depict detachment and separation of pvACef in (**B**) and psACef in (**D**). Note that this detachment and separation creates a perivascular space (PVS) (**B**) and a perisynaptic space (**D**) that may continue to become enlarged with dysfunctional dystroglycan (dysDG) and dysfunctional aquaporin4 (dysAQP4). Note that the cyan green line denoting the glia limitans in (**B**) is not present in (**D**). Images in A and B are reproduced courtesy of CC 4.0 [7]. Scale bars = 0.5 µm in (**A**,**B**) and100 nm (**D**,**E**). BBB = blood–brain barrier; CL = capillary lumen; dys = dysfunctional; DG = dystroglycans; EC = brain endothelial cell; NVU = neurovascular unit; PcP = pericyte process; PSD = post synaptic density; TJ/AJ = tight and adherens junctions.

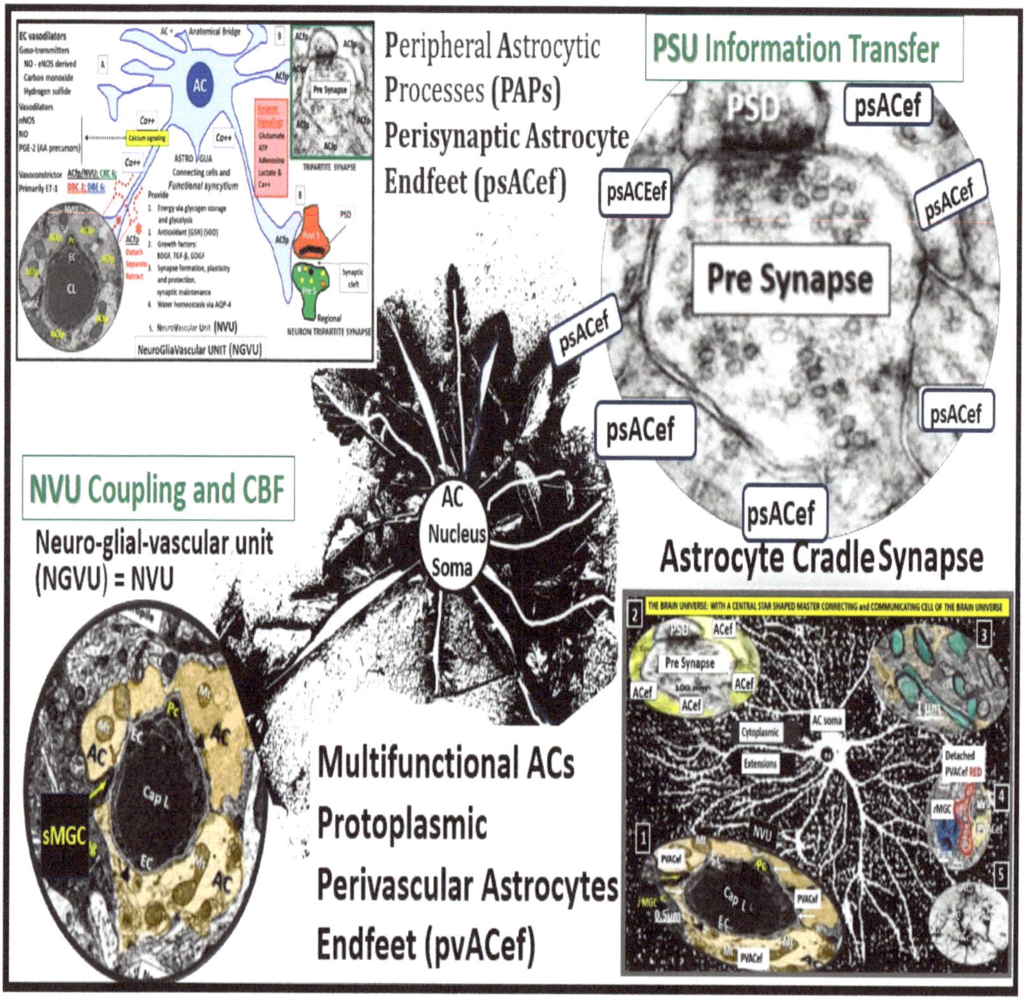

Figure 20. Protoplasmic astrocytes are multifunctional and may be perivascular (perivascular astrocytes endfeet (pvACef)), perisynaptic (perisynaptic astrocyte endfeet (psACef)), or both. This image illustrates a central astrocyte (AC) with its nucleus and soma with multiple protoplasmic extensions. Lower-left image illustrates a transmission electron microscopic (TEM) image of the control true capillary neurovascular unit/neuro-glial-vascular unit (NVU) with a protoplasmic extension connecting with the golden yellow pvACef. Upper-right image illustrates a perisynaptic unit (PSU) with cradling psACef with a protoplasmic AC extension connecting to the psACef that cradle the synapse. Upper-left image illustrates a similar central AC connection to both a NVU and a PSU (image supplied by 4.0 [94]). Lower-right image also illustrates a similar AC connective morphologic image with the central AC connecting to both a NVU and a PSU. Upper-left and lower-left and -right provided by CC 4.0 [94].

In summary, the detachment–retraction of pvACef and psACef can both have widespread effects on the NVU and PSU, respectively, and can potentially lead to disruptions in BBB integrity, altered blood flow regulation, compromised metabolic support to neurons, and disturbances in synaptic function. These changes may contribute to neurological disorders and impair overall brain function via neuroinflammation, impaired glymphatic system efflux associated with EPVS, compromised metabolic support to neurons, impaired synaptic function, plasticity, synaptogenesis, and impaired synaptic transmission with impaired transmission of information [9,89,91,94,97].

8. The Venular Side of the Perivascular Unit (PVU) and Enlarged Perivascular Spaces (EPVS)

All small vessels including the postcapillary venules, venules, and veins are almost completely covered by astrocytic endfeet that provide for a relative barrier, since there are known to be up to 20 nm clefts/gaps between some pvACef [102]. pvACef ensheathment of cortical postcapillary venules, venules, and veins play an essential role in the transcellular trafficking of metabolic solutes, ions, and water as they diffuse bidirectionally into and out of the interstitial neuronal parenchymal spaces at the level of the PVS within the PVUs and subsequently empty into the SAS and CSF to the systemic circulation (Figure 2) [63,102,103]. Also, the polarized AQP4 water channels located at the plasmalemma of pvACef/psACef are responsible for bidirectional water flow and are essential for proper water and volume homeostasis including the ISF within the PVS and the interstitial spaces of the parenchyma, CNS, CSF, PVS within the PVU, and the astrocyte itself [9,89]. These important functions of the pvACef are necessary and essential for the proper removal of waste by the glymphatic system to the SAS and CSF and systemic circulation previously discussed in Section 7.

The response to injury wound-healing mechanisms due to any brain injury is known to result in the loss of polarization of AQP4 water channels and dysfunction or loss of DG (especially if it is ongoing or chronic as in obesity, MetS, and T2DM due to metainflammation, or age-related disease such as LOAD), which results in dysfunctional regulation of water [9,88,89]. Dysfunction of the AQP4 water channel would allow for the development of both interstitial space edema and enlargement of the PVS with resultant EPVS. In turn, the EPVS would result in increased stalling and stasis within the PVS with the accumulation of neurotoxins, proinflammatory leukocytes, and metabolic debris with the accumulation of neurotoxins that could be delivered to the interstitial space of the neuronal parenchyma. Additionally, this would result in damage to neurons and synapses with the development of impaired cognition and neurodegeneration and the eventual development of various dementias depending on the specific instigating injurious mechanisms. For example, venular pathology has been shown to contribute to vascular dysfunction in LOAD, which results in WMHs and microthrombosis, infarcts, and microbleeds as in Figures 8 and 9 to result in regional ischemia [104–106]. Also, single venule blockade in mice models resulted in impaired cerebrovascular structure and function [107–110].

Even though histopathological documentation of venular accumulation of amyloid fragments in both human and animal models has been identified [103,109–116], the role of the venous network and venous dysfunction induced by amyloid accumulation in SVD and CAA will need to be further studied [103].

Notably, Hartmann et al. [107] and Duvernoy et al. [117] have previously described how penetrating venules formed "units" that were surrounded by rings of penetrating arterioles [107]. Exact ratios were not specified in their studies; however, a typical penetrating venule appeared to drain blood supplied by ~4–5 penetrating arterioles (Figure 21) [63,103].

Therefore, should the findings of Hartmann et al. [107] and Duvernoy et al. [117] be supported by others, this would help to explain the importance of venular neurovasculome [118] with obstruction–thrombosis to be studied with greater detail since the venular systems have not yet been studied to the degree that the arteriole neurovasculome systems have been studied to date. Notably, pre-clinical studies have confirmed that venular occlusion causes microinfarcts that are remarkably similar to those found in clinical–pathological human studies [119]. Recently, blockage of a single venule in mice increased microinfarcts and vastly impaired cerebrovascular structure and function [107,108]. Importantly, as we continue to study and explore the glymphatic system, we will also advance our understanding of the venular system.

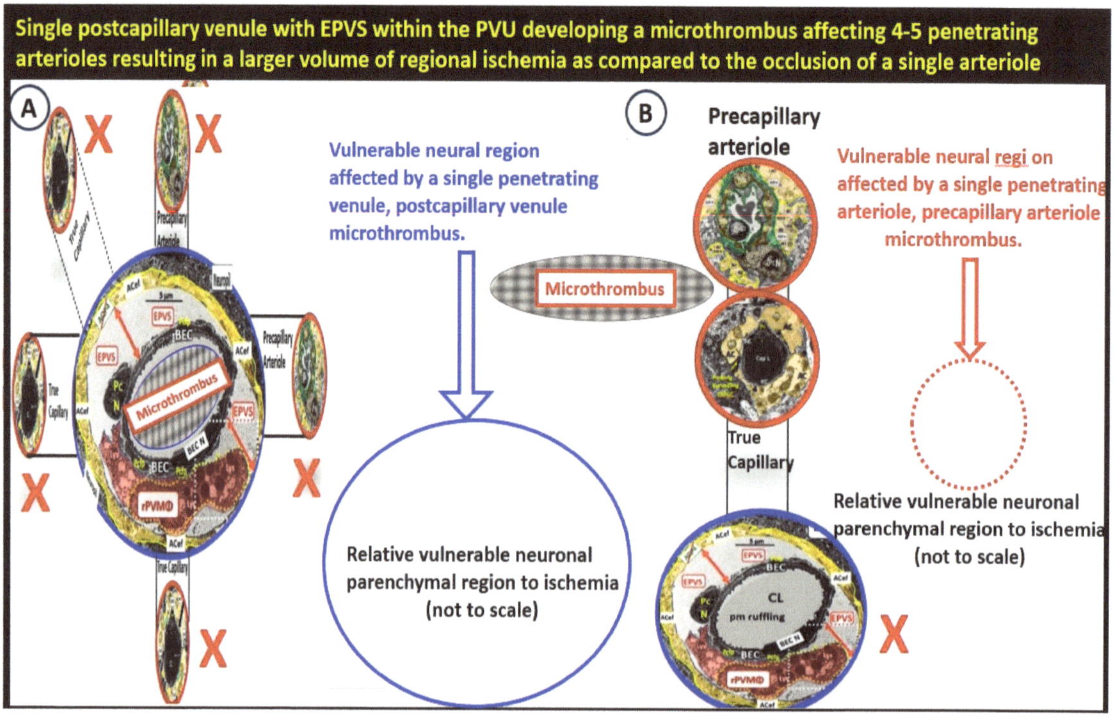

Figure 21. Comparison of a single postcapillary venule microthrombus with enlarged perivascular spaces (EPVS) of the perivascular unit (PVU) compared with a single precapillary arteriole/true capillary microthrombus. (**A**) A single microvessel thrombosis in a postcapillary PVU venule that may affect 4–5 penetrating precapillary arterioles or true capillaries and a much larger vulnerable neuronal region due to decreased cerebral blood flow and ischemia. This also may result in increased microinfarcts, neuronal dysfunction, and even neurodegeneration as compared to the microthrombosis of a single penetrating arteriole as illustrated in (**B**). (**B**) A single precapillary arteriole/true capillary microthrombosis may affect a much smaller vulnerable neuronal parenchymal regional volume as compared to a single venular microthrombosis as in (**A**).

9. Conclusions

PVUs with their normal PVS and pathologic remodeled EPVS play essential roles in the development of neuroinflammation, cerebrovascular disease, and neurodegeneration as summarized in Sections 9.1–9.3.

9.1. In Regards to Neuroinflammation

Postcapillary venular EPVS and alterations in the perivascular unit can trigger inflammatory responses in the brain. Disruption of the NVU BBB integrity and increased permeability can allow immune cells and peripheral cytokines/chemokines to infiltrate the brain PVS (Figures 5 and 11–13), which completes step 1 of Owens' 2-step process and contributes to neuroinflammation [10]. Once the glia limitans is breeched (step 2 of Owens' 2-step process) [10], CNS neuroinflammation can exacerbate neuronal damage and contribute to the progression of various neuropathologic disorders. Further, the PVU and its EPVS can release inflammatory mediators that further amplify the neuroinflammatory response via its resident PVMΦs. Thus, the PVU allows a crossroad for extensive cellular crosstalk (Figures 5 and 11–13) [1].

9.2. In regards to Cerebrovascular Disease

Postcapillary venular EPVS are excessive fluid-filled spaces surrounding blood vessels in the brain and are associated with impaired drainage of interstitial fluid, contributing to changes in cerebral blood flow regulation. Impaired perivascular clearance mechanisms lead to the accumulation of toxic substances (neurotoxins) and contribute to the development of cerebrovascular diseases such as SVD, ischemic and hemorrhagic stroke, and venular thromboses, which are emerging as a risk for the development of microinfarcts, and possibly even accelerated microinfarctions and microhemorrhages or microbleeds that are known to be increased in obesity, MetS, and T2DM (Figures 8 and 9) [5,39]. Rotta et al. found that in human individuals with SVD and LOAD that cerebral microbleeds were common [103]. Further, these microbleeds were associated with the venous vasculature in human individuals with higher resolution 7TMRI [103]. CMBs are a common finding in patients with VSD, CAA, and LOAD [103,120]. Additionally, van Veluw et al. were able to demonstrate that cerebral microbleeds and infarcts commonly co-occurred in CAA utilizing high-resolution 7T MRI that were not identified with either 1.5 or 3T MRI and further identified more microbleeds with 7T [120]. Thus, as we begin to use newer high-resolution MRI, we will begin to advance the knowledge in the field of venular pathology [120]. Also, Morrone et al. proposed that venular amyloid is an important part of both CAA and AD or LOAD pathology [121]. Further, numerous mechanisms related to the pathophysiology of veins, which include EPVS, venous impaired cerebrovascular pulsatility due to the vascular stiffness of aging, and continued studies would allow for further insight into cerebral SVD, CAA, LOAD, and cerebrovascular dysfunction as an associated early remodeling change associated with multiple brain injuries that result in the response to injury wound-healing mechanisms to add further insight [64,120,121] and may lend further insight into the cerebrovascular dysfunction in SVD, CAA, and LOAD [120,121].

9.3. In Regards to Neurodegeneration

Postcapillary venular EPVS have been linked to neurodegenerative diseases like LOAD, PD, and MS, and have recently been described to associate with migraine headaches. The associated compromised perivascular drainage may result in the accumulation of beta-amyloid plaques and other neurotoxic substances, which are known to contribute to the progression of neurodegeneration. Accordingly, the PVU, which consists of blood microvessels, pvACef, Pcs with their endfeet, and resident PVMΦs play a crucial role in maintaining the microenvironment for neuronal health. Dysfunction within the PVU with its pathologic EPVS can lead to impaired nutrient supply and waste removal that are known to contribute to neurodegenerative mechanisms. The above three major headings, neuroinflammation, cerebrovascular disease, and neurodegeneration are equivalent to original description of the PVUs by Troili et al. [1] as a result of being the key anatomical and functional substrate for the interaction between immune, vascular, and neuronal mechanisms associated with multiple types of brain injury.

Advancing science in novel areas typically grows on the shoulders of preexisting scientific discoveries and concepts along with advancing technology, such as high-resolution 7T MRI. Since the glymphatic system is totally dependent on the anatomic structure of the fluid-filled perivascular spaces as its conduit, we can expect the growth of knowledge regarding the PVS, EPVS, and the PVU to grow along with this exciting, novel topic of research regarding the GS; and vice versa, as the knowledge expands in the PVS, EPVS, and the PVU field of studies, we may come to understand more about the GS.

A better understanding of the intricate relationships between EPVS, the PVU, and neurological disorders are crucial for developing targeted therapeutic interventions. Strategies aimed at preserving perivascular space function and the promotion of efficient clearance mechanisms hold great promise in mitigating the impact on neuroinflammation, cerebrovascular disease, and neurodegeneration that are associated with each of the clinical diseases mentioned throughout this review.

In summary, it is hoped that this narrative review will help to increase the understanding regarding the importance of not only the microvascular arterial system including its PVS, but also the venular system. The venular system with its PVU and PVS are essential for waste removal including the effective removal of neurotoxins via the glymphatic system that is dependent on the normal functioning of the postcapillary venular PVS as it is known to be the structural conduit for waste removal within the PVU. The normal postcapillary venular PVS are effective in the removal of neurotoxins including soluble oligomeric amyloid beta, tau, other neurotoxic misfolded proteins, and metabolic waste to slow or prevent neurovascular, neuroinflammatory, and neurodegenerative diseases discussed in this review. EPVS identified by MRI are a biomarker of GS stasis and stalled or obstructed waste removal, and are associated with multiple neurodegenerative diseases. Therefore, we must continue to strive to better understand the microvascular venular systems in addition to the arterial systems. Importantly, EPVS can now be identified and quantitated via algorithms that have been created for deep machine learning, which reduce time and effort, and increase specificity in contrast to earlier visual quantifications.

Funding: This research received no external funding.

Institutional Review Board Statement: The tissues provided for the representative electron microscopic images utilized in this manuscript were all approved in advance by the University of Missouri Institutional Animal Care and Use Committee (No. 190). The animals were cared for in accordance with National Institutes of Health guidelines and by the Institutional Animal Care and Use Committees at the Harry S. Truman Memorial Veterans Hospital and the University of Missouri, Columbia, MO, USA, which conformed to the Guide for the Care and Use of Laboratory Animals published by the National Institutes of Health (NIH).

Informed Consent Statement: Not applicable.

Data Availability Statement: The data and materials can be provided upon reasonable request.

Acknowledgments: The author would like to acknowledge Tatyana Shulyatnikova for the contribution of many artistic illustrations and her editing of this manuscript. The author would also like to acknowledge DeAna Grant, Research Specialist of the Electron Microscopy Core Facility at the Roy Blunt NextGen Precision Health Research Center, University of Missouri, Columbia, Missouri. The author also acknowledges the kind support of the William A. Banks Lab at the VA Medical Center, Seattle, Washington.

Conflicts of Interest: The author declares no conflict of interest.

Abbreviations

AC: astrocyte: ACef, astrocyte endfeet; AQP4, aquaporin-4; ATIII, BBB, blood–brain barrier; BEC(s), brain endothelial cell(s); BECact/dys, brain endothelial cell activation/dysfunction; BG, basal ganglia; BM(s), basement membranes; CAA, cerebral amyloid angiopathy; CBF, cerebral blood flow; CID, cognitive impairment and dysfunction; CL, capillary lumen; CMB(s), cerebral microbleed(s); CSF, cerebrospinal fluid; CSO, central semiovale; EPVS, enlarged perivascular spaces; DG, dystroglycan; EC, brain endothelial cell; EPVS, enlarged perivascular spaces; GS, glymphatic space; IPAD, intramural periarterial drainage; ISF, interstitial fluid; ISS, interstitial space; LAN, lanthanum nitrate; LOAD, late-onset Alzheimer's disease; LPS, lipopolysaccharide; MetS, metabolic syndrome; MGCs, microglia cells; MMP-2,-9, matrix metalloproteinase-2,-9; MRI, magnetic resonance imaging; MS, multiple sclerosis; PD, Parkinson's disease; NVU, neurovascular unit-neuro-glia-vascular unit; Pc, pericyte; Pcfp, pericyte foot process; pvACef; perivascular astrocyte endfeet; psACef, perisynaptic astrocyte endfeet; PVS, perivascular spaces; PVS/EPVS, perivascular space/enlarged perivascular space; rPVMΦ, resident perivascular macrophages; SAS, subarachnoid space; rPVMΦ, reactive perivascular macrophage; SVD, cerebral small vessel disease; T, Tesla; T2DM, type 2 diabetes mellitus; TEM, transmission electron microscopy; TI/AJs, tight and adherens junctions; VAD, vascular dementia; VAT, visceral adipose tissue; VCID, vascular contributions to cognitive impairment and dementia; WMH, white matter hyperintensities.

References

1. Troili, F.; Cipollini, V.; Moci, M.; Morena, E.; Palotai, M.; Rinaldi, V.; Romano, C.; Ristori, G.; Giubilei, F.; Salvetti, M.; et al. Perivascular Unit: This Must Be the Place. The Anatomical Crossroad between the Immune, Vascular and Nervous System. *Front. Neuroanat.* **2020**, *14*, 17. [CrossRef]
2. Shulyatnikova, T.; Hayden, M.R. Why Are Perivascular Spaces Important? *Medicina* **2023**, *59*, 917. [CrossRef]
3. Presa, J.L.; Saravia, F.; Bagi, Z.; Filosa, J.A. Vasculo-Neuronal Coupling and Neurovascular Coupling at the Neurovascular Unit: Impact of Hypertension. *Front. Physiol.* **2020**, *11*, 584135. [CrossRef]
4. Iadecola, C. The Neurovascular Unit Coming of Age: A Journey through Neurovascular Coupling in Health and Disease. *Neuron* **2017**, *96*, 17–42. [CrossRef]
5. Hayden, M.R.; Grant, D.G.; Aroor, A.R.; DeMarco, V.G. Ultrastructural Remodeling of the Neurovascular Unit in the Female Diabetic db/db Model—Part I: Astrocyte. *Neuroglia* **2018**, *1*, 220–244. [CrossRef]
6. Szczygielski, J.; Kopańska, M.; Wysocka, A.; Oertel, J. Cerebral Microcirculation, Perivascular Unit, and Glymphatic System: Role of Aquaporin-4 as the Gatekeeper for Water Homeostasis. *Front. Neurol.* **2021**, *12*, 767470. [CrossRef]
7. Hayden, M.R. Pericytes and Resident Perivascular Macrophages Play a Key Role in the Development of Enlarged Perivascular Spaces in Obesity, Metabolic Syndrome and Type 2 Diabetes Mellitus. *Preprints* **2023**, 2023070461. [CrossRef]
8. Hayden, M.R. The Brain Endothelial Cell Glycocalyx Plays a Crucial Role in the Development of Enlarged Perivascular Spaces in Obesity, Metabolic Syndrome, and Type 2 Diabetes Mellitus. *Life* **2023**, *13*, 1955. [CrossRef]
9. Hayden, M.R. Protoplasmic Perivascular Astrocytes Play a Crucial Role in the Development of Enlarged Perivascular Spaces in Obesity, Metabolic Syndrome, and Type 2 Diabetes Mellitus. *Neuroglia* **2023**, *4*, 307–328. [CrossRef]
10. Owens, T.; Bechmann, I.; Engelhardt, B. Perivascular Spaces and the Two Steps to Neuroinflammation. *J. Neuropathol. Exp. Neurol.* **2008**, *67*, 1113–1121. [CrossRef]
11. Iliff, J.J.; Wang, M.; Liao, Y.; Plogg, B.A.; Peng, W.; Gundersen, G.A.; Benveniste, H.; Vates, G.E.; Deane, R.; Goldman, S.A.; et al. A Paravascular Pathway Facilitates CSF Flow Through the Brain Parenchyma and the Clearance of Interstitial Solutes, Including Amyloid β. *Sci. Transl. Med.* **2012**, *4*, 147ra111. [CrossRef]
12. Mestre, H.; Kostrikov, S.; Mehta, R.I.; Nedergaard, M. Perivascular spaces, glymphatic dysfunction, and small vessel disease. *Clin. Sci.* **2017**, *131*, 2257–2274. [CrossRef]
13. Kutuzov, N.; Flyvbjerg, H.; Lauritzen, M. Contributions of the glycocalyx, endothelium, and extravascular compartment to the blood–brain barrier. *Proc. Natl. Acad. Sci. USA* **2018**, *115*, E9429–E9438. [CrossRef]
14. Horng, S.; Therattil, A.; Moyon, S.; Gordon, A.; Kim, K.; Argaw, A.T.; Hara, Y.; Mariani, J.N.; Sawai, S.; Flodby, P.; et al. Astrocytic tight junctions control inflammatory CNS lesion pathogenesis. *J. Clin. Investig.* **2017**, *127*, 3136–3151. [CrossRef]
15. Diem, A.K.; Carare, R.O.; Weller, R.O.; Bressloff, N.W. A control mechanism for intra-mural peri-arterial drainage via astrocytes: How neuronal activity could improve waste clearance from the brain. *PLoS ONE* **2018**, *13*, e0205276. [CrossRef]
16. Carare, R.O.; Aldea, R.; Agarwal, N.; Bacskai, B.J.; Bechman, I.; Boche, D.; Bu, G.; Bulters, D.; Clemens, A.; Counts, S.E.; et al. Clearance of interstitial fluid (ISF) and CSF (CLIC) group—Part of Vascular Professional Interest Area (PIA): Cerebrovascular disease and the failure of elimination of Amyloid-β from the brain and retina with age and Alzheimer's disease-Opportunities for Therapy. *Alzheimer's Dement. Diagn. Assess. Dis. Monit.* **2020**, *12*, e12053. [CrossRef]
17. Brinker, T.; Stopa, E.; Morrison, J.; Klinge, P. A new look at cerebrospinal fluid circulation. *Fluids Barriers CNS* **2014**, *11*, 10. [CrossRef]
18. Kaplan, L.; Chow, B.W.; Gu, C. Neuronal regulation of the blood–brain barrier and neurovascular coupling. *Nat. Rev. Neurosci.* **2020**, *21*, 416–432. [CrossRef]
19. Hayden, M.R. New Section 2. Type 2 Diabetes Mellitus Increases the Risk of Late-Onset Alzheimer's Disease Increases The Risk of Late-Onset Alzheimer's Disease: Ultrastructural Remodeling of the Neurovascular Unit and Diabetic Gliopathy. *Brain Sci.* **2019**, *9*, 262. [CrossRef]
20. Qi, Y.; Lin, M.; Yang, Y.; Li, Y. Relationship of Visceral Adipose Tissue with Dilated Perivascular Spaces. *Front. Neurosci.* **2021**, *14*, 583557. [CrossRef]
21. Lee, T.H.-Y.; Yau, S.-Y. From Obesity to Hippocampal Neurodegeneration: Pathogenesis and Non-Pharmacological Interventions. *Int. J. Mol. Sci.* **2021**, *22*, 201. [CrossRef] [PubMed]
22. Wu, D.; Yang, X.; Zhong, P.; Ye, X.; Li, C.; Liu, X. Insulin Resistance Is Independently Associated With Enlarged Perivascular Space in the Basal Ganglia in Nondiabetic Healthy Elderly Population. *Am. J. Alzheimer's Dis. Other Dementiasr* **2020**, *35*. [CrossRef] [PubMed]
23. Cai, Y.; Chen, B.; Zeng, X.; Xie, M.; Wei, X.; Cai, J. The Triglyceride Glucose Index Is a Risk Factor for Enlarged Perivascular Space. *Front. Neurol.* **2022**, *13*, 782286. [CrossRef] [PubMed]
24. Hayden, M.R. Islet amyloid, metabolic syndrome, and the natural progressive history of type 2 diabetes mellitus. *JOP* **2002**, *3*, 126–138. [PubMed]
25. Jiang, Q.; Zhang, L.; Ding, G.; Davoodi-Bojd, E.; Li, Q.; Li, L.; Sadry, N.; Nedergaard, M.; Chopp, M.; Zhang, Z. Impairment of the glymphatic system after diabetes. *J. Cereb. Blood Flow Metab.* **2017**, *37*, 1326–1337. [CrossRef]
26. Tiehuis, A.M.; van der Graaf, Y.; Mali, W.P.; Vincken, K.; Muller, M.; Geerlings, M.I.; SMART Study Group. Metabolic Syndrome, Prediabetes, and Brain Abnormalities on MRI in Patients with Manifest Arterial Disease: The SMART-MR Study. *Diabetes Care* **2014**, *37*, 2515–2521. [CrossRef]

27. Munis, B. Association of Type 2 Diabetes Mellitus with Perivascular Spaces and Cerebral Amyloid Angiopathy in Alzheimer's Disease: Insights from MRI Imaging. *Dement. Neurocognitive Disord.* **2023**, *22*, 87–99. [CrossRef]
28. Choi, E.Y.; Park, Y.W.; Lee, M.; Kim, M.; Lee, C.S.; Ahn, S.S.; Kim, J.; Lee, S.-K. Magnetic Resonance Imaging-Visible Perivascular Spaces in the Basal Ganglia Are Associated with the Diabetic Retinopathy Stage and Cognitive Decline in Patients with Type 2 Diabetes. *Front. Aging Neurosci.* **2021**, *13*, 666495. [CrossRef]
29. Liu, J.-C.; Pan, S.-N.; Zhao, H.; Wang, F.; Luo, G.-H.; Lei, H.; Peng, F.; Ren, Q.-P.; Chen, W.; Wu, Y.-F.; et al. Assessment of structural brain changes in patients with type 2 diabetes mellitus using the MRI-based brain atrophy and lesion index. *Neural Regen. Res.* **2022**, *17*, 618–624. [CrossRef]
30. Craft, S. The Role of Metabolic Disorders in Alzheimer Disease and Vascular Dementia: Two roads converged. *Arch. Neurol.* **2009**, *66*, 300–305. [CrossRef]
31. Hayden, M.R. Overview and New Insights into the Metabolic Syndrome: Risk Factors and Emerging Variables in the Development of Type 2 Diabetes and Cerebrocardiovascular Disease. *Medicina* **2023**, *59*, 561. [CrossRef] [PubMed]
32. Hayden, M.; Banks, W.; Shah, G.; Gu, Z.; Sowers, J. Cardiorenal Metabolic Syndrome and Diabetic Cognopathy. *Cardiorenal Med.* **2013**, *3*, 265–282. [CrossRef] [PubMed]
33. Zebarth, J.; Kamal, R.; Perlman, G.; Ouk, M.; Xiong, L.Y.; Yu, D.; Lin, W.Z.; Ramirez, J.; Masellis, M.; Goubran, M.; et al. Perivascular spaces mediate a relationship between diabetes and other cerebral small vessel disease markers in cerebrovascular and neurodegenerative diseases. *J. Stroke Cerebrovasc. Dis.* **2023**, *32*, 107273. [CrossRef] [PubMed]
34. Biessels, G.J.; Staekenborg, S.; Brunner, E.; Brayne, C.; Scheltens, P. Risk of dementia in diabetes mellitus: A systematic review. *Lancet Neurol.* **2006**, *5*, 64–74. [CrossRef] [PubMed]
35. Ott, A.; Stolk, R.P.; van Harskamp, F.; Pols, H.A.; Hofman, A.; Breteler, M.M.B. Diabetes mellitus and risk of dementia: The Rotterdam study. *Neurology* **1999**, *53*, 1037–1942. [CrossRef]
36. Cukierman, T.; Gerstein, H.C.; Williamson, J.D. Cognitive decline and dementia in diabetes—Systematic overview of prospective observational studies. *Diabetologia* **2005**, *48*, 2460–2469. [CrossRef]
37. Diabetes Diabetes Control and Complications Trial/Epidemiology of Diabetes Interventions and Complications Study Research Group; Jacobson, A.M.; Musen, G.; Ryan, C.M.; Silvers, N.; Cleary, P.; Waberski, B.; Burwood, A.; Weinger, K.; Bayless, M.; et al. Long-Term Effect of Diabetes and Its Treatment on Cognitive Function. *N. Engl. J. Med.* **2007**, *356*, 1842–1852. [CrossRef]
38. McCrimmon, R.J.; Ryan, C.M.; Frier, B.M. Diabetes and cognitive dysfunction. *Lancet* **2012**, *379*, 2291–2299. [CrossRef]
39. Fulop, G.A.; Tarantini, S.; Yabluchanskiy, A.; Molnar, A.; Prodan, C.I.; Kiss, T.; Csipo, T.; Lipecz, A.; Balasubramanian, P.; Farkas, E.; et al. Role of age-related alterations of the cerebral venous circulation in the pathogenesis of vascular cognitive impairment. *Am. J. Physiol. Heart Circ. Physiol.* **2019**, *316*, H1124–H1140. [CrossRef]
40. Ungvari, Z.; Tarantini, S.; Kirkpatrick, A.C.; Csiszar, A.; Prodan, C.I. Cerebral microhemorrhages: Mechanisms, consequences, and prevention. *Am. J. Physiol. Heart Circ. Physiol.* **2017**, *312*, H1128–H1143. [CrossRef]
41. Paavonsalo, S.; Hariharan, S.; Lackman, M.H.; Karaman, S. Capillary Rarefaction in Obesity and Metabolic Diseases—Organ-Specificity and Possible Mechanisms. *Cells* **2020**, *9*, 2683. [CrossRef] [PubMed]
42. Chantler, P.D.; Shrader, C.D.; Tabone, L.E.; D'Audiffret, A.C.; Huseynova, K.; Brooks, S.D.; Branyan, K.W.; Grogg, K.A.; Frisbee, J.C. Cerebral Cortical Microvascular Rarefaction in Metabolic Syndrome is Dependent on Insulin Resistance and Loss of Nitric Oxide Bioavailability. *Microcirculation* **2015**, *22*, 435–445. [CrossRef] [PubMed]
43. Estato, V.; Nascimento, A.; Antunes, B.; Gomes, F.; Coelho, L.; Rangel, R.; Garzoni, L.; Daliry, A.; Bousquet, P.; Tibiriçá, E.; et al. Cerebral Microvascular Dysfunction and Inflammation Are Improved by Centrally Acting Antihypertensive Drugs in Metabolic Syndrome. *Metab. Syndr. Relat. Disord.* **2017**, *15*, 26–35. [CrossRef] [PubMed]
44. Tucsek, Z.; Toth, P.; Tarantini, S.; Sosnowska, D.; Gautam, T.; Warrington, J.P.; Giles, C.B.; Wren, J.D.; Koller, A.; Ballabh, P.; et al. Aging exacerbates obesity-induced cerebro-microvascular rarefaction, neurovascular uncoupling, and cognitive decline in mice. *J. Gerontol. A Biol. Sci. Med. Sci.* **2014**, *69*, 1339–1352. [CrossRef] [PubMed]
45. Grotta, J.C. Stroke Progress Review Group: Summary of successes and lack of progress. incorporating neurons and the adjacent vasculature. *Stroke* **2013**, *44* (Suppl. 1), S111–S113. [CrossRef] [PubMed]
46. Lo, E.H.; Rosenberg, G.A. The Neurovascular Unit in Health and Disease: Introduction. *Stroke* **2009**, *40* (Suppl. 3), S2–S3. [CrossRef] [PubMed]
47. Zlokovic, B.V. Neurovascular pathways to neurodegeneration in Alzheimer's disease and other disorders. *Nat. Rev. Neurosci.* **2011**, *12*, 723–738. [CrossRef]
48. Alkayed, N.J.; Cipolla, M.J. The Ever-Evolving Concept of the Neurovascular Unit. *Stroke* **2023**, *54*, 2178–2180. [CrossRef]
49. Faraco, G.; Park, L.; Anrather, J.; Iadecola, C. Brain perivascular macrophages: Characterization and functional roles in health and disease. *J. Mol. Med.* **2017**, *95*, 1143–1152. [CrossRef]
50. Whitmer, R.A.; Gustafson, D.R.; Barrett-Connor, E.; Haan, M.N.; Gunderson, E.P.; Yaffe, K. Central obesity and increased risk of dementia more than three decades later. *Neurology* **2008**, *71*, 1057–1064. [CrossRef]
51. Luchsinger, J.A.; Roberts, R. Relation of diabetes to mild cognitive impairment. *Arch Neurol.* **2007**, *64*, 570–575. [CrossRef] [PubMed]
52. Knopman, D.S.; Roberts, R. Vascular Risk Factors: Imaging and Neuropathologic Correlates. *J. Alzheimer's Dis.* **2010**, *20*, 699–709. [CrossRef] [PubMed]
53. Iadecola, C.; Davisson, R.L. Hypertension and Cerebrovascular Dysfunction. *Cell Metab.* **2008**, *7*, 476–484. [CrossRef] [PubMed]

54. Jellinger, K.A. Prevalence and Impact of Cerebrovascular Lesions in Alzheimer and Lewy Body Diseases. *Neurodegener. Dis.* **2010**, *7*, 112–115. [CrossRef] [PubMed]
55. Marchesi, V.T. Alzheimer's dementia begins as a disease of small blood vessels, damaged by oxidative-induced inflammation and dysregulated amyloid metabolism: Implications for early detection and therapy. *FASEB J.* **2011**, *25*, 5–13. [CrossRef] [PubMed]
56. Ruitenberg, A.; den Heijer, T.; Bakker, S.L.; van Swieten, J.C.; Koudstaal, P.J.; Hofman, A. Cerebral hypoperfusion and clinical onset of dementia: The Rotterdam Study. *Ann. Neurol. Off. J. Am. Neurol. Assoc. Child Neurol. Soc.* **2005**, *57*, 789–794. [CrossRef]
57. Vermeer, S.E.; Prins, N.D.; Heijer, T.D.; Hofman, A.; Koudstaal, P.J.; Breteler, M.M. Silent Brain Infarcts and the Risk of Dementia and Cognitive Decline. *N. Engl. J. Med.* **2003**, *348*, 1215–1222. [CrossRef]
58. Snowdon, D.A.; Greiner, L.H.; Mortimer, J.A.; KP Riley, K.P.; Greiner, P.A.; Markesbery, W.R. Brain infarction and the clinical ex-pression of Alzheimer disease. The Nun Study. *JAMA* **1997**, *277*, 813–817. [CrossRef]
59. Hayden, M.R. Hypothesis: Neuroglia Activation Due to Increased Peripheral and CNS Proinflammatory Cytokines/Chemokines with Neuroinflammation May Result in Long COVID. *Neuroglia* **2021**, *2*, 7–35. [CrossRef]
60. González-Castro, V.; Hernández, M.d.C.V.; Chappell, F.M.; Armitage, P.A.; Makin, S.; Wardlaw, J.M. Reliability of an automatic classifier for brain enlarged perivascular spaces burden and comparison with human performance. *Clin. Sci.* **2017**, *131*, 1465–1481. [CrossRef]
61. Iliff, J.J.; Lee, H.; Yu, M.; Feng, T.; Logan, J.; Nedergaard, M.; Benveniste, H. Brain-wide pathway for waste clearance captured by contrast-enhanced MRI. *J. Clin. Investig.* **2013**, *123*, 1299–1309. [CrossRef] [PubMed]
62. Rudie, J.D.; Rauschecker, A.M.; Nabavizadeh, S.A.; Mohan, S. Neuroimaging of Dilated Perivascular Spaces: From Benign and Pathologic Causes to Mimics. *J. Neuroimaging* **2018**, *28*, 139–149. [CrossRef] [PubMed]
63. Zedde, M.; Grisendi, I.; Assenza, F.; Vandelli, G.; Napoli, M.; Moratti, C.; Lochner, P.; Seiffge, D.J.; Piazza, F.; Valzania, F.; et al. The Venular Side of Cerebral Amyloid Angiopathy: Proof of Concept of a Neglected Issue. *Biomedicines* **2023**, *11*, 2663. [CrossRef] [PubMed]
64. Hayden, M.R. Brain Injury: Response to Injury Wound-Healing Mechanisms and Enlarged Perivascular Spaces in Obesity, Metabolic Syndrome, and Type 2 Diabetes Mellitus. *Medicina* **2023**, *59*, 1337. [CrossRef] [PubMed]
65. Wardlaw, J.M.; Smith, E.E.; Biessels, G.J.; Cordonnier, C.; Fazekas, F.; Frayne, R.; Lindley, R.I.; O'Brien, J.T.; Barkhof, F.; Benavente, O.R.; et al. STandards for ReportIng Vascular changes on nEuroimaging (STRIVE v1). *Lancet Neurol.* **2013**, *12*, 822–838. [CrossRef] [PubMed]
66. Brown, R.; Benveniste, H.; Black, S.E.; Charpak, S.; Dichgans, M.; Joutel, A.; Nedergaard, M.; Smith, K.J.; Zlokovic, B.V.; Wardlaw, J.M. Understanding the role of the perivascular space in cerebral small vessel disease. *Cardiovasc. Res.* **2018**, *114*, 1462–1473. [CrossRef]
67. Bown, C.W.; Carare, R.O.; Schrag, M.S.; Jefferson, A.L. Physiology and Clinical Relevance of Enlarged Perivascular Spaces in the Aging Brain. *Neurology* **2022**, *98*, 107–117. [CrossRef]
68. Doubal, F.N.; MacLullich, A.M.; Ferguson, K.J.; Dennis, M.S.; Wardlaw, J.M. Enlarged Perivascular Spaces on MRI Are a Feature of Cerebral Small Vessel Disease. *Stroke* **2010**, *41*, 450–454. [CrossRef]
69. Paradise, M.; Crawford, J.D.; Lam, B.C.; Wen, W.; Kochan, N.A.; Makkar, S.; Dawes, L.; Trollor, J.; Draper, B.; Brodaty, H.; et al. Association of Dilated Perivascular Spaces With Cognitive Decline and Incident Dementia. *Neurology* **2021**, *96*, e1501–e1511. [CrossRef]
70. Ramirez, J.; Berezuk, C.; McNeely, A.A.; Scott, C.J.M.; Gao, F.; Black, S.E. Visible Virchow-Robin spaces on magnetic resonance imaging of Alzheimer's disease patients and normal elderly from the Sunnybrook Dementia Study. *J. Alzheimer Dis.* **2015**, *43*, 415–424. [CrossRef]
71. Banerjee, G.; Kim, H.J.; Fox, Z.; Jäger, H.R.; Wilson, D.; Charidimou, A.; Na, H.K.; Na, D.L.; Seo, S.W.; Werring, D.J. MRI-visible perivascular space location is associated with Alzheimer's disease independently of amyloid burden. *Brain* **2017**, *140*, 1107–1116. [CrossRef] [PubMed]
72. Bohr, T.; Hjorth, P.G.; Holst, S.C.; Hrabětová, S.; Kiviniemi, V.; Lilius, T.; Lundgaard, I.; Mardal, K.-A.; Martens, E.A.; Mori, Y.; et al. The glymphatic system: Current understanding and modeling. *iScience* **2022**, *25*, 104987. [CrossRef] [PubMed]
73. Rasmussen, M.K.; Mestre, H.; Nedergaard, M. Fluid transport in the brain. *Physiol. Rev.* **2022**, *102*, 1025–1151. [CrossRef] [PubMed]
74. Kress, B.T.; Iliff, J.J.; Xia, M.; Wang, M.; Wei, H.S.; Zeppenfeld, D.; Xie, L.; Kang, H.; Xu, Q.; Liew, J.A.; et al. Impairment of paravascular clearance pathways in the aging brain. *Ann. Neurol.* **2014**, *76*, 845–861. [CrossRef] [PubMed]
75. Hablitz, L.M.; Plá, V.; Giannetto, M.; Vinitsky, H.S.; Stæger, F.F.; Metcalfe, T.; Nguyen, R.; Benrais, A.; Nedergaard, M. Circadian control of brain glymphatic and lymphatic fluid flow. *Nat. Commun.* **2020**, *11*, 4411. [CrossRef]
76. Loos, C.M.J.; Klarenbeek, P.; van Oostenbrugge, R.J.; Staals, J. Association between Perivascular Spaces and Progression of White Matter Hyperintensities in Lacunar Stroke Patients. *PLoS ONE* **2015**, *10*, e0137323. [CrossRef]
77. Pham, W.; Lynch, M.; Spitz, G.; O'brien, T.; Vivash, L.; Sinclair, B.; Law, M. A critical guide to the automated quantification of perivascular spaces in magnetic resonance imaging. *Front. Neurosci.* **2022**, *16*, 1021311. [CrossRef]
78. Wang, Y.; Huang, C.; Guo, Q.; Chu, H. Aquaporin-4 and Cognitive Disorders. *Aging Dis.* **2022**, *13*, 61–72. [CrossRef]
79. Mader, S.; Brimberg, L. Aquaporin-4 Water Channel in the Brain and Its Implication for Health and Disease. *Cells* **2019**, *8*, 90. [CrossRef]
80. Benveniste, H.; Nedergaard, M. Cerebral small vessel disease: A glymphopathy? *Curr. Opin. Neurobiol.* **2022**, *72*, 15–21. [CrossRef]

81. Papadopoulos, M.C.; Verkman, A. Aquaporin 4 and neuromyelitis optica. *Lancet Neurol.* **2012**, *11*, 535–544. [CrossRef] [PubMed]
82. Jarius, S.; Paul, F.; Weinshenker, B.G.; Levy, M.; Kim, H.J.; Wildemann, B. Neuromyelitis Optica. *Nat. Rev. Dis. Primers* **2020**, *6*, 85. [CrossRef] [PubMed]
83. Huang, T.-L.; Wang, J.-K.; Chang, P.-Y.; Hsu, Y.-R.; Lin, C.-H.; Lin, K.-H.; Tsai, R.-K. Neuromyelitis Optica Spectrum Disorder: From Basic Research to Clinical Perspectives. *Int. J. Mol. Sci.* **2022**, *23*, 7908. [CrossRef] [PubMed]
84. Huda, S.; Whittam, D.; Bhojak, M.; Chamberlain, J.; Noonan, C.; Jacob, A.; Kneen, R. Neuromyelitis optica spectrum disorders. *Clin. Med.* **2019**, *19*, 169–176. [CrossRef] [PubMed]
85. Verkman, A.; Binder, D.K.; Bloch, O.; Auguste, K.; Papadopoulos, M.C. Three distinct roles of aquaporin-4 in brain function revealed by knockout mice. *Biochim. Biophys. Acta* **2006**, *1758*, 1085–1093. [CrossRef]
86. Nielsen, S.; Nagelhus, E.A.; Amiry-Moghaddam, M.; Bourque, C.; Agre, P.; Ottersen, O.P. Specialized Membrane Domains for Water Transport in Glial Cells: High-Resolution Immunogold Cytochemistry of Aquaporin-4 in Rat Brain. *J. Neurosci.* **1997**, *17*, 171–180. [CrossRef]
87. Rasmussen, M.K.; Mestre, H.; Nedergaard, M. The glymphatic pathway in neurological disorders. *Lancet Neurol.* **2018**, *17*, 1016–1024. [CrossRef]
88. Verkhratsky, A. Astrocytes: The Housekeepers and Guardians of the CNS. *Adv. Neurobiol.* **2021**, *26*, 21–53. [CrossRef]
89. Verkhratsky, A.; Butt, A.M. *Neuroglia: Function and Pathology*, 1st ed.; Academic Press: London, UK, 2023.
90. Scemes, E.; Spray, D.C. *Chapter: The Astrocytic Syncytium In Non-Neural Cells in the Nervous System: Function and Dysfunction*; Hertz, L., Ed.; Elsevier: New York, NY, USA, 2004; Volume 31, pp. 165–179.
91. Verkhratsky, A.; Nedergaard, M. Astroglial cradle in the life of the synapse. *Philos. Trans. R. Soc. B Biol. Sci.* **2014**, *369*, 20130595. [CrossRef]
92. Cashion, J.M.; Young, K.M.; Sutherland, B.A. How does neurovascular unit dysfunction contribute to multiple sclerosis? *Neurobiol. Dis.* **2013**, *178*, 106028. [CrossRef]
93. Díaz-Castro, B.; Robel, S.; Mishra, A. Astrocyte Endfeet in Brain Function and Pathology: Open Questions. *Annu. Rev. Neurosci.* **2023**, *46*, 101–121. [CrossRef] [PubMed]
94. Hayden, M.R. Hypothesis: Astrocyte Foot Processes Detachment from the Neurovascular Unit in Female Diabetic Mice May Impair Modulation of Information Processing—Six Degrees of Separation. *Brain Sci.* **2019**, *9*, 83. [CrossRef] [PubMed]
95. Milner, R.; Hung, S.; Wang, X.; Spatz, M.; del Zoppo, G.J. The Rapid Decrease in Astrocyte-Associated Dystroglycan Expression by Focal Cerebral Ischemia is Protease-Dependent. *J. Cereb. Blood Flow Metab.* **2008**, *28*, 812–823. [CrossRef]
96. Scharfman, H.E.; Binder, D.K. Aquaporin-4 water channels and synaptic plasticity in the hippocampus. *Neurochem. Int.* **2013**, *63*, 702–711. [CrossRef] [PubMed]
97. Nedergaard, M.; Verkhratsky, A. Artifact versus reality—How astrocytes contribute to synaptic events. *Glia* **2012**, *60*, 1013–1023. [CrossRef]
98. Skucas, V.A.; Mathews, I.B.; Yang, J.; Cheng, Q.; Treister, A.; Duffy, A.M.; Verkman, A.S.; Hempstead, B.L.; Wood, M.A.; Binder, D.K.; et al. Impairment of Select Forms of Spatial Memory and Neurotrophin-Dependent Synaptic Plasticity by Deletion of Glial Aquaporin-4. *J. Neurosci.* **2011**, *31*, 6392–6397. [CrossRef]
99. González-Marrero, I.; Hernández-Abad, L.G.; González-Gómez, M.; Soto-Viera, M.; Carmona-Calero, E.M.; Castañeyra-Ruiz, L.; Castañeyra-Perdomo, A. Altered Expression of AQP1 and AQP4 in Brain Barriers and Cerebrospinal Fluid May Affect Cerebral Water Balance during Chronic Hypertension. *Int. J. Mol. Sci.* **2022**, *23*, 12277. [CrossRef]
100. Li, Y.K.; Wang, F.; Wang, W.; Luo, Y.; Wu, P.F.; Xiao, J.L.; Hu, Z.L.; Jin, Y.; Hu, G.; Chen, J.G. Aquaporin-4 deficiency impairs synaptic plasticity and associative fear memory in the lateral amygdala: Involvement of downregulation of glutamate transporter-1 expression. *Neuropsycopharmacolgy* **2012**, *37*, 1867–1878. [CrossRef]
101. Chelini, G.; Pantazopoulos, H.; Durning, P.; Berretta, S. The tetrapartite synapse: A key concept in the pathophysiology of schizophrenia. *Eur. Psychiatry* **2018**, *50*, 60–69. [CrossRef]
102. Mathiisen, T.M.; Lehre, K.P.; Danbolt, N.C.; Ottersen, O.P. The perivascular astroglial sheath provides a complete covering of the brain microvessels: An electron microscopic 3D reconstruction. *Glia* **2010**, *58*, 1094–1103. [CrossRef]
103. Rotta, J.; Perosa, V.; Yakupov, R.; Kuijf, H.J.; Schreiber, F.; Dobisch, L.; Oltmer, J.; Assmann, A.; Speck, O.; Heinze, H.-J.; et al. Detection of Cerebral Microbleeds With Venous Connection at 7-Tesla MRI. *Neurology* **2021**, *96*, e2048–e2057. [CrossRef] [PubMed]
104. Schmid, F.; Barrett, M.J.; Jenny, P.; Weber, B. Vascular density and distribution in neocortex. *NeuroImage* **2019**, *197*, 792–805. [CrossRef] [PubMed]
105. Lai, A.Y.; Dorr, A.; Thomason, L.A.M.; Koletar, M.M.; Sled, J.G.; Stefanovic, B.; McLaurin, J. Venular degeneration leads to vascular dysfunction in a transgenic model of Alzheimer's disease. *Brain* **2015**, *138*, 1046–1058. [CrossRef] [PubMed]
106. Keith, J.; Gao, F.; Noor, R.; Kiss, A.; Balasubramaniam, G.; Au, K.; Rogaeva, E.; Masellis, M.; Black, S.E. Collagenosis of the Deep Medul-lary Veins: An Underrecognized Pathologic Correlate of White Matter Hyperintensities and Periventricular Infarction? *J. Neuropathol. Exp. Neurol.* **2017**, *76*, 299–312. [CrossRef] [PubMed]
107. Hartmann, D.A.; Hyacinth, H.I.; Liao, F.; Shih, A.Y. Does pathology of small venules contribute to cerebral microinfarcts and dementia? *J. Neurochem.* **2018**, *144*, 517–526. [CrossRef]
108. Shih, A.Y.; Rühlmann, C.; Blinder, P.; Devor, A.; Drew, P.J.; Friedman, B.; Knutsen, P.M.; Lyden, P.D.; Matéo, C.; Mellander, L.; et al. Robust and Fragile Aspects of Cortical Blood Flow in Relation to the Underlying Angioarchitecture. *Microcirculation* **2015**, *22*, 204–218. [CrossRef]

109. Klakotskaia, D.; Agca, C.; Richardson, R.A.; Stopa, E.G.; Schachtman, T.R.; Agca, Y. Memory deficiency, cerebral amyloid angiopathy, and amyloid-β plaques in APP+PS1 double transgenic rat model of Alzheimer's disease. *PLoS ONE* **2018**, *13*, e0195469. [CrossRef]
110. Cohen, R.M.; Rezai-Zadeh, K.; Weitz, T.M.; Rentsendorj, A.; Gate, D.; Spivak, I.; Bholat, Y.; Vasilevko, V.; Glabe, C.G.; Breunig, J.J.; et al. A Transgenic Alzheimer Rat with Plaques, Tau Pathology, Behavioral Impairment, Oligomeric Aβ, and Frank Neuronal Loss. *J. Neurosci.* **2013**, *33*, 6245–6256. [CrossRef]
111. Joo, I.L.; Lai, A.Y.; Bazzigaluppi, P.; Koletar, M.M.; Dorr, A.; Brown, M.E.; Thomason, L.A.M.; Sled, J.G.; McLaurin, J.; Stefanovic, B. Early neurovascular dysfunction in a transgenic rat model of Alzheimer's disease. *Sci. Rep.* **2017**, *7*, srep46427. [CrossRef]
112. Revesz, T.; Holton, J.L.; Lashley, T.; Plant, G.; Rostagno, A.; Ghiso, J.; Frangione, B. Sporadic and Familial Cerebral Amyloid Angiopa-thies. *Brain Pathol.* **2002**, *12*, 343–357. [CrossRef]
113. Michaud, J.-P.; Bellavance, M.-A.; Préfontaine, P.; Rivest, S. Real-Time In Vivo Imaging Reveals the Ability of Monocytes to Clear Vascular Amyloid Beta. *Cell Rep.* **2013**, *5*, 646–653. [CrossRef] [PubMed]
114. Dorr, A.; Sahota, B.; Chinta, L.V.; Brown, M.E.; Lai, A.Y.; Ma, K.; Hawkes, C.A.; McLaurin, J.; Stefanovic, B. Amyloid-β-dependent compromise of microvascular structure and function in a model of Alzheimer's disease. *Brain* **2012**, *135*, 3039–3050. [CrossRef] [PubMed]
115. Weller, R.O.; Boche, D.; Nicoll, J.A.R. Microvasculature changes and cerebral amyloid angiopathy in Alzheimer's disease and their potential impact on therapy. *Acta Neuropathol.* **2009**, *118*, 87–102. [CrossRef] [PubMed]
116. Weller, R.O.; Massey, A.; Newman, T.A.; Hutchings, M.; Kuo, Y.-M.; Roher, A.E. Cerebral Amyloid Angiopathy: Amyloid β Accumulates in Putative Interstitial Fluid Drainage Pathways in Alzheimer's Disease. *Am. J. Pathol.* **1998**, *153*, 725–733. [CrossRef] [PubMed]
117. Duvernoy, H.; Delon, S.; Vannson, J. Cortical blood vessels of the human brain. *Brain Res. Bull.* **1981**, *7*, 519–579. [CrossRef]
118. Iadecola, C.; Smith, E.E.; Anrather, J.; Gu, C.; Mishra, A.; Misra, S.; Perez-Pinzon, M.A.; Shih, A.Y.; Sorond, F.A.; van Veluw, S.J.; et al. The Neurovasculome: Key Roles in Brain Health and Cognitive Impairment: A Scientific Statement from the American Heart Association/American Stroke Association. *Stroke* **2023**, *54*, E251–E271. [CrossRef]
119. E Smith, E.; A Schneider, J.; Wardlaw, J.M.; Greenberg, S.M. Cerebral microinfarcts: The invisible lesions. *Lancet Neurol.* **2012**, *11*, 272–282. [CrossRef]
120. van Veluw, S.J.; Charidimou, A.; van der Kouwe, A.J.; Lauer, A.; Reijmer, Y.D.; Costantino, I.; Gurol, M.E.; Biessels, G.J.; Frosch, M.P.; Viswanathan, A.; et al. Microbleed and microinfarct detection in amyloid angiopathy: A high-resolution MRI-histopathology study. *Brain* **2016**, *139*, 3151–3162. [CrossRef]
121. Morrone, C.D.; Bishay, J.; McLaurin, J. Potential Role of Venular Amyloid in Alzheimer's Disease Pathogenesis. *Int. J. Mol. Sci.* **2020**, *21*, 1985. [CrossRef]

biomedicines

MDPI

Review

A Review of Fetal Development in Pregnancies with Maternal Type 2 Diabetes Mellitus (T2DM)-Associated Hypothalamic-Pituitary-Adrenal (HPA) Axis Dysregulation: Possible Links to Pregestational Prediabetes

Mathuli Ngema, Nombuso D. Xulu, Phikelelani S. Ngubane and Andile Khathi *

School of Laboratory Medicine & Medical Sciences, University of KwaZulu-Natal, Private Bag X54001, Durban 4001, South Africa; 218022309@stu.ukzn.ac.za (M.N.); 215019278@stu.ukzn.ac.za (N.D.X.); ngubanep1@ukzn.ac.za (P.S.N.)
* Correspondence: khathia@ukzn.ac.za; Tel.: +27-31-260-7585

Abstract: Research has identified fetal risk factors for adult diseases, forming the basis for the Developmental Origins of Health and Disease (DOHaD) hypothesis. DOHaD suggests that maternal insults during pregnancy cause structural and functional changes in fetal organs, increasing the risk of chronic diseases like type 2 diabetes mellitus (T2DM) in adulthood. It is proposed that altered maternal physiology, such as increased glucocorticoid (GC) levels associated with a dysregulated hypothalamic-pituitary-adrenal (HPA) axis in maternal stress and T2DM during pregnancy, exposes the fetus to excess GC. Prenatal glucocorticoid exposure reduces fetal growth and programs the fetal HPA axis, permanently altering its activity into adulthood. This programmed HPA axis is linked to increased risks of hypertension, cardiovascular diseases, and mental disorders in adulthood. With the global rise in T2DM, particularly among young adults of reproductive age, it is crucial to prevent its onset. T2DM is often preceded by a prediabetic state, a condition that does not show any symptoms, causing many to unknowingly progress to T2DM. Studying prediabetes is essential, as it is a reversible stage that may help prevent T2DM-related pregnancy complications. The existing literature focuses on HPA axis dysregulation in T2DM pregnancies and its link to fetal programming. However, the effects of prediabetes on HPA axis function, specifically glucocorticoid in pregnancy and fetal outcomes, are not well understood. This review consolidates research on T2DM during pregnancy, its impact on fetal programming via the HPA axis, and possible links with pregestational prediabetes.

Keywords: type 2 diabetes mellitus; prediabetes; pregnancy; maternal HPA axis; fetal HPA axis; programming; fetal development; placenta; glucocorticoids; metabolic diseases

Citation: Ngema, M.; Xulu, N.D.; Ngubane, P.S.; Khathi, A. A Review of Fetal Development in Pregnancies with Maternal Type 2 Diabetes Mellitus (T2DM)-Associated Hypothalamic-Pituitary-Adrenal (HPA) Axis Dysregulation: Possible Links to Pregestational Prediabetes. *Biomedicines* **2024**, *12*, 1372. https://doi.org/10.3390/biomedicines12061372

Academic Editors: Ana Dascalu and Dragos Serban

Received: 30 April 2024
Revised: 15 June 2024
Accepted: 17 June 2024
Published: 20 June 2024

1. Introduction

Fetal programming occurs during embryonic and fetal development, a vital stage during which tissues and organs are formed [1–3]. Many environmental cues, such as excess glucocorticoid exposure in utero, can contribute to various changes that include changes in molecular biological functions, such as receptor cell density or sensitivity, as well as alterations in metabolism or responses to physiological stressors [4,5]. Essentially, fetal programming refers to the process of sustaining or affecting a stimulus or impairment that occurs at a crucial point in its development [6,7]. Studies show that maternal diabetes, particularly type 2 diabetes mellitus (T2DM), with increased glucocorticoid (GC) levels, may be one of the common mechanisms through which glucocorticoid insults exert their programming effects [8,9]. Rapid economic development and urbanization, sedentary lifestyles, and the Westernized diet have led to a rising burden of 463 million (aged 20–79 years) adults living with T2DM in many parts of the world, especially in developing countries [10]. Although the weights of infants of diabetic mothers are generally skewed into the upper range, intrauterine growth restriction (IUGR), commonly diagnosed as low birth weight, occurs with concerning frequency in diabetic women, especially those with underlying hypertension, uncontrolled blood glucose levels, and vascular diseases [11,12]. T2DM has been shown to account for 30–50% of cases of pregestational diabetes during pregnancy [13].

Glucocorticoids (GC), such as cortisol in humans and corticosterone in rodents, are well-known for their role in glucose homeostasis in adult life [14]. The HPA axis regulates GC production through a feedback loop involving glucocorticoid receptors (GR) and mineralocorticoid receptors (MR) [14]. During pregnancy, this feedback mechanism ensures that GC levels are kept within a range that supports pregnancy while avoiding the adverse effects of hypercortisolism [15]. GC in pregnancy has also been shown to be essential in fetal maturation [5,16]. However, fetal GC load is usually regulated by 11β-hydroxysteroid dehydrogenase type-2 (11β-HSD2), a placental enzyme that inactivates GCs [17,18]. Increased maternal GC levels observed in T2DM and attenuating 11β-HSD2 expression potentially increase fetal exposure to GCs, slowing fetal growth and altering the gestational period [19–21]. Excessive glucocorticoid exposure in utero goes as far as altering the set-point and development of the offspring's HPA axis that alternately reprograms the HPA axis, thus compromising its function after birth [22–24]. In addition, excess maternal or fetal corticosterone causes the downregulation of fetal GR and MR and impairs the feedback regulation of the HPA axis in both infancy and adulthood [25,26]. Cross-sectional research has also indicated a connection between lower birth weights, elevated cortisol levels, catch-up growth in the neonatal period, and adult obesity, which may be an indication of unfavorable adaptive responses until birth [27,28]. The association with low birth weight was first reported as a several-fold increase in the incidence of glucose intolerance and T2DM in adult men compared with those born with normal birth weight [29,30]. Furthermore, studies demonstrate that low birth weight has been associated with high risks of other non-communicable diseases (NCDs) such as hypertension, cardiovascular diseases, and mental disorders in adulthood, correlating with the concept of the Developmental Origins of Health and Disease (DOHaD) hypothesis [31,32]. The DOHaD, which emerged as a broadening of the "Barker hypothesis" and was named after epidemiologist David Barker, explains the scenario in which in utero maternal insults cause structural and functional alterations in fetal organs, extending postnatal life and increasing susceptibility to chronic disease in adulthood [33,34].

Prediabetes is characterized by impaired glucose metabolism, with glucose concentration above the optimal value but still below the diagnostic levels for T2DM [35]. In 2019, the prevalence of prediabetes was 373.9 million, with 15.3% undiagnosed, and it is expected to increase to 453.8 million by 2030 in parallel with increasing T2DM prevalence [36,37]. Studies show that prediabetes precedes T2DM, and it has been suggested that the onset of complications associated with T2DM begins during the prediabetic state, including myocardial injury, renal dysfunction, hormonal dysfunction, and dysregulation in the HPA axis function, among others [38–41]. The literature primarily reports alterations that occur in pre-existing T2DM pregnancies and fetal programming, while the changes in maternal pregestational prediabetes HPA axis function, specifically glucocorticoid and its influence on fetal outcomes, have not yet been explored. Therefore, this review consolidates research on T2DM during pregnancy, its impact on fetal programming via the HPA axis, and its possible links with pregestational prediabetes. The following section describes fetal programming, theories, and associated diseases.

2. Fetal Programming

According to Barker (1995), the early life environment affects fetal growth and adds to disease susceptibility [42,43]. The developing baby adapts to an insult in utero, leading to long-term changes in form, physiology, and metabolism that are beneficial for survival [33,34]. Gluckman discovered that mismatches between early and later life circumstances might cause maladaptive alterations that raise the risk of a variety of cardiometabolic and psychiatric disorders as well as vulnerability factors, pertaining to the phenomena known as fetal programming [44,45]. Fetal programming occurs when the normal pattern of fetal development is disrupted by an abnormal stimulus or 'insult' applied at a critical point in in utero development [46–48]. According to the evidence for the Developmental Origins of Health and Disease (DOHaD) hypothesis, the antenatal period is a particularly vulnerable period of development in which exposure to adverse environments, such as glucocorticoid exposure, can have long-term or permanent effects on the offspring's health trajectory [49,50]. Studies have shown that maternal HPA axis is crucial during fetal development [51,52]. However, maternal dysregulation in the HPA axis during pregnancy or before pregnancy has been shown to exert its programming effect, especially in the brain, notably in the HPA axis [53–55]. The following section details the physiological role of the maternal HPA axis in pregnancy, and its role in fetal development.

3. Role of the Maternal Hypothalamic–Pituitary–Adrenal (HPA) Axis in Pregnancy

The hypothalamic–pituitary–adrenal axis is a complex system of neuroendocrine pathways and feedback loops that functionally maintain physiological homeostasis through a synthesis of glucocorticoids (GCs). Active GCs are known as cortisol in humans and corticosterone in rodents [56]. The maternal HPA axis adapts during pregnancy, and regulates stress-related deleterious effects on the mother and offspring [57,58]. A non-diabetic pregnancy is a state of hyperactivity in the HPA axis and is also a state of hypercortisolism, especially towards late gestation [59–61]. The increased cortisol in late gestation is regulated by the placenta, an important source secreting the corticotropin-releasing hormone (CRH), which further enters the maternal pituitary gland via the hypophyseal portal circulation and enhances adrenocorticotropin (ACTH) synthesis and secretion into the peripheral circulation (Figure 1) [62,63]. ACTH increases glucocorticoid synthesis and secretion through the adrenal cortex in the kidney into the bloodstream in the course of pregnancy [64,65]. GC levels influence the hypothalamic CRH in a negative feedback loop, while the placental CRH is strongly stimulated by GC in a mechanism of a positive feedback loop [66,67]. In addition, studies show that high GC in pregnancy also plays a primary role in regulating fuel homeostasis. After the uptake of free cortisol from the circulation, cortisol increases the availability of potential fuel substrates by the mobilization of glucose, free fatty acids, and amino acids through the enhancement of hepatic gluconeogenesis and glycogenolysis [68,69]. Hence, research shows that GC contributes to insulin resistance, which is necessary to ensure that an adequate amount of glucose reaches the fetus for its growth and development [70].

Research shows that there are two types of corticosteroid receptors in the brain, namely glucocorticoid (GR) and mineralocorticoid (MR) receptors involved in the feedback regulation of the HPA axis [15,71]. The proper functioning of these receptors ensures that glucocorticoid levels remain within a range that supports pregnancy without causing undue stress to the mother or the fetus [72]. In the brain, MR binds to endogenous glucocorticoid with a higher affinity than GR, and, at basal concentrations of cortisol and corticosterone, MR is occupied while the GR remains largely unoccupied [73,74]. During times of elevated plasma glucocorticoid levels, such as during stress, increased occupation of GR helps to reduce the release of CRH and ACTH, ultimately lowering glucocorticoid production, and thereby regulating the function of the HPA axis. [75,76]. In pregnancy, MR mRNA expression in the hippocampus is unaltered, and GR gene expression is only modestly increased, which promotes negative feedback, maintaining HPA axis activity (Figure 1) and, hence, the diurnal secretion of cortisol is maintained throughout pregnancy [16,77]. Furthermore, studies show that late pregnancy (the last week, in the rat) is associated with a substantial reduction in HPA axis responses to both psychological and physical stressors in several species [77–79]. This adaptation is considered to buffer the impact of stress by reducing fetal exposure to maternal glucocorticoid, thus minimizing the risk of detrimental glucocorticoid programming [16,80].

Furthermore, studies show that glucocorticoids are lipophilic and can readily pass through the placental barrier by simple diffusion [81,82]. While glucocorticoid's most well-known function is to stimulate differentiation and functional development of the lungs, glucocorticoids also play crucial roles in the development of several other organ systems, including the HPA axis [51,52]. However, 11β-hydroxysteroid dehydrogenase type 2 (11β-HSD2) acts as a barrier enzyme to control the passage of glucocorticoids from the mother to the fetus and protects the fetus from the much higher maternal levels of glucocorticoids [83,84]. This enzyme is found on both placental sides of the syncytiotrophoblast [84,85]. It metabolizes active glucocorticoids (cortisol in humans, and corticosterone in rats) into inactive glucocorticoids, thereby shielding the fetus against excessive glucocorticoid exposure from the mother, as shown in (Figure 1) [85,86]. Although the placenta metabolizes a significant proportion of cortisol (80–90% during gestation), excess cortisol may reach the fetus, and the 'barrier' can be further weakened by maternal high maternal glucocorticoid or placental dysfunction, which is commonly caused by increased oxidative stress, resulting in hypoxia, allowing for the increased transfer of glucocorticoids from the mother to the fetus [16,54,87]. The following section describes T2DM, prevalences, pregestational consequences associated with fetal programming, and associated diseases in adulthood.

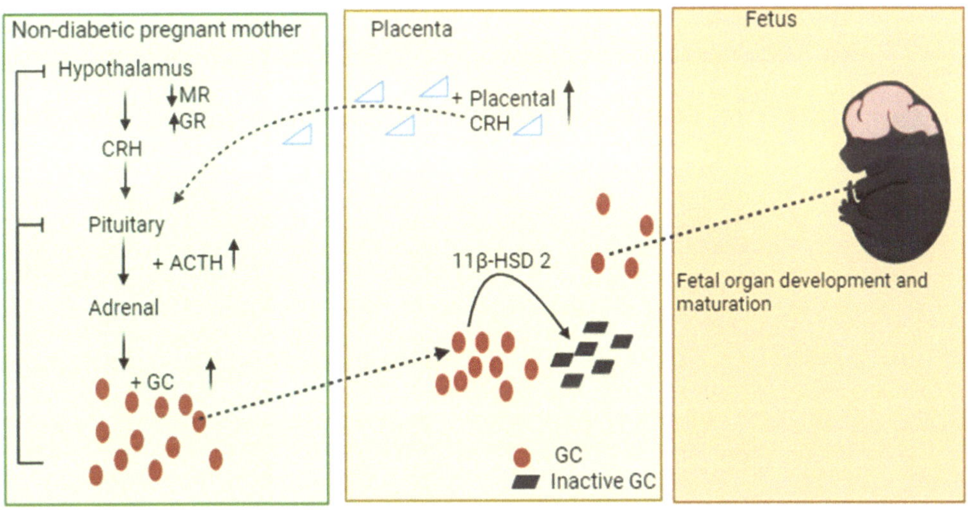

Figure 1. A schematic presentation of maternal HPA axis and GC signalling between mother, placenta, and fetus. Glucocorticoids tightly control HPA axis activity through glucocorticoid receptors (GR) and mineralocorticoid receptors (MR) in the pituitary and hypothalamus to inhibit CRH release, ACTH, and its own secretion [83,88]. In pregnancy, the placenta secretes large quantities of CRH into the maternal bloodstream as the pregnancy progresses, which promotes the production of GC [16,77]. Increased placental CRH secretion and GC also increase GR, promoting negative feedback and, therefore, maintaining the HPA axis activity in pregnancy. Nevertheless, the fetus is shielded from excess maternal GC exposure by the increased activity of 11β-hydroxysteroid dehydrogenase type 2 (11β-HSD2). The minimal transfer of GC from the placenta to the fetal compartment plays a vital role in the development of fetal organs, particularly the brain's HPA axis and the maturation of the lungs.

4. Changes in the HPA Axis in T2DM Pregnancies Associated with Fetal Programming

T2DM accounts for 90–95% of all diagnosed diabetes mellitus (DM) cases and is regarded as a complicated and multifaceted illness caused by a mix of genetic and environmental risk factors [89]. T2DM is characterized by insulin resistance and inadequate β-cell responsiveness to glucose stimulation [90]. Globally, the International Diabetes Federation (IDF) estimated that by 2045, 629 million are expected to have T2DM aged 20–70 years [91].

Swift urbanization, marked by the uptake of unhealthy, calorie-rich diets and sedentary lifestyles, has played a role in the progressively rising prevalence of T2DM, particularly among females compared to males, and the prevalence rises with age [92,93]. Research suggests that, while T2DM is often associated with macro-and microvascular complications, individuals with poor management of everyday stress is also associated, in diabetic patients, with the constant activation and disrupted regulation of the HPA axis, showing a similar resemblance to maternal stress, accompanied by high levels of glucocorticoids [94]. Champaneri et al. similarly found high cortisol (hypercortisolism) levels throughout the day in diabetic women [95]. Established diabetes mellitus, either type 1 or 2, is the most common pre-existing medical condition in pregnant women at younger ages, resulting in an increasing proportion of pregnancies complicated by diabetes [96–98]. In some areas, pregnant women with T2DM now outnumber those with type 1 diabetes (T1DM) [99]. Research indicates that pregnant women with T2DM exhibit comparable patterns to those observed in maternal obesity and in depressed and stressed pregnant women [100,101]. These patterns involve the prolonged activation and dysregulated function of the HPA axis with elevated glucocorticoid levels [100,101].

The pathophysiology of fetal growth in the context of T2DM pregnancy is intricate and multifaceted [102,103]. However, the complications of diabetes affecting the mother and fetus are well-known [104–106]. Maternal complications include preterm labour, nephropathy, vascular diseases, caesarean section, postoperative wound complications, uncontrolled hyperglycemia, and increased oxidative stress, among others [107]. Fetal complications include fetal wastage from early pregnancy loss or congenital anomalies, macrosomia, shoulder dystocia, stillbirth, and intrauterine growth restriction (IUGR), among others [108,109].

Approximately 20% of pregnant women with diabetes experience gestational hypertension and/or preeclampsia [110]. The individuals most susceptible to these conditions are those who have pre-existing microvascular complications such as microangiopathy, hypertension, or inadequate control of blood glucose levels, which also contribute to endothelial dysfunction [110,111]. These complications have been shown to induce a reduction in trophoblast proliferation, delaying placental growth and development, particularly in the first few weeks of gestation [112]. This mechanism suggests the presence of dysregulation of trophoblast invasion by the diabetic environment, leading to decreased placental perfusion, which results in placental dysfunction [113].

Studies show that placental dysfunction is associated with relatively low placental 11β-HSD2 activity, therefore increasing active maternal GC to the fetus's bloodstream [114,115]. Overexposure to glucocorticoids during fetal development causes modifications in the expression of various cytostructural proteins, receptors, enzymes, ion channels, and growth factors [116]. These modifications result in changes in tissue structure, biochemical composition, metabolism, and hormone responsiveness, impacting the functionality of several fetal organ systems [117,118]. A study has discovered a correlation between maternal diabetes accompanied by elevated cortisol levels and alterations in the development of the brain's cortical neuroendocrine system by reducing the number of hippocampal neurons [119]. Nevertheless, the precise molecular and cellular process via which diabetes during pregnancy impacts brain development remains unknown [120]. Consequently, glucocorticoids trigger physiological processes that have little or insignificant roles in utero but which become crucial at birth, such as the HPA axis [121]. The HPA axis and its key limbic regulator, the hippocampus, are particularly sensitive to glucocorticoids and their perinatal programming actions [52,122]. Previous studies show that glucocorticoid excess exposure during fetal development programs has specific effects on the brain, notably the HPA axis [123,124]. This exposure changes its development, sensitivity, and activity in utero, relatively stressing its growth as the HPA axis begins to develop during the embryonic stage and continues to mature throughout pregnancy [25,125]. As a result, studies show that prenatal glucocorticoid exposure permanently increases basal plasma corticosterone levels in adult rats [122,126]. This was because the density of both types of corticosteroid receptors, GRs and MRs, are permanently reduced in the hippocampus, changes which are anticipated to attenuate HPA axis feedback sensitivity from maternal stress shown in Figure 2 [122,126,127].

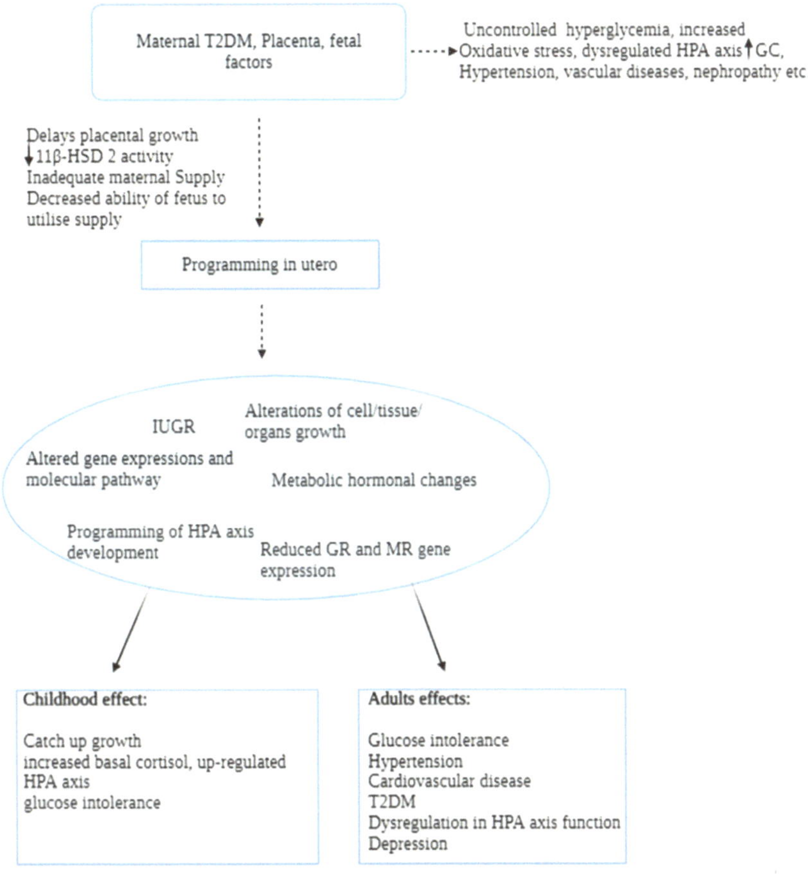

Figure 2. The schematic diagram presents the summary of maternal T2DM pregnancy complications, leading to fetal consequences in utero that persist until adulthood.

In addition, studies show that fetal exposure to excess maternal GC relative to early increases in the fetal GC concentration also triggers tissue differentiation and reduces accretion in the fetus [5]. As a result, the overall rate of maturation and growth declines as GC concentrations rise in the fetus toward term and in response to adverse intrauterine conditions, resulting in growth-retarded fetuses recognized as having IUGR [12,128]. The term intrauterine growth restriction (IUGR) refers to neonates whose birth weight and length fall below the tenth percentile for their gestational age [129,130]. IUGR is a common antenatal diagnosis; nevertheless, some of these fetuses, particularly those who were not checked during pregnancy, may be discovered only after birth [129,131]. The primary diagnostic criteria for IUGR include low birth weights (LWs), a surrogate marker of an adverse intrauterine environment, and subsequent cardio-metabolic disease and mental health problems [132,133].

On the other hand, the brain is heavily reliant on glucose for energy, and mammals have redundant systems for controlling glucose production [134]. As a result, it is possible that altered hypothalamus function may promote the dysregulation of peripheral glucose metabolism, leading to insulin resistance or T2DM in adulthood [135]. Research conducted by Hales et al. revealed a several-fold higher incidence of glucose intolerance and T2DM in adult men who were born with low birth weight as opposed to those who were born with normal birth weight, which established the first link between low birth weight and the development of T2DM [29]. A study in rats showed that the smallest fetuses with the largest placentas had lower placental 11β-HSD2 activity and were projected to have the highest adult blood pressures [17,136]. Heightened HPA axis activity, particularly with increased ACTH and high plasma glucocorticoid levels, is seen in children and adults who were born underweight [137,138]. In addition, previous research has indicated that infants born with lower birth weights undergo catch-up growth within the initial two years of life [139,140]. This process is seen as a means to offset their genetically predetermined growth patterns [140]. Catch-up growth is also observed in other aspects of growth, such as changes in body weight and body composition [141]. As per the theory of DOHaD, the rapid catch-up growth experienced by low-birth-weight infants in their early years is associated with various metabolic conditions such as obesity, hypertension, cardiovascular diseases, metabolic syndrome, and endothelial dysfunction later in adulthood shown in Figure 2 above [142,143]. Men were found to be more likely to develop cardiovascular disorders than women born with IUGR due to hormonal differences, as men had lower levels of estrogen, which has protective effects on the circulatory system and may contribute to women's decreased risk of cardiovascular disease [144,145]. A prior study discovered that a combination of placental weight and birth weight predicts the risk of high blood pressure and hypertension in men and women around the age of 50 [137,146]. People who were babies with large placentas had the highest blood pressure and a higher risk of hypertension [147]. Both the Barker and DOHaD hypotheses support these theories.

5. Prediabetes

Prediabetes is a condition in which blood glucose levels are abnormally high, but do not meet the diagnostic criteria for type 2 diabetes mellitus (T2DM) [148]. Prediabetes can be identified by at least two of these characteristics: impaired fasting glucose (IFG), impaired glucose tolerance (IGT), or high glycated hemoglobin A1c (HbA1c) [149]. IFG patients have significant hepatic insulin resistance with normal skeletal muscle values and poor glucose suppression, which causes hyperglycemia during fasting due to impaired insulin secretion or reduced sensitivity of β-cells to glucose stimulation [150]. IGT mainly impacts muscle insulin resistance, with minimal effects on liver insulin sensitivity. The reduced glucose absorption observed in individuals with impaired glucose tolerance (IGT) contributes to postprandial hyperglycemia [151,152]. This is primarily due to pancreatic β-cell dysfunction, resulting in inadequate secretion of insulin to counteract elevated glucose levels and stimulate a response in insulin-targeted peripheral tissues [153,154]. Lastly, IFG Individuals have a poor early insulin response during the oral glucose tolerance test (OGTT), but improve insulin secretion during the second phase, which is one of the reasons why prediabetes is frequently undetectable [155]. As a result, the American Diabetes Association (ADA) recommendations were changed in 2003 to identify patients with prediabetes based on the following values. Fasting plasma glucose levels range from 5.6 mmol/L to 6.9 mmol/L, whereas IGT values recorded after OGTT range from 7.8 mmol/L to 11.0 mmol/L [156,157]. Glycated haemoglobin (HbA1c) levels between 5.7% and 6.4% are used as an additional diagnostic criterion for prediabetes [158,159]. Prior to the diagnosis of pre-diabetes, there is a presence of insulin resistance and malfunctioning of pancreatic β-cells [153,154]. Studies show that a diet high in saturated fats, high in carbohydrates, or high in fructose contributes to the development of intermediate hyperglycemia [160,161]. In addition, studies also show that these caloric foods lead to elevated triglycerides and increased release of free fatty acids (FFA) from adipocytes into circulation, which is accompanied by decreased FFA uptake by adipocytes in insulin-dependent tissues, promoting insulin resistance and dyslipidemia [162,163]. These actions result in increased circulating FFA levels and FFA flux to the liver, which stimulates increased production and secretion of atherogenic very-low-density lipoprotein and small dense low-density lipoprotein particles, and reduced high-density lipoprotein cholesterol (HDL-C) levels, increasing the risk of microvascular and macrovascular diseases [162,164]. In addition, in the condition of insulin resistance, normal levels of insulin in the blood would be unable to elicit a reaction in the peripheral tissues that are targeted by insulin due to a decrease in the number of insulin receptors on the surface of cells, including muscle cells [165,166]. With fewer receptors available, the cells become less responsive to insulin, reducing their ability to take up glucose [90]. Consequently, the β-cells of the pancreas react by producing additional insulin to counterbalance the increased glucose levels [167]. When the β-cells fail to secrete sufficient insulin to counteract insulin resistance, the blood glucose levels commence fluctuating, leading to intermediate hyperglycemia and hyperinsulinemia, leading to further alterations in β-cell function [168].

5.1. Prevalence of Prediabetes

The prevalence of prediabetes has grown worldwide, and in 2019, the International Diabetes Federation estimated the worldwide prevalence of prediabetes to be 373.9 million, with 15.3% undiagnosed according to studies [36,37]. It is also projected that by 2030, approximately 453.8 million people will have prediabetes [37,169]. The prevalence of prediabetes is anticipated to increase to 8.3% of the global adult population, equivalent to an estimated 587 million individuals by the year 2045 [170]. Studies show that prediabetes is frequently undetected due to its often-asymptomatic nature in its early stages; hence, most humans tend to unknowingly bypass the prediabetes stage to overt T2DM [159,171]. In addition, studies show that the increase in prediabetes prevalence is due to rapid urbanization, increasingly sedentary lifestyles, and unhealthy eating habits [172,173]. As a result, a retrospective study in a rapidly urbanizing area such as Durban, South Africa, indicated that 68% of the individuals are prediabetic in the sample population between the ages of 20–45 years, with 51.0% of the study population being women [174]. This suggests that women of childbearing age are also affected by the global rise in prediabetes [175]. Furthermore, studies show that people with prediabetes have a two-fold increased likelihood of developing T2DM [176,177]. Moreover, studies show that complications associated with T2DM are already evident in some people with prediabetes; these complications include myocardial injury, renal dysfunction, hormonal dysfunction, and dysregulation in HPA axis function, among others [38–41].

5.2. HPA Axis Function in Prediabetes

Animal models have been observed to mirror human disease conditions, making them extensively utilizable for studying physiological systems and human disease states [178,179]. A high-fat, high carbohydrate and 15% fructose diet-induced animal model of prediabetes has been found to mimic the human condition [38]. In addition, this animal model showed dysregulation in the functioning of the HPA axis in the prediabetic state, as shown by elevated basal corticosterone and impaired regulation of their glucocorticoid receptor (GR) and mineralocorticoid receptor (MR) in male prediabetic rats [39]. At present, there have been no investigations to show whether this phenomenon also exists in female prediabetic rats. Furthermore, if present, this raised the question of whether maternal basal corticosterone and ACTH levels in prediabetic dams may impact fetal HPA axis development.

6. Possible Links of Prediabetes with T2DM Pregnancies in Association with the HPA Axis

Both T2DM and prediabetes result from insulin resistance, leading to impaired glucose tolerance and chronic hyperglycemia [154,180]. Several studies have shown that during pregnancy, increased maternal serum glucocorticoids (GC), observed in T2DM and maternal stress, may cross the placenta, overwhelming the protective placental barrier [11,30]. Conversely, research shows that pre-existing metabolic disorders associated with T2DM, such as hypertension, renal disease, and maternal microangiopathy during pregnancy, decrease trophoblast proliferation [110,111]. This delays placental growth and its ability to supply the fetus with enough nutrients and oxygen, resulting in fetal hypoxia and inadequate nutrition supply [112]. Decreased placental function is linked to placental dysfunction and relatively low activity of placental 11β-HSD2 [115,118]. These complications have been associated with increasing the vulnerability of the fetus during this period to unwanted programming effects such as increased transfer of active maternal cortisol to the fetal compartments [115,118]. Early maternal GC exposure to the fetus has been associated with alterations to the balance of both GR and MR development in the fetal brain, leading to changes in gene expression patterns and neural circuit formation, evidenced by low GR and MR expression in offspring after birth, even in both basal and stressed animals [126]. Moreover, studies show that excessive GC exposure during critical periods of development could program the fetal HPA axis to be hyper-responsive, beginning in utero and persisting later in life, leading to outcomes such as the prolonged increases in ACTH and GC seen in adulthood, potentially predisposing the offspring to diseases such as depression, cardiometabolic disorders, or T2DM [53]. Furthermore, studies suggest that excessive maternal GC exposure or reduced placental function during pregnancy is associated with reduced fetal growth, contributing to intrauterine growth restriction (IUGR), commonly diagnosed at birth as low birth weight [12,128]. Low-birth-weight offspring have been associated with HPA axis hyperactivity, glucose intolerance, hypertension, obesity, and greater risks for developing depression, anxiety, T2DM, and cardiovascular diseases in adulthood, especially when there was catch-up growth in the first 2 years, as supported by the DOHaD hypotheses [49,50].

There are various possible causes that contribute to IUGR or an unfavourable hostile environment during pregnancy, including preeclampsia and hypertensive disorders that have also been associated with greater risk in T2DM pregnancies [110]. Prediabetes, which often precedes the onset of T2DM, has been shown by various studies to be the genesis of complications associated with T2DM, including myocardial injury, renal dysfunction, hormonal dysfunction, and dysregulation in HPA axis function, among others [38–41]. A recent study by Ludidi and colleagues showed that prediabetes is a risk factor for developing pre-eclampsia, similar to T2DM pregnancy [181]. In addition, since prediabetes often goes undiagnosed, this suggests that there is a population of people unaware of their elevated risk of developing hypertension and preeclampsia [181]. Therefore, the presence of prediabetes in pregnancy might increase the likelihood of IUGR and impaired glucose tolerance in offspring. With the increasing prevalence of prediabetes, especially in women of childbearing age, there is an increased possibility of pre-gestational, gestational, and fetal outcome consequences. However, there have been no studies that have looked at how maternal prediabetes affects the HPA axis along with fetal outcomes (summarised in Table 1 below). Conducting preliminary investigations, generating hypotheses, carrying out invasive procedures, collecting tissue samples, and understanding the fundamental mechanism of pregnancy-related disorders would be ethically or practically challenging in humans. As a result, most primary studies are recommended to begin with animal models; hence, this study recommends that future studies focus on pregestational prediabetes, detailing its effects on maternal HPA axis function and its influences on fetal outcomes in Sprague Dawley rats.

Table 1. Summarizes possible links of prediabetes with T2DM pregnancies in association with the HPA axis.

Prediabetes	T2DM
• Prediabetes is characterized by the concurrent occurrence of insulin resistance and β-cell dysfunction, which are abnormalities that occur prior to the detection of changes in glucose levels [150,182].	• T2DM, a condition marked by deficient insulin secretion by pancreatic islet β-cells, tissue insulin resistance (IR), and an inadequate compensatory insulin secretory response [183].
• Studies show that young adults from 25–45 are diagnosed with prediabetes. • The number of individuals with prediabetes is expected to rise to around 587 million by 2045 [170,174].	• It is projected that, by 2045, the number of people with type 2 diabetes mellitus (T2DM) will reach 629 million people aged 20–79 years, respectively [91].
• This study shows that a prediabetic diet-induced animal model showed dysregulation in the functioning of the HPA axis, associated with elevated basal corticosterone and impaired regulation of the glucocorticoid receptor (GR) and mineralocorticoid receptor (MR) in male prediabetic rats animals [39].	• Patients with diabetes present alterations of the HPA axis negative feedback, suggestive of the impairment of corticosteroid receptor sensitivity, which is associated with high levels of glucocorticoids and ACTH [94].
• Pre-eclampsia has been demonstrated to be associated with an increased risk in those with prediabetes [181]. Studies show that prediabetes is also linked to the early development of complications shown in T2DM. Therefore, the occurrence of prediabetes during pregnancy may increase the likelihood of exacerbated maternal dysregulation in HPA axis function, leading to fetal adverse consequences.	• Research indicates that T2DM pregnancy with uncontrolled hyperglycemia, hypertension, elevated glucocorticoids, and an increased risk of pre-eclampsia is associated with higher rates of complications for both the mother and the baby [110,111]. • The complications include IUGR, altered gene expression, HPA axis programming, glucose intolerance, increased risk of developing T2DM, and cardiovascular diseases, among others [67,137,184–186].

Nevertheless, no research has been conducted to investigate the impact of maternal prediabetes on the HPA axis in relation to fetal outcomes in pregnancy.

7. Conclusions

This review paper highlights the significant impact of the dysregulation of the maternal HPA axis during pregnancy, particularly elevated glucocorticoid levels, on fetal growth and programming, with potential implications for HPA axis development in the fetus seen in T2DM. It discusses the possible links between prediabetes and T2DM pregnancies relative to impaired HPA axis function. Women with type 2 diabetes represent high-risk groups during pregnancy. As the incidence of diagnosed diabetes and prediabetes continues to increase, especially at young ages, the number of women with diabetes or prediabetes in pregnancy may also continue to increase. However, further research is needed to understand the effects of pregestational prediabetes on the maternal HPA axis and its impact on fetal outcomes. This could further underscore the importance of a continued investigation into the complex interplay between maternal metabolic health, HPA axis regulation, and fetal development to inform clinical management and improve pregnancy outcomes.

Author Contributions: Conceptualization, M.N., A.K., N.D.X. and P.S.N.; writing—original draft preparation, M.N.; writing—review and editing, A.K. and P.S.N.; visualization, M.N. and A.K.; supervision, A.K., N.D.X. and P.S.N. All authors have read and agreed to the published version of the manuscript.

Funding: This research received no external funding.

Conflicts of Interest: The authors declare no conflicts of interest.

References

1. Lemley, C.O.; Littlejohn, B.P.; Burnett, D.D. Fetal programming. In *Bovine Reproduction*; John Wiley & Sons, Inc.: Hoboken, NJ, USA, 2021; pp. 339–346.
2. Öztürk, H.N.O.; Türker, P.F. Fetal programming: Could intrauterin life affect health status in adulthood? *Obstet. Gynecol. Sci.* **2021**, *64*, 473–483. [CrossRef] [PubMed]
3. Seneviratne, S.N.; Rajindrajith, S. Fetal programming of obesity and type 2 diabetes. *World J. Diabetes* **2022**, *13*, 482. [CrossRef] [PubMed]
4. Hong, J.Y. Developmental Programming by Perinatal Glucocorticoids. *Mol. Cells* **2022**, *45*, 685. [CrossRef] [PubMed]
5. Fowden, A.L.; Vaughan, O.R.; Murray, A.J.; Forhead, A.J. Metabolic consequences of glucocorticoid exposure before birth. *Nutrients* **2022**, *14*, 2304. [CrossRef] [PubMed]
6. Piro, E.; Serra, G.; Schierz, I.A.M.; Giuffrè, M.; Corsello, G. Fetal growth restriction: A growth pattern with fetal, neonatal and long-term consequences. *Euromediterr. Biomed. J.* **2019**, *14*, 038–044.
7. Raja, G.L.; Subhashree, K.D.; Kantayya, K.E. In utero exposure to endocrine disruptors and developmental neurotoxicity: Implications for behavioural and neurological disorders in adult life. *Environ. Res.* **2022**, *203*, 111829. [CrossRef] [PubMed]
8. Bashir, M.; Dabbous, Z.; Baagar, K.; Elkhatib, F.; Ibrahim, A.; Brich, S.-A.; Abdel-Rahman, M.E.; Konje, J.C.; Abou-Samra, A.-B. Type 2 diabetes mellitus in pregnancy: The impact of maternal weight and early glycaemic control on outcomes. *Eur. J. Obstet. Gynecol. Reprod. Biol.* **2019**, *233*, 53–57. [CrossRef] [PubMed]
9. Vickers, M.H. Early life nutrition and its effect on the development of obesity and type-2 diabetes. In *Early Nutrition and Long-Term Health*; Elsevier: Amsterdam, The Netherlands, 2022; pp. 281–307.
10. Aschner, P.; Karuranga, S.; James, S.; Simmons, D.; Basit, A.; Shaw, J.E.; Wild, S.H.; Ogurtsova, K.; Saeedi, P. The International Diabetes Federation's guide for diabetes epidemiological studies. *Diabetes Res. Clin. Pract.* **2021**, *172*, 108630. [CrossRef]
11. Cleary, E.M.; Thung, S.F.; Buschur, E.O. Pregestational Diabetes Mellitus. In *Endotext [Internet]*; MDText.com, Inc.: South Dartmouth, MA, USA, 2021.
12. Bedell, S.; Hutson, J.; de Vrijer, B.; Eastabrook, G. Effects of maternal obesity and gestational diabetes mellitus on the placenta: Current knowledge and targets for therapeutic interventions. *Curr. Vasc. Pharmacol.* **2021**, *19*, 176–192. [CrossRef]
13. Raets, L.; Ingelbrecht, A.; Benhalima, K. Management of type 2 diabetes in pregnancy: A narrative review. *Front. Endocrinol.* **2023**, *14*, 1193271. [CrossRef]
14. Bhaumik, S.; Lockett, J.; Cuffe, J.; Clifton, V.L. Glucocorticoids and Their Receptor Isoforms: Roles in Female Reproduction, Pregnancy, and Foetal Development. *Biology* **2023**, *12*, 1104. [CrossRef] [PubMed]
15. Pofi, R.; Tomlinson, J.W. Glucocorticoids in pregnancy. *Obstet. Med.* **2020**, *13*, 62–69. [CrossRef] [PubMed]
16. Sheng, J.A.; Bales, N.J.; Myers, S.A.; Bautista, A.I.; Roueinfar, M.; Hale, T.M.; Handa, R.J. The hypothalamic-pituitary-adrenal axis: Development, programming actions of hormones, and maternal-fetal interactions. *Front. Behav. Neurosci.* **2021**, *14*, 256. [CrossRef] [PubMed]
17. Yu, P.; Zhou, J.; Ge, C.; Fang, M.; Zhang, Y.; Wang, H. Differential expression of placental 11β-HSD2 induced by high maternal glucocorticoid exposure mediates sex differences in placental and fetal development. *Sci. Total Environ.* **2022**, *827*, 154396. [CrossRef] [PubMed]
18. Rensel, M.A.; Schlinger, B.A. 11ß hydroxysteroid dehydrogenases regulate circulating glucocorticoids but not central gene expression. *Gen. Comp. Endocrinol.* **2021**, *305*, 113734. [CrossRef] [PubMed]
19. Chavey, A.; Kioon, M.-D.A.; Bailbé, D.; Movassat, J.; Portha, B. Maternal diabetes, programming of beta-cell disorders and intergenerational risk of type 2 diabetes. *Diabetes Metab.* **2014**, *40*, 323–330. [CrossRef] [PubMed]
20. Kadayifci, F.Z.; Haggard, S.; Jeon, S.; Ranard, K.; Tao, D.; Pan, Y.-X. Early-life programming of type 2 diabetes mellitus: Understanding the association between epigenetics/genetics and environmental factors. *Curr. Genom.* **2019**, *20*, 453–463. [CrossRef]
21. Kong, L.; Chen, X.; Gissler, M.; Lavebratt, C. Relationship of prenatal maternal obesity and diabetes to offspring neurodevelopmental and psychiatric disorders: A narrative review. *Int. J. Obes.* **2020**, *44*, 1981–2000. [CrossRef]
22. Reynolds, R.M. Glucocorticoid excess and the developmental origins of disease: Two decades of testing the hypothesis–2012 Curt Richter Award Winner. *Psychoneuroendocrinology* **2013**, *38*, 1–11. [CrossRef]
23. Seckl, J.R. Glucocorticoids and Fetal Programming; Necessary and Sufficient? In *Hormones, Intrauterine Health and Programming*; Springer: Cham, Switzerland, 2014; pp. 1–15.
24. Rizzo, H.E.; Escaname, E.N.; Alana, N.B.; Lavender, E.; Gelfond, J.; Fernandez, R.; Hibbs, M.A.; King, J.M.; Carr, N.R.; Blanco, C.L. Maternal diabetes and obesity influence the fetal epigenome in a largely Hispanic population. *Clin. Epigenet.* **2020**, *12*, 34. [CrossRef]
25. Cottrell, E.C.; Seckl, J. Prenatal stress, glucocorticoids and the programming of adult disease. *Front. Behav. Neurosci.* **2009**, *3*, 707. [CrossRef] [PubMed]
26. Miranda, A.; Sousa, N. Maternal hormonal milieu influence on fetal brain development. *Brain Behav.* **2018**, *8*, e00920. [CrossRef] [PubMed]
27. Grace, C.E.; Kim, S.J.; Rogers, J.M. Maternal influences on epigenetic programming of the developing hypothalamic-pituitary-adrenal axis. *Birth Defects Res. Part A Clin. Mol. Teratol.* **2011**, *91*, 797–805. [CrossRef] [PubMed]

28. de Mendonça, E.L.S.S.; de Lima Macêna, M.; Bueno, N.B.; de Oliveira, A.C.M.; Mello, C.S. Premature birth, low birth weight, small for gestational age and chronic non-communicable diseases in adult life: A systematic review with meta-analysis. *Early Hum. Dev.* **2020**, *149*, 105154. [CrossRef] [PubMed]
29. Hales, C.N.; Barker, D.J.; Clark, P.M.; Cox, L.J.; Fall, C.; Osmond, C.; Winter, P. Fetal and infant growth and impaired glucose tolerance at age 64. *BMJ Br. Med. J.* **1991**, *303*, 1019. [CrossRef] [PubMed]
30. Weiss, U.; Cervar, M.; Puerstner, P.; Schmut, O.; Haas, J.; Mauschitz, R.; Arikan, G.; Desoye, G. Hyperglycaemia in vitro alters the proliferation and mitochondrial activity of the choriocarcinoma cell lines BeWo, JAR and JEG-3 as models for human first-trimester trophoblast. *Diabetologia* **2001**, *44*, 209–219. [CrossRef] [PubMed]
31. Starikov, R.; Has, P.; Wu, R.; Nelson, D.M.; He, M. Small-for-gestational age placentas associate with an increased risk of adverse outcomes in pregnancies complicated by either type I or type II pre-gestational diabetes mellitus. *J. Matern.-Fetal Neonatal Med.* **2022**, *35*, 1677–1682. [CrossRef]
32. Baird, J.; Jacob, C.; Barker, M.; Fall, C.H.; Hanson, M.; Harvey, N.C.; Inskip, H.M.; Kumaran, K.; Cooper, C. Developmental origins of health and disease: A lifecourse approach to the prevention of non-communicable diseases. *Healthcare* **2017**, *5*, 14. [CrossRef] [PubMed]
33. Lampl, M.; Mummert, A.; Schoen, M. Auxological perspectives on 'growth' in DOHaD. *J. Dev. Orig. Health Dis.* **2015**, *6*, 390–398. [CrossRef]
34. Ventura, N.M. Developmental Origins of Cardiovascular Disease: Alterations in Gestational Hypertension and Stroke Outcome in Adult Offspring. Ph.D. Thesis, Queen's University, Kingston, ON, Canada, 2015.
35. Baranowska-Jurkun, A.; Matuszewski, W.; Bandurska-Stankiewicz, E. Chronic microvascular complications in prediabetic states—An overview. *J. Clin. Med.* **2020**, *9*, 3289. [CrossRef]
36. Saeedi, P.; Petersohn, I.; Salpea, P.; Malanda, B.; Karuranga, S.; Unwin, N.; Colagiuri, S.; Guariguata, L.; Motala, A.A.; Ogurtsova, K. Global and regional diabetes prevalence estimates for 2019 and projections for 2030 and 2045: Results from the International Diabetes Federation Diabetes Atlas. *Diabetes Res. Clin. Pract.* **2019**, *157*, 107843. [CrossRef] [PubMed]
37. Sosibo, A.M.; Mzimela, N.C.; Ngubane, P.S.; Khathi, A. Prevalence and correlates of pre-diabetes in adults of mixed ethnicities in the South African population: A systematic review and meta-analysis. *PLoS ONE* **2022**, *17*, e0278347. [CrossRef] [PubMed]
38. Luvuno, M.; Mabandla, M.; Khathi, A. Voluntary ingestion of a high-fat high-carbohydrate diet: A model for prediabetes. *Ponte Int. Sci. Res. J.* **2018**, *74*, 119–143.
39. Mosili, P.; Mkhize, B.C.; Ngubane, P.; Sibiya, N.; Khathi, A. The dysregulation of the hypothalamic–pituitary–adrenal axis in diet-induced prediabetic male Sprague Dawley rats. *Nutr. Metab.* **2020**, *17*, 104. [CrossRef] [PubMed]
40. Mkhize, B.C.; Mosili, P.; Ngubane, P.S.; Sibiya, N.H.; Khathi, A. Diet-induced prediabetes: Effects on the activity of the renin–angiotensin–aldosterone system in selected organs. *J. Diabetes Investig.* **2022**, *13*, 768–780. [CrossRef] [PubMed]
41. Dimba, N.R.; Mzimela, N.; Mosili, P.; Ngubane, P.S.; Khathi, A. Investigating the association between diet-induced "leaky gut" and the development of prediabetes. *Exp. Clin. Endocrinol. Diabetes* **2023**, *131*, 569–576. [CrossRef] [PubMed]
42. Barker, D.J.; Clark, P.M. Fetal undernutrition and disease in later life. *Rev. Reprod.* **1997**, *2*, 105–112. [CrossRef] [PubMed]
43. Monahan, B.; Farland, L.V.; Shadyab, A.H.; Hankinson, S.E.; Manson, J.E.; Spracklen, C.N. Birthweight and subsequent risk for thyroid and autoimmune conditions in postmenopausal women. *J. Dev. Orig. Health Dis.* **2022**, *13*, 463–470. [CrossRef]
44. Gluckman, P.D.; Buklijas, T.; Hanson, M.A. The developmental origins of health and disease (DOHaD) concept: Past, present, and future. In *The Epigenome and Developmental Origins of Health and Disease*; Elsevier: Amsterdam, The Netherlands, 2016; pp. 1–15.
45. Poston, L.; Godfrey, K.; Gluckman, P.; Hanson, M. *Developmental Origins of Health and Disease*; Cambridge University Press: Cambridge, UK, 2022.
46. Lemley, C. Fetal programming: Maternal-fetal interactions and postnatal performance. *Clin. Theriogenology* **2020**, *12*, 252–267.
47. Cerritelli, F.; Frasch, M.G.; Antonelli, M.C.; Viglione, C.; Vecchi, S.; Chiera, M.; Manzotti, A. A review on the vagus nerve and autonomic nervous system during fetal development: Searching for critical windows. *Front. Neurosci.* **2021**, *15*, 721605. [CrossRef]
48. Gundacker, A.; Cuenca Rico, L.; Stoehrmann, P.; Tillmann, K.E.; Weber-Stadlbauer, U.; Pollak, D.D. Interaction of the pre-and postnatal environment in the maternal immune activation model. *Discov. Ment. Health* **2023**, *3*, 15. [CrossRef] [PubMed]
49. Wolford, E. Developmental Origins of Mental Health: Human Observational Studies of Preterm Birth, Antenatal Synthetic Glucocorticoid Exposure, and Maternal Depressive Symptoms during Pregnancy. Ph.D. Thesis, University of Helsinki, Helsinki, Finland, 2018.
50. Hoffman, D.J.; Powell, T.L.; Barrett, E.S.; Hardy, D.B. Developmental origins of metabolic diseases. *Physiol. Rev.* **2021**, *101*, 739–795. [CrossRef] [PubMed]
51. Morsi, A.; DeFranco, D.; Witchel, S.F. The hypothalamic-pituitary-adrenal axis and the fetus. *Horm. Res. Paediatr.* **2018**, *89*, 380–387. [CrossRef]
52. Drake, A.J.; Tang, J.I.; Nyirenda, M.J. Mechanisms underlying the role of glucocorticoids in the early life programming of adult disease. *Clin. Sci.* **2007**, *113*, 219–232. [CrossRef]
53. Braun, T.; Challis, J.R.; Newnham, J.P.; Sloboda, D.M. Early-life glucocorticoid exposure: The hypothalamic-pituitary-adrenal axis, placental function, and long-term disease risk. *Endocr. Rev.* **2013**, *34*, 885–916. [CrossRef] [PubMed]
54. Duthie, L.; Reynolds, R.M. Changes in the maternal hypothalamic-pituitary-adrenal axis in pregnancy and postpartum: Influences on maternal and fetal outcomes. *Neuroendocrinology* **2013**, *98*, 106–115. [CrossRef] [PubMed]

55. Xiong, F.; Zhang, L. Role of the hypothalamic–pituitary–adrenal axis in developmental programming of health and disease. *Front. Neuroendocrinol.* **2013**, *34*, 27–46. [CrossRef]
56. Matthews, S.G.; McGowan, P.O. Developmental programming of the HPA axis and related behaviours: Epigenetic mechanisms. *J. Endocrinol.* **2019**, *242*, T69–T79. [CrossRef]
57. Maniam, J.; Antoniadis, C.; Morris, M.J. Early-life stress, HPA axis adaptation, and mechanisms contributing to later health outcomes. *Front. Endocrinol.* **2014**, *5*, 73. [CrossRef]
58. Brunton, P.J. Effects of maternal exposure to social stress during pregnancy: Consequences for mother and offspring. *Reproduction* **2013**, *146*, R175–R189. [CrossRef]
59. Tal, R.; Taylor, H.S. Endocrinology of pregnancy. In *Endotext [Internet]*; MDText.com, Inc.: South Dartmouth, MA, USA, 2021.
60. O'Reilly, J.R. Effects of Maternal Stress and Obesity on Human Feto-Placental Glucocorticoid Exposure. Ph.D. Thesis, The University of Edinburgh, Edinburgh, UK, 2014.
61. Ying, S. Investigating the Mechanisms Mediating the Outcomes of Prenatal Stress. Ph.D. Thesis, The University of Edinburgh, Edinburgh, UK, 2020.
62. Erhuma, A.M. The interaction between maternal and fetal hypothalamic–pituitary–adrenal axes. In *Corticosteroids—A Paradigmatic Drug Class*; IntechOpen Limited: London, UK, 2021; Volume 6.
63. Valsamakis, G.; Chrousos, G.; Mastorakos, G. Stress, female reproduction and pregnancy. *Psychoneuroendocrinology* **2019**, *100*, 48–57. [CrossRef] [PubMed]
64. Opichka, M.A.; Livergood, M.C.; Grobe, J.L.; McIntosh, J.J. Cardiovascular Neuroendocrinology of Pregnancy. In *Cardiovascular Neuroendocrinology*; Springer: Cham, Switzerland, 2023; pp. 111–135.
65. Gonzalez-Iglesias, A.E.; Freeman, M.E. Brain control over pituitary gland hormones. In *Neuroscience in the 21st Century: From Basic to Clinical*; Springer: Cham, Switzerland, 2022; pp. 2291–2344.
66. Kota, S.K.; Gayatri, K.; Jammula, S.; Kota, S.K.; Krishna, S.V.; Meher, L.K.; Modi, K.D. Endocrinology of parturition. *Indian J. Endocrinol. Metab.* **2013**, *17*, 50–59. [CrossRef] [PubMed]
67. Gans, I.M.; Coffman, J.A. Glucocorticoid-mediated developmental programming of vertebrate stress responsivity. *Front. Physiol.* **2021**, *12*, 812195. [CrossRef] [PubMed]
68. Kuo, T.; McQueen, A.; Chen, T.-C.; Wang, J.-C. Regulation of glucose homeostasis by glucocorticoids. In *Glucocorticoid Signaling: From Molecules to Mice to Man*; Springer: New York, NY, USA, 2015; pp. 99–126.
69. Mourtzi, N.; Sertedaki, A.; Charmandari, E. Glucocorticoid signaling and epigenetic alterations in stress-related disorders. *Int. J. Mol. Sci.* **2021**, *22*, 5964. [CrossRef]
70. Grilo, L.F.; Tocantins, C.; Diniz, M.S.; Gomes, R.M.; Oliveira, P.J.; Matafome, P.; Pereira, S.P. Metabolic disease programming: From mitochondria to epigenetics, glucocorticoid signalling and beyond. *Eur. J. Clin. Investig.* **2021**, *51*, e13625. [CrossRef] [PubMed]
71. Künzel, R.G.; Elgazzar, M.; Bain, P.A.; Kirschbaum, C.; Papatheodorou, S.; Gelaye, B. The Association Between Maternal Prenatal Hair Cortisol Concentration and Preterm Birth: A Systematic Review and Meta-Analysis. *Psychoneuroendocrinology* **2024**, *165*, 107041. [CrossRef] [PubMed]
72. Solano, M.E.; Arck, P.C. Steroids, pregnancy and fetal development. *Front. Immunol.* **2020**, *10*, 477454. [CrossRef] [PubMed]
73. Groeneweg, F.L.; Karst, H.; de Kloet, E.R.; Joëls, M. Rapid non-genomic effects of corticosteroids through the membrane-associated MR and GR and their role in the central stress. *J. Endocrinol.* **2011**, *2*, 153–167. [CrossRef] [PubMed]
74. Vyas, S.; Maatouk, L. Contribution of glucocorticoids and glucocorticoid receptors to the regulation of neurodegenerative processes. *CNS Neurol. Disord. Drug Targets* **2013**, *12*, 1175–1193. [CrossRef]
75. de Kloet, E.R.; Schmidt, M.; Meijer, O.C. Corticosteroid receptors and HPA-axis regulation. In *Techniques in the Behavioral and Neural Sciences*; Elsevier: Amsterdam, The Netherlands, 2005; Volume 15, pp. 265–294.
76. de Kloet, E.R.; Joëls, M. Mineralocorticoid receptors and glucocorticoid receptors in HPA stress responses during coping and adaptation. In *Oxford Research Encyclopedia of Neuroscience*; Oxford University Press: Oxford, UK, 2020.
77. Brunton, P. Resetting the dynamic range of hypothalamic-pituitary-adrenal axis stress responses through pregnancy. *J. Neuroendocrinol.* **2010**, *22*, 1198–1213. [CrossRef]
78. Brunton, P.; Russell, J.; Douglas, A. Adaptive responses of the maternal hypothalamic-pituitary-adrenal axis during pregnancy and lactation. *J. Neuroendocrinol.* **2008**, *20*, 764–776. [CrossRef]
79. Azizi, N.; Heidari, M.; Saboory, E.; Abdollahzade, N.; Roshan-Milani, S. Investigating the effect of parental pre-gestational stress on ethological parameters in male rat offspring. *J. Vet. Behav.* **2024**, *73*, 31–40. [CrossRef]
80. Ruffaner-Hanson, C.; Noor, S.; Sun, M.S.; Solomon, E.; Marquez, L.E.; Rodriguez, D.E.; Allan, A.M.; Caldwell, K.K.; Bakhireva, L.N.; Milligan, E.D. The maternal-placental-fetal interface: Adaptations of the HPA axis and immune mediators following maternal stress and prenatal alcohol exposure. *Exp. Neurol.* **2022**, *355*, 114121. [CrossRef]
81. Stirrat, L.I.; Sengers, B.G.; Norman, J.E.; Homer, N.Z.; Andrew, R.; Lewis, R.M.; Reynolds, R.M. Transfer and metabolism of cortisol by the isolated perfused human placenta. *J. Clin. Endocrinol. Metab.* **2018**, *103*, 640–648. [CrossRef]
82. Fowden, A.L.; Forhead, A.J. Glucocorticoids as regulatory signals during intrauterine development. *Exp. Physiol.* **2015**, *100*, 1477–1487. [CrossRef]
83. Chapman, K.; Holmes, M.; Seckl, J. 11β-hydroxysteroid dehydrogenases: Intracellular gate-keepers of tissue glucocorticoid action. *Physiol. Rev.* **2013**, *93*, 1139–1206. [CrossRef]

84. Zhu, P.; Wang, W.; Zuo, R.; Sun, K. Mechanisms for establishment of the placental glucocorticoid barrier, a guard for life. *Cell. Mol. Life Sci.* **2019**, *76*, 13–26. [CrossRef]
85. Shallie, P.D.; Naicker, T. The placenta as a window to the brain: A review on the role of placental markers in prenatal programming of neurodevelopment. *Int. J. Dev. Neurosci.* **2019**, *73*, 41–49. [CrossRef]
86. Waldorf, K.M.A. Maternal-fetal immunology. In *Obstetrics: Normal and Problem Pregnancies*; Elsevier: Amsterdam, The Netherlands, 2016; Volume 7, pp. 64–82.
87. Volqvartz, T.; Andersen, H.H.B.; Pedersen, L.H.; Larsen, A. Obesity in pregnancy—Long-term effects on offspring hypothalamic-pituitary-adrenal axis and associations with placental cortisol metabolism: A systematic review. *Eur. J. Neurosci.* **2023**, *58*, 4393–4422. [CrossRef]
88. Koorneef, L.L.; Viho, E.M.; Wahl, L.F.; Meijer, O.C. Do corticosteroid receptor mRNA levels predict the expression of their target genes? *J. Endocr. Soc.* **2023**, *7*, bvac188. [CrossRef]
89. Lima, J.E.; Moreira, N.C.; Sakamoto-Hojo, E.T. Mechanisms underlying the pathophysiology of type 2 diabetes: From risk factors to oxidative stress, metabolic dysfunction, and hyperglycemia. *Mutat. Res./Genet. Toxicol. Environ. Mutagen.* **2022**, *874*, 503437. [CrossRef]
90. Rachdaoui, N. Insulin: The friend and the foe in the development of type 2 diabetes mellitus. *Int. J. Mol. Sci.* **2020**, *21*, 1770. [CrossRef]
91. Cho, N.H.; Shaw, J.; Karuranga, S.; Huang, Y.; da Rocha Fernandes, J.; Ohlrogge, A.; Malanda, B. IDF Diabetes Atlas: Global estimates of diabetes prevalence for 2017 and projections for 2045. *Diabetes Res. Clin. Pract.* **2018**, *138*, 271–281. [CrossRef]
92. Goedecke, J.H.; Mtintsilana, A.; Dlamini, S.N.; Kengne, A.P. Type 2 diabetes mellitus in African women. *Diabetes Res. Clin. Pract.* **2017**, *123*, 87–96. [CrossRef]
93. Nanditha, A.; Chamukuttan, S.; Raghavan, A.; Ramachandran, A. Global Epidemic of Type 2 Diabetes Mellitus: An Epidemiologist's Perspective. In *Current Trends in Diabetes*; JP Medical Publishers: Ashland, OH, USA, 2020; p. 36.
94. Kivimäki, M.; Bartolomucci, A.; Kawachi, I. The multiple roles of life stress in metabolic disorders. *Nat. Rev. Endocrinol.* **2023**, *19*, 10–27. [CrossRef]
95. Champaneri, S.; Xu, X.; Carnethon, M.R.; Bertoni, A.G.; Seeman, T.; Roux, A.D.; Golden, S.H. Diurnal salivary cortisol and urinary catecholamines are associated with diabetes mellitus: The Multi-Ethnic Study of Atherosclerosis. *Metabolism* **2012**, *61*, 986–995. [CrossRef]
96. Kitzmiller, J.L.; Ferrara, A.; Peng, T.; Cissell, M.A.; Kim, C. Preexisting diabetes and pregnancy. In *Diabetes in America*, 3rd ed.; National Institutes of Health: Bethesda, MD, USA, 2018.
97. Bapayeva, G.; Terzic, S.; Dotlic, J.; Togyzbayeva, K.; Bugibaeva, U.; Mustafinova, M.; Alisheva, A.; Garzon, S.; Terzic, M.; Laganà, A.S. Pregnancy outcomes in women with diabetes mellitus–the impact of diabetes type and treatment. *Menopause Rev./Przegląd Menopauzalny* **2022**, *21*, 37–46. [CrossRef]
98. Chivese, T.; Hoegfeldt, C.A.; Werfalli, M.; Yuen, L.; Sun, H.; Karuranga, S.; Li, N.; Gupta, A.; Immanuel, J.; Divakar, H. IDF Diabetes Atlas: The prevalence of pre-existing diabetes in pregnancy–A systematic review and meta-analysis of studies published during 2010–2020. *Diabetes Res. Clin. Pract.* **2022**, *183*, 109049. [CrossRef]
99. Thong, E.P.; Codner, E.; Laven, J.S.; Teede, H. Diabetes: A metabolic and reproductive disorder in women. *Lancet Diabetes Endocrinol.* **2020**, *8*, 134–149. [CrossRef]
100. Johns, E.C.; Denison, F.C.; Reynolds, R.M. The impact of maternal obesity in pregnancy on placental glucocorticoid and macronutrient transport and metabolism. *Biochim. Biophys. Acta (BBA)-Mol. Basis Dis.* **2020**, *1866*, 165374. [CrossRef]
101. Valsamakis, G.; Papatheodorou, D.; Chalarakis, N.; Manolikaki, M.; Margeli, A.; Papassotiriou, I.; Barber, T.M.; Kumar, S.; Kalantaridou, S.; Mastorakos, G. Maternal chronic stress correlates with serum levels of cortisol, glucose and C-peptide in the fetus, and maternal non chronic stress with fetal growth. *Psychoneuroendocrinology* **2020**, *114*, 104591. [CrossRef]
102. Basu, M.; Garg, V. Maternal hyperglycemia and fetal cardiac development: Clinical impact and underlying mechanisms. *Birth Defects Res.* **2018**, *110*, 1504–1516. [CrossRef]
103. Mukerji, G.; Bacon, S.; Feig, D.S. Gestational diabetes and type 2 diabetes during pregnancy. In *Maternal-Fetal and Neonatal Endocrinology*; Elsevier: Amsterdam, The Netherlands, 2020; pp. 371–388.
104. Vrachnis, N.; Antonakopoulos, N.; Iliodromiti, Z.; Dafopoulos, K.; Siristatidis, C.; Pappa, K.I.; Deligeoroglou, E.; Vitoratos, N. Impact of maternal diabetes on epigenetic modifications leading to diseases in the offspring. *J. Diabetes Res.* **2012**, *2012*, 538474. [CrossRef]
105. Ornoy, A.; Reece, E.A.; Pavlinkova, G.; Kappen, C.; Miller, R.K. Effect of maternal diabetes on the embryo, fetus, and children: Congenital anomalies, genetic and epigenetic changes and developmental outcomes. *Birth Defects Res. Part C Embryo Today Rev.* **2015**, *105*, 53–72. [CrossRef]
106. Ornoy, A.; Becker, M.; Weinstein-Fudim, L.; Ergaz, Z. Diabetes during pregnancy: A maternal disease complicating the course of pregnancy with long-term deleterious effects on the offspring. a clinical review. *Int. J. Mol. Sci.* **2021**, *22*, 2965. [CrossRef]
107. Aboud, F.; Torjman, F.; Sultan, M.; Benjama, A. Histopathological changes of umbilical cord blood vessels in diabetic pregnancies. *Tripolitan Med. J.* **2018**, *7*, 1–6.
108. Suma, M. Pregnancy Outcome in Gestational Diabetes Mellitus. Master's Thesis, Rajiv Gandhi University of Health Sciences, Bengaluru, India, 2011.

109. Mackin, S.T. Vascular Function in Hyperglycaemic Pregnancy: Studies into Potential Mechanisms for Adverse Obstetric Outcomes in Diabetes. Ph.D. Thesis, University of Glasgow, Glasgow, UK, 2021.
110. Sullivan, S.D.; Umans, J.G.; Ratner, R. Hypertension complicating diabetic pregnancies: Pathophysiology, management, and controversies. *J. Clin. Hypertens.* **2011**, *13*, 275–284. [CrossRef]
111. Mota, R.I.; Morgan, S.E.; Bahnson, E.M. Diabetic vasculopathy: Macro and microvascular injury. *Curr. Pathobiol. Rep.* **2020**, *8*, 1–14. [CrossRef]
112. Biesenbach, G.; Grafinger, P.; Zazgornik, J.; Stöger, H. Perinatal complications and three-year follow up of infants of diabetic mothers with diabetic nephropathy stage IV. *Ren. Fail.* **2000**, *22*, 573–580. [CrossRef]
113. Fowden, A.; Valenzuela, O.; Vaughan, O.; Jellyman, J.; Forhead, A. Glucocorticoid programming of intrauterine development. *Domest. Anim. Endocrinol.* **2016**, *56*, S121–S132. [CrossRef]
114. Lu, M.; Sferruzzi-Perri, A.N. Placental mitochondrial function in response to gestational exposures. *Placenta* **2021**, *104*, 124–137. [CrossRef]
115. Nugent, J.L. Effects of Glucocorticoids on Placental Development and Function: Implications for Fetal Growth Restriction. Ph.D. Thesis, The University of Manchester, Manchester, UK, 2012.
116. Cheong, J.N.; Wlodek, M.E.; Moritz, K.M.; Cuffe, J.S. Programming of maternal and offspring disease: Impact of growth restriction, fetal sex and transmission across generations. *J. Physiol.* **2016**, *594*, 4727–4740. [CrossRef]
117. Agorastos, A.; Pervanidou, P.; Chrousos, G.P.; Baker, D.G. Developmental trajectories of early life stress and trauma: A narrative review on neurobiological aspects beyond stress system dysregulation. *Front. Psychiatry* **2019**, *10*, 377300. [CrossRef]
118. Korgun, E.T.; Ozmen, A.; Unek, G.; Mendilcioglu, I. The effects of glucocorticoids on fetal and placental development. In *Glucocorticoids—New Recognition of Our Familiar Friend*; IntechOpen Limited: London, UK, 2012.
119. Vafaei-Nezhad, S.; Hami, J.; Sadeghi, A.; Ghaemi, K.; Hosseini, M.; Abedini, M.; Haghir, H. The impacts of diabetes in pregnancy on hippocampal synaptogenesis in rat neonates. *Neuroscience* **2016**, *318*, 122–133. [CrossRef]
120. Sadeghi, A.; Asghari, H.; Hami, J.; Roodi, M.M.; Mostafaee, H.; Karimipour, M.; Namavar, M.; Idoon, F. Volumetric investigation of the hippocampus in rat offspring due to diabetes in pregnancy—A stereological study. *J. Chem. Neuroanat.* **2019**, *101*, 101669. [CrossRef]
121. Rakers, F.; Rupprecht, S.; Dreiling, M.; Bergmeier, C.; Witte, O.W.; Schwab, M. Transfer of maternal psychosocial stress to the fetus. *Neurosci. Biobehav. Rev.* **2020**, *117*, 185–197. [CrossRef]
122. McGowan, P.O.; Matthews, S.G. Prenatal stress, glucocorticoids, and developmental programming of the stress response. *Endocrinology* **2018**, *159*, 69–82. [CrossRef]
123. Van den Bergh, B.R.; van den Heuvel, M.I.; Lahti, M.; Braeken, M.; de Rooij, S.R.; Entringer, S.; Hoyer, D.; Roseboom, T.; Räikkönen, K.; King, S. Prenatal developmental origins of behavior and mental health: The influence of maternal stress in pregnancy. *Neurosci. Biobehav. Rev.* **2020**, *117*, 26–64. [CrossRef]
124. Manojlović-Stojanoski, M.; Nestorović, N.; Milošević, V. Prenatal glucocorticoids: Short-term benefits and long-term risks. In *Glucocorticoids—New Recognition of Our Familiar Friend*; Tech Open Access Publisher: Rijeka, Croatia, 2012; pp. 337–390.
125. Rooij, S.R.d. Prenatal Diet and Stress Responsiveness. In *Handbook of Behavior, Food and Nutrition*; Springer: New York, NY, USA, 2011; pp. 2023–2039.
126. Moisiadis, V.G.; Matthews, S.G. Glucocorticoids and fetal programming part 2: Mechanisms. *Nat. Rev. Endocrinol.* **2014**, *10*, 403–411. [CrossRef]
127. de Kloet, E.R.; Claessens, S.E.; Kentrop, J. Context modulates outcome of perinatal glucocorticoid action in the brain. *Front. Endocrinol.* **2014**, *5*, 100. [CrossRef] [PubMed]
128. Sallam, N.A.; Palmgren, V.A.; Singh, R.D.; John, C.M.; Thompson, J.A. Programming of vascular dysfunction in the intrauterine milieu of diabetic pregnancies. *Int. J. Mol. Sci.* **2018**, *19*, 3665. [CrossRef]
129. Sharma, D.; Shastri, S.; Sharma, P. Intrauterine growth restriction: Antenatal and postnatal aspects. *Clin. Med. Insights Pediatr.* **2016**, *10*, 67–83. [CrossRef]
130. Kesavan, K.; Devaskar, S.U. Intrauterine growth restriction: Postnatal monitoring and outcomes. *Pediatr. Clin.* **2019**, *66*, 403–423.
131. Parikh, N.I.; Gonzalez, J.M.; Anderson, C.A.; Judd, S.E.; Rexrode, K.M.; Hlatky, M.A.; Gunderson, E.P.; Stuart, J.J.; Vaidya, D.; American Heart Association Council on Epidemiology and Prevention; et al. Adverse pregnancy outcomes and cardiovascular disease risk: Unique opportunities for cardiovascular disease prevention in women: A scientific statement from the American Heart Association. *Circulation* **2021**, *143*, e902–e916. [CrossRef]
132. Burd, I.; Baschat, A.; Costantine, M. *Prenatal Beginnings for Better Health*; Frontiers Media SA: Lausanne, Switzerland, 2018.
133. Robinson, N. Early Life Risk Factors and Epigenetic Biomarkers of Obesity across the Life Course. Ph.D. Thesis, Newcastle University, Newcastle upon Tyne, UK, 2020.
134. López-Gambero, A.; Martínez, F.; Salazar, K.; Cifuentes, M.; Nualart, F. Brain glucose-sensing mechanism and energy homeostasis. *Mol. Neurobiol.* **2019**, *56*, 769–796. [CrossRef]
135. Nimgampalle, M.; Chakravarthy, H.; Devanathan, V. Glucose metabolism in the brain: An update. In *Recent Developments in Applied Microbiology and Biochemistry*; Elsevier: Amsterdam, The Netherlands, 2021; pp. 77–88.
136. Tartour, A.; Chivese, T.; Eltayeb, S.; Elamin, F.M.; Fthenou, E.; Ahmed, M.S.; Babu, G.R. Prenatal Psychological Distress and 11β-HSD2 Gene Expression in Human Placentas: Systematic Review and Meta-analysis. *Psychoneuroendocrinology* **2024**, *166*, 107060. [CrossRef] [PubMed]

137. Reynolds, R.M.; Walker, B.R.; Syddall, H.E.; Andrew, R.; Wood, P.J.; Whorwood, C.B.; Phillips, D.I. Altered control of cortisol secretion in adult men with low birth weight and cardiovascular risk factors. *J. Clin. Endocrinol. Metab.* **2001**, *86*, 245–250. [CrossRef] [PubMed]
138. Dunlop, K.; Sarr, O.; Stachura, N.; Zhao, L.; Nygard, K.; Thompson, J.A.; Hadway, J.; Richardson, B.S.; Bureau, Y.; Borradaile, N. Differential and synergistic effects of low birth weight and western diet on skeletal muscle vasculature, mitochondrial lipid metabolism and insulin signaling in male Guinea pigs. *Nutrients* **2021**, *13*, 4315. [CrossRef]
139. Singhal, A. Should we promote catch-up growth or growth acceleration in low-birthweight infants? In *Low-Birthweight Baby: Born too Soon or too Small*; Karger: Basel, Switzerland, 2015; Volume 81, pp. 51–60.
140. Möllers, L.S.; Yousuf, E.I.; Hamatschek, C.; Morrison, K.M.; Hermanussen, M.; Fusch, C.; Rochow, N. Metabolic-endocrine disruption due to preterm birth impacts growth, body composition, and neonatal outcome. *Pediatr. Res.* **2022**, *91*, 1350–1360. [CrossRef]
141. Marcovecchio, M.L.; Gorman, S.; Watson, L.P.; Dunger, D.B.; Beardsall, K. Catch-up growth in children born small for gestational age related to body composition and metabolic risk at six years of age in the UK. *Horm. Res. Paediatr.* **2020**, *93*, 119–127. [CrossRef]
142. Dulloo, A.G. Thrifty energy metabolism in catch-up growth trajectories to insulin and leptin resistance. *Best Pract. Res. Clin. Endocrinol. Metab.* **2008**, *22*, 155–171. [CrossRef] [PubMed]
143. Kelishadi, R.; Haghdoost, A.A.; Jamshidi, F.; Aliramezany, M.; Moosazadeh, M. Low birthweight or rapid catch-up growth: Which is more associated with cardiovascular disease and its risk factors in later life? A systematic review and cryptanalysis. *Paediatr. Int. Child Health* **2015**, *35*, 110–123. [CrossRef] [PubMed]
144. Dasinger, J.H.; Alexander, B.T. Gender differences in developmental programming of cardiovascular diseases. *Clin. Sci.* **2016**, *130*, 337–348. [CrossRef] [PubMed]
145. Tarnovski, L.; Brinar, I.V.; Kirhmajer, M.V.; Vrkic, T.Z.; Laganovic, M. Sex Differences in Cardiovascular Risk Factors and Renal Function among Young Adults after Intrauterine Growth Restriction. *Acta Clin. Croat.* **2021**, *60*, 164–173.
146. Andersson, S.W.; Lapidus, L.; Niklasson, A.; Hallberg, L.; Bengtsson, C.; Hulthén, L. Blood pressure and hypertension in middle-aged women in relation to weight and length at birth: A follow-up study. *J. Hypertens.* **2000**, *18*, 1753–1761. [CrossRef] [PubMed]
147. Barker, D.; Bull, A.R.; Osmond, C.; Simmonds, S.J. Fetal and placental size and risk of hypertension in adult life. *Br. Med. J.* **1990**, *301*, 259–262. [CrossRef] [PubMed]
148. Brar, P.C. Update on the current modalities used to screen high risk youth for prediabetes and/or type 2 diabetes mellitus. *Ann. Pediatr. Endocrinol. Metab.* **2019**, *24*, 71. [CrossRef] [PubMed]
149. Tura, A.; Grespan, E.; Göbl, C.S.; Koivula, R.W.; Franks, P.W.; Pearson, E.R.; Walker, M.; Forgie, I.M.; Giordano, G.N.; Pavo, I. Profiles of glucose metabolism in different prediabetes phenotypes, classified by fasting glycemia, 2-hour OGTT, glycated hemoglobin, and 1-hour OGTT: An IMI DIRECT study. *Diabetes* **2021**, *70*, 2092–2106. [CrossRef] [PubMed]
150. Yip, W.C.; Sequeira, I.R.; Plank, L.D.; Poppitt, S.D. Prevalence of pre-diabetes across ethnicities: A review of impaired fasting glucose (IFG) and impaired glucose tolerance (IGT) for classification of dysglycaemia. *Nutrients* **2017**, *9*, 1273. [CrossRef]
151. Succurro, E.; Pedace, E.; Andreozzi, F.; Papa, A.; Vizza, P.; Fiorentino, T.V.; Perticone, F.; Veltri, P.; Cascini, G.L.; Sesti, G. Reduction in global myocardial glucose metabolism in subjects with 1-hour postload hyperglycemia and impaired glucose tolerance. *Diabetes Care* **2020**, *43*, 669–676. [CrossRef]
152. Penhaligan, J.; Sequeira-Bisson, I.R.; Miles-Chan, J.L. The role of postprandial thermogenesis in the development of impaired glucose tolerance and type II diabetes. *Am. J. Physiol.-Endocrinol. Metab.* **2023**, *325*, E171–E179. [CrossRef] [PubMed]
153. Mittendorfer, B.; Patterson, B.W.; Smith, G.I.; Yoshino, M.; Klein, S. β cell function and plasma insulin clearance in people with obesity and different glycemic status. *J. Clin. Investig.* **2022**, *132*, e154068. [CrossRef] [PubMed]
154. Papakonstantinou, E.; Oikonomou, C.; Nychas, G.; Dimitriadis, G.D. Effects of diet, lifestyle, chrononutrition and alternative dietary interventions on postprandial glycemia and insulin resistance. *Nutrients* **2022**, *14*, 823. [CrossRef] [PubMed]
155. Roncero-Ramos, I.; Alcala-Diaz, J.F.; Rangel-Zuñiga, O.A.; Gomez-Delgado, F.; Jimenez-Lucena, R.; García-Rios, A.; Vals-Delgado, C.; Romero-Baldonado, C.; Luque, R.M.; Ordovas, J.M. Prediabetes diagnosis criteria, type 2 diabetes risk and dietary modulation: The CORDIOPREV study. *Clin. Nutr.* **2020**, *39*, 492–500. [CrossRef] [PubMed]
156. Bergman, M.; Manco, M.; Sesti, G.; Dankner, R.; Pareek, M.; Jagannathan, R.; Chetrit, A.; Abdul-Ghani, M.; Buysschaert, M.; Olsen, M.H. Petition to replace current OGTT criteria for diagnosing prediabetes with the 1-hour post-load plasma glucose $\geq$ 155 mg/dL (8.6 mmol/L). *Diabetes Res. Clin. Pract.* **2018**, *146*, 18–33. [CrossRef] [PubMed]
157. Herman, W.H. Prediabetes Diagnosis and Management. *JAMA* **2023**, *329*, 1157–1159. [CrossRef] [PubMed]
158. Basit, A.; Fawwad, A.; Basit, K.A.; Waris, N.; Tahir, B.; Siddiqui, I.A. Glycated hemoglobin (HbA1c) as diagnostic criteria for diabetes: The optimal cut-off points values for the Pakistani population; a study from second National Diabetes Survey of Pakistan (NDSP) 2016–2017. *BMJ Open Diabetes Res. Care* **2020**, *8*, e001058. [CrossRef] [PubMed]
159. Rett, K.; Gottwald-Hostalek, U. Understanding prediabetes: Definition, prevalence, burden and treatment options for an emerging disease. *Curr. Med. Res. Opin.* **2019**, *35*, 1529–1534. [CrossRef]
160. Ekta; Gupta, M.; Kaur, A.; Singh, T.G.; Bedi, O. Pathobiological and molecular connections involved in the high fructose and high fat diet induced diabetes associated nonalcoholic fatty liver disease. *Inflamm. Res.* **2020**, *69*, 851–867. [CrossRef]

161. Aimaretti, E.; Chimienti, G.; Rubeo, C.; Di Lorenzo, R.; Trisolini, L.; Dal Bello, F.; Moradi, A.; Collino, M.; Lezza, A.M.S.; Aragno, M. Different Effects of High-Fat/High-Sucrose and High-Fructose Diets on Advanced Glycation End-Product Accumulation and on Mitochondrial Involvement in Heart and Skeletal Muscle in Mice. *Nutrients* **2023**, *15*, 4874. [CrossRef]
162. Sangrós, F.J.; Torrecilla, J.; Giraldez-Garcia, C.; Carrillo, L.; Mancera, J.; Mur, T.; Franch, J.; Diez, J.; Goday, A.; Serrano, R. Association of general and abdominal obesity with hypertension, dyslipidemia and prediabetes in the PREDAPS Study. *Rev. Esp. Cardiol.* **2018**, *71*, 170–177. [CrossRef]
163. Hirano, T. Pathophysiology of diabetic dyslipidemia. *J. Atheroscler. Thromb.* **2018**, *25*, 771–782. [CrossRef]
164. Chi, J.H.; Lee, B.J. Risk factors for hypertension and diabetes comorbidity in a Korean population: A cross-sectional study. *PLoS ONE* **2022**, *17*, e0262757. [CrossRef]
165. Lee, S.-H.; Park, S.-Y.; Choi, C.S. Insulin resistance: From mechanisms to therapeutic strategies. *Diabetes Metab. J.* **2022**, *46*, 15. [CrossRef] [PubMed]
166. Cignarelli, A.; Genchi, V.A.; Perrini, S.; Natalicchio, A.; Laviola, L.; Giorgino, F. Insulin and insulin receptors in adipose tissue development. *Int. J. Mol. Sci.* **2019**, *20*, 759. [CrossRef] [PubMed]
167. Losada-Barragán, M. Physiological effects of nutrients on insulin release by pancreatic beta cells. *Mol. Cell. Biochem.* **2021**, *476*, 3127–3139. [CrossRef]
168. Gil-Rivera, M.; Medina-Gali, R.M.; Martínez-Pinna, J.; Soriano, S. Physiology of pancreatic β-cells: Ion channels and molecular mechanisms implicated in stimulus-secretion coupling. *Int. Rev. Cell Mol. Biol.* **2021**, *359*, 287–323. [PubMed]
169. Jing, J.; Li, J.; Yan, N.; Li, N.; Liu, X.; Li, X.; Zhang, J.; Wang, Q.; Yang, C.; Qiu, J. Increased TG Levels and HOMA-IR Score Are Associated With a High Risk of Prediabetes: A Prospective Study. *Asia Pac. J. Public Health* **2023**, *35*, 413–419. [CrossRef] [PubMed]
170. Hostalek, U. Global epidemiology of prediabetes-present and future perspectives. *Clin. Diabetes Endocrinol.* **2019**, *5*, 5. [CrossRef]
171. Carmichael, J.; Fadavi, H.; Ishibashi, F.; Shore, A.C.; Tavakoli, M. Advances in screening, early diagnosis and accurate staging of diabetic neuropathy. *Front. Endocrinol.* **2021**, *12*, 671257. [CrossRef]
172. Hill, J.O.; Galloway, J.M.; Goley, A.; Marrero, D.G.; Minners, R.; Montgomery, B.; Peterson, G.E.; Ratner, R.E.; Sanchez, E.; Aroda, V.R. Scientific statement: Socioecological determinants of prediabetes and type 2 diabetes. *Diabetes Care* **2013**, *36*, 2430–2439. [CrossRef]
173. Mohan, V.; Sudha, V.; Shobana, S.; Gayathri, R.; Krishnaswamy, K. Are unhealthy diets contributing to the rapid rise of type 2 diabetes in India? *J. Nutr.* **2023**, *153*, 940–948. [CrossRef]
174. Sosibo, A.M.; Mzimela, N.C.; Ngubane, P.S.; Khathi, A. Prevalence of pre-diabetes in adults aged 25–45 years in a Durban-based clinical setting, South Africa: A retrospective study. *Prim. Care Diabetes* **2023**, *17*, 650–654. [CrossRef]
175. Al-Rifai, R.H.; Majeed, M.; Qambar, M.A.; Ibrahim, A.; AlYammahi, K.M.; Aziz, F. Type 2 diabetes and pre-diabetes mellitus: A systematic review and meta-analysis of prevalence studies in women of childbearing age in the Middle East and North Africa, 2000–2018. *Syst. Rev.* **2019**, *8*, 268. [CrossRef]
176. Andes, L.J.; Cheng, Y.J.; Rolka, D.B.; Gregg, E.W.; Imperatore, G. Prevalence of prediabetes among adolescents and young adults in the United States, 2005-2016. *JAMA Pediatr.* **2020**, *174*, e194498. [CrossRef]
177. Bansal, N. Prediabetes diagnosis and treatment: A review. *World J. Diabetes* **2015**, *6*, 296. [CrossRef] [PubMed]
178. Mzimela, N.C.; Ngubane, P.S.; Khathi, A. The changes in immune cell concentration during the progression of pre-diabetes to type 2 diabetes in a high-fat high-carbohydrate diet-induced pre-diabetic rat model. *Autoimmunity* **2019**, *52*, 27–36. [CrossRef] [PubMed]
179. Naidoo, K.; Ngubane, P.S.; Khathi, A. Investigating the Effects of Diet-Induced Pre-Diabetes on the Functioning of Calcium-Regulating Organs in Male Sprague Dawley Rats: Effects on Selected Markers. *Front. Endocrinol.* **2022**, *13*, 914189. [CrossRef]
180. Chakraborty, S.; Verma, A.; Garg, R.; Singh, J.; Verma, H. Cardiometabolic Risk Factors Associated with Type 2 Diabetes Mellitus: A Mechanistic Insight. *Clin. Med. Insights Endocrinol. Diabetes* **2023**, *16*, 11795514231220780. [CrossRef] [PubMed]
181. Ludidi, A.; Siboto, A.; Nkosi, A.; Xulu, N.D.; Khathi, A.; Sibiya, N.H.; Ngubane, P.S. High-fat, high-carbohydrate diet-induced prediabetes preconception in Sprague–Dawley rats as a risk factor for the development of preeclampsia: Assessing changes in placental metabolic insults. *Front. Nutr.* **2023**, *10*, 1241785. [CrossRef]
182. Alatrach, M.; Agyin, C.; Adams, J.; DeFronzo, R.A.; Abdul-Ghani, M.A. Decreased basal hepatic glucose uptake in impaired fasting glucose. *Diabetologia* **2017**, *60*, 1325–1332. [CrossRef] [PubMed]
183. Hudish, L.I.; Reusch, J.E.; Sussel, L. β Cell dysfunction during progression of metabolic syndrome to type 2 diabetes. *J. Clin. Investig.* **2019**, *129*, 4001–4008. [CrossRef] [PubMed]
184. Katugampola, H.; Gevers, E.F.; Dattani, M.T. Fetal Endocrinology. In *Brook's Clinical Pediatric Endocrinology*; John Wiley & Sons Ltd.: Hoboken, NJ, USA, 2019; pp. 47–104.
185. Hoffman, D.J.; Reynolds, R.M.; Hardy, D.B. Developmental origins of health and disease: Current knowledge and potential mechanisms. *Nutr. Rev.* **2017**, *75*, 951–970. [CrossRef] [PubMed]
186. Zygula, A.; Kosinski, P.; Wielgos, M. Saliva, hair, tears, and other biological materials obtained non-invasively for diagnosis in pregnancy: A literature review. *Ginekol. Pol.* **2019**, *90*, 475–481. [CrossRef] [PubMed]

biomedicines

MDPI

Article

Associations between Skin Autofluorescence Levels with Cardiovascular Risk and Diabetes Complications in Patients with Type 2 Diabetes

Delia Reurean-Pintilei [1,2,3], Anca Pantea Stoian [4], Teodor Salmen [5,*], Roxana-Adriana Stoica [5,*], Liliana Mititelu-Tartau [6], Sandra Lazăr [2,7,8] and Bogdan Timar [2,9,10]

1 Doctoral School of Medicine and Pharmacy, "Victor Babes" University of Medicine and Pharmacy, 300041 Timisoara, Romania; drdeliapintilei@gmail.com
2 Centre for Molecular Research in Nephrology and Vascular Disease, "Victor Babes" University of Medicine and Pharmacy, 300041 Timisoara, Romania
3 Department of Diabetes, Nutrition and Metabolic Diseases, Consultmed Medical Centre, 700544 Iasi, Romania
4 Diabetes, Nutrition and Metabolic Diseases Department, "Carol Davila" University of Medicine and Pharmacy, 050474 Bucharest, Romania
5 Doctoral School of "Carol Davila", University of Medicine and Pharmacy, 050474 Bucharest, Romania
6 Department of Pharmacology, Faculty of Medicine, "Grigore T. Popa" University of Medicine and Pharmacy, 700115 Iasi, Romania
7 First Department of Internal Medicine, "Victor Babes" University of Medicine and Pharmacy, 300041 Timisoara, Romania
8 Department of Hematology, Emergency Municipal Hospital Timisoara, 300041 Timisoara, Romania
9 Second Department of Internal Medicine, "Victor Babes" University of Medicine and Pharmacy, 300041 Timisoara, Romania
10 Department of Diabetes, Nutrition and Metabolic Diseases, "Pius Brinzeu" Emergency Hospital, 300723 Timisoara, Romania
* Correspondence: teodor.salmen@drd.umfcd.ro (T.S.); roxana.stoica@drd.umfcd.ro (R.-A.S.)

Citation: Reurean-Pintilei, D.; Pantea Stoian, A.; Salmen, T.; Stoica, R.-A.; Mititelu-Tartau, L.; Lazăr, S.; Timar, B. Associations between Skin Autofluorescence Levels with Cardiovascular Risk and Diabetes Complications in Patients with Type 2 Diabetes. *Biomedicines* **2024**, *12*, 890. https://doi.org/10.3390/biomedicines12040890

Academic Editor: Juergen Schrezenmeir

Received: 24 March 2024
Revised: 10 April 2024
Accepted: 16 April 2024
Published: 17 April 2024

Abstract: Advanced Glycation End Products (AGEs) contribute to the pathophysiology of type 2 diabetes mellitus (T2DM) and cardiovascular (CV) diseases (CVDs), making their non-invasive assessment through skin autofluorescence (SAF) increasingly important. This study aims to investigate the relationship between SAF levels, cardiovascular risk, and diabetic complications in T2DM patients. We conducted a single-center, cross-sectional study at Consultmed Hospital in Iasi, Romania, including 885 T2DM patients. The assessment of SAF levels was performed with the AGE Reader™, (Diagnoptics, Groningen, The Netherlands). CVD prevalence was 13.9%, and according to CV risk category distribution, 6.1% fell into the moderate-risk, 1.13% into the high-risk, and 92.77% into the very-high-risk category. The duration of DM averaged 9.0 ± 4.4 years and the mean HbA1c was 7.1% ± 1.3. After adjusting for age and eGFR, HbA1c values showed a correlation with SAF levels in the multivariate regression model, where a 1 SD increase in HbA1c was associated with a 0.105 SD increase in SAF levels (Nagelkerke R^2 = 0.110; $p < 0.001$). For predicting very high risk with an SAF cut-off of 2.35, sensitivity was 67.7% and specificity was 56.2%, with an AUC of 0.634 (95% CI 0.560–0.709, $p = 0.001$). In T2DM, elevated SAF levels were associated with higher CV risk and HbA1c values, with 2.35 identified as the optimal SAF cut-off for very high CV risk.

Keywords: Advanced Glycation End Products (AGEs); skin autofluorescence; type 2 diabetes mellitus (T2DM); cardiovascular risk

1. Introduction

Advanced Glycation End Products (AGEs) are detrimental byproducts of hyperglycemia in diabetes mellitus (DM) and aging in the non-diabetic population, with their accumulation exacerbating glucose metabolism imbalances, metabolic syndrome (MetS), and chronic diseases, including cardiovascular (CV) disease, chronic kidney disease (CKD), and neurodegenerative disorders [1–9]. These products are formed through the non-enzymatic glycation of biomolecules, such as proteins, lipids, and nucleic acids, with an alternative pathway involving the reaction with α-dicarbonyl compounds, which are highly reactive and able to establish covalent bonds [3,10,11]. Upon binding to the receptor for AGEs (RAGE), a cascade of cellular oxidative stress and inflammatory pathways is triggered [12]. AGEs possess the ability to create durable chemical bonds with nearby proteins, especially through cross-linking, altering their structure and function in a process that bypasses enzymatic catalysis [11]. This increase in cross-links, induced by AGEs, leads to a reduction in tissue elasticity and flexibility, thus contributing to the pathogenesis of atherosclerosis, elevated vascular resistance, and CV events (CVEs) [13,14].

In recent years, the emphasis on non-invasive assessment techniques has grown, with skin autofluorescence (SAF) emerging as a pivotal tool. Utilizing the fluorescent properties of certain AGEs, SAF has been recognized as a reliable method for quantifying AGE levels in both diabetic and non-diabetic populations. This approach has gained attention for its ability to deliver precise and reproducible results without necessitating invasive procedures [15–18]. SAF has been identified as a practical, straightforward method facilitating DM screening in populations without a prior DM diagnosis [18], and has been shown to refine the FINDRISC model, enhancing DM detection strategies [19]. Moreover, its predictive capacity for CV disease (CVD) and mortality has been documented, positioning SAF as an important tool in expanding the understanding of DM screening and CV risk assessment [20–22].

The link between SAF and CV mortality in individuals with CKD, DM, or peripheral artery disease (PAD) further emphasizes AGEs' contribution to vascular complications in CVD through inflammatory mechanisms. Compared with established CV measures, like pulse wave velocity and intima-media thickness, SAF's simplicity and effectiveness offer a comparative advantage, facilitating vascular health screenings [23,24]. Moreover, SAF's ability to detect subclinical atherosclerosis, independent of conventional CV risk factors, underscores its significance in identifying early signs of arterial damage, as illustrated in studies by Pan et al. [25]. The association of SAF with adverse CVEs in heart failure (HF) patients also underscores its role in risk stratification [26]. Although reference values for SAF exist for Caucasian populations, significant differences have been noted between various European groups [27]. Calls for more detailed clinical analysis and the creation of tailored reference data for specific demographics have been made since 2014, aiming to improve the test's utility and accuracy for a wide range of populations [28]. However, it is significant to note that, as of now, no study has specifically assessed the utility of SAF across a broader Eastern European context or the Romanian population.

This study aimed to evaluate the associations between SAF levels, CV risk, and DM chronic complications, within a large cohort of patients with T2DM.

2. Materials and Methods

2.1. Study Design, Protocol, and Patients

In a single-center, cross-sectional, non-interventional, and consecutive case population-based enrollment study, a total of 896 patients with T2DM were included. All participants were recruited during their regularly scheduled clinical visits between January and July 2019 at Consultmed Hospital in Iasi, Romania.

The inclusion criteria were as follows: patients aged 18 years or older with a confirmed diagnosis of T2DM who had received standard stable management for associated conditions for a minimum of three months, as per the medical guidelines in effect at that time.

Exclusion criteria for this study included patient refusal, individuals under 18 years of age, patients diagnosed with other forms of DM (secondary, type 1, latent autoimmune), those with mental illness, pregnant or breastfeeding women, and patients with an identified presence of autoimmune markers (GADA, ICA, IA-2A, and ZnT8-Ab antibodies).

Utilizing statistical information from the Iasi County population, which was reported as 944,074 inhabitants, with a margin of error of 3.29% and a confidence level of 95%, we calculated a representative sample consisting of at least 885 patients. The remaining patients were included in the analysis.

2.2. Demographic, Anthropometric, Clinical, and Laboratory Data Collection

This study involved an analysis of hospitalized patients' medical records, concentrating on sociodemographic and anthropometric characteristics, comorbidities, laboratory results, DM complications, and treatment plans. The data collected included variables such as residency, age, sex, weight, height, body mass index (BMI), high blood pressure (HBP), atherosclerotic CVD (ASCVD), dyslipidemia, HbA1c levels, estimated glomerular filtration rate (eGFR), uric acid, total cholesterol (total-C) and its fractions, SAF levels, DM complications, and prescribed medications.

2.3. Cardiovascular Risk Profile Assessment

The CV risk profile of the patients was assessed according to the ESC/EAS 2019 classification, which was applicable at the time of this study [29]. These guidelines divide patients into moderate-, high-, and very-high-risk categories according to their CV risk levels. Considering that all participants had T2DM, the low-risk category was deemed irrelevant, resulting in their classification into at least the moderate-risk group by default.

Patients under 50 years old with a T2DM duration of fewer than 10 years and no additional risk factors were categorized as moderate risk. The high-risk category included patients who had been living with T2DM for at least 10 years but showed no signs of target organ damage (TOD) (microalbuminuria, retinopathy, or neuropathy). This category also encompassed patients with significantly elevated individual risk factors, as specified by the ESC/EAS guidelines. Those with an eGFR between 30 and 59 mL/min/1.73 m^2 were classified as high risk as well [29].

Within the very-high-risk group, individuals meeting the following criteria were considered: T2DM diagnosis accompanied by TOD or those with at least three major risk factors. Additionally, this group included individuals with established ASCVD, as determined by the ESC/EAS guidelines. Patients with severe CKD, marked by an eGFR lower than 30 mL/min/1.73 m^2, were also placed in the very-high-risk category [29].

2.4. Measurements of SAF Levels

The assessment of SAF was performed non-invasively by placing intact skin on the ventral side of the forearm, following the manufacturer's instructions provided with the AGE Reader™, developed by DiagnOptics Technologies in Groningen, The Netherlands. This device measures the accumulation of AGEs within approximately 1 mm beneath the epidermis by utilizing the autofluorescent properties of specific AGEs. It features a 4 cm^2 aperture that allows light emission through a UV-A lamp. Upon contact, this emitted light reflects off the epidermis before being detected by a spectrometer within a wavelength range of 300 to 600 nanometers and is processed using specialized software. The results are then presented in arbitrary units (AUs). This methodology has undergone rigorous testing with additional technical specifications detailed in previous publications [15,17,18].

2.5. Diabetes Complication Assessment

T2DM complications, such as nephropathy, microalbuminuria, retinopathy, and diabetic sensory peripheral neuropathy (DSPN), were identified and included in this study.

The diagnosis of DSPN was confirmed either by a previously documented diagnosis of DSPN or by clinical examination incorporating at least two of the following tests: evaluation of thermal or pinprick sensibility (assessing small-fiber function) and assessment of vibration sensing using a 128 Hz tuning fork (evaluating large-fiber function) and the 10 g monofilament test. Cardiac autonomic neuropathy (CAN) was evaluated with Sudoscan®. The Sudoscan® device, developed by Impeto Medical in Paris, France, and FDA approved, offers a non-invasive method for assessing sudomotor function. It measures the electrochemical skin conductance (ESC) of sweat glands using electrodes for the hands and feet. The device assesses the release of chloride ions by sweat glands under low voltage, conducting a brief assessment that lasts approximately 2 min and does not require any special conditions or preparations. Sudoscan's algorithms combine ESC with demographic and health data to generate scores for CAN Risk (CANRS) and kidney dysfunction risk, providing insights into the risks of CAN and CKD [30–32]. The CAN score and nephropathy (NEPHRO) score were reported for each patient.

Identification of diabetic retinopathy (DR) was achieved either through a documented previous diagnosis in the patient's medical record or via an ophthalmologic examination conducted at our medical center, which included a fundoscopic evaluation.

CKD was defined as either a pre-existing diagnosis or persistent evidence of a urine albumin/creatinine ratio (UACR) equal to or greater than 30 mg/g, and/or a consistently low eGFR below 60 mL/min/1.73 m^2, and/or the presence of structural abnormalities or hematuria, as outlined in the guidelines [33].

2.6. Statistical Analysis

The information was structured into Excel databases and subjected to analysis using both the Excel program for Windows 11 and Gnu PSPP 1.4.1 software, enabling a thorough statistical assessment of the gathered parameters. Results are presented as mean (standard deviation—SD) or median (interquartile range—IQR). The distribution of data was tested using the Kolmogorov–Smirnov test. In univariate correlation analysis, Pearson or Spearman coefficients were selected depending on variables' distribution. First, a receiver operating curve (ROC) was used to identify the cut-off for SAF determination. The cut-off value was determined using Youden's method as the value in which the sum of sensitivity (Sn) and specificity (Spe) is maximized. Second, using logistic regression, we estimated the probability of the binary outcome based on the SAF cut-off. Additional multivariate regression models for predicting SAF as a continuous variable were generated. A p-value less than 0.05 was considered statistically significant.

2.7. Ethical Aspects of the Research

The investigation adhered to the principles outlined in the Declaration of Helsinki and received approval from the Institutional Ethics Committee of Consultmed Hospital in Iași County, Romania (protocol number CMD102018006, dated 18 October 2018). The objectives and methods of this study were explained to the patients, who consented to the utilization and publication of the data collected, and informed consent was obtained.

3. Results

During their regularly scheduled appointments at the Consultmed Hospital, the initial cohort of 1000 patients who met the inclusion criteria were invited to participate in this study. During the enrolment visit, medical records were collected without any subsequent study-related appointments being scheduled. Among the 896 patients who provided informed consent, 11 were excluded due to extreme laboratory results indicating compromised blood samples or errors in laboratory analysis (Figure 1).

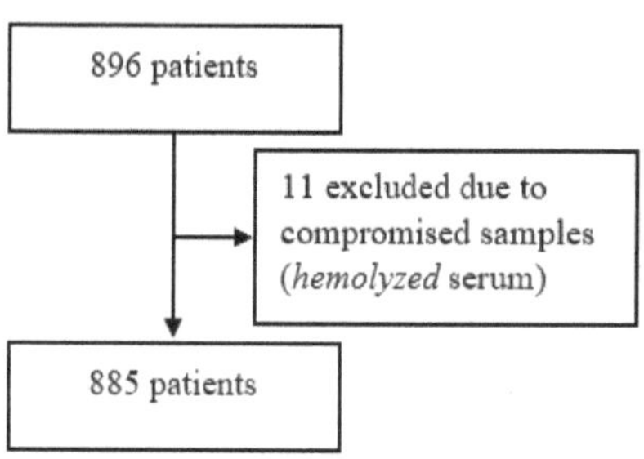

Figure 1. Patients inclusion flowchart.

Our study cohort demonstrated a relatively balanced distribution, consisting of 53.7% females and 46.3% males, with no statistically significant differences between genders. The mean age of the analyzed patient sample was 62.9 ± 7.7 years, with 74.69% residing in urban areas. Anthropometric characteristics included a mean BMI of 32.3 ± 5.3 kg/m^2, a mean waist circumference of 104.74 ± 11.7 cm, a mean hip circumference of 109.34 ± 9.94 cm, and a mean waist–hip ratio of 0.96 ± 0.074, as detailed in Table 1.

Table 1. Patients' characteristics.

Characteristic	*n* = 885
Demographics	
Urban, %, *n*	74.69% (661)
Age (years), mean (SD)	62.9 ± 7.7
Women, %, (*n*)	53.7% (475)
Anthropometrics	
BMI (kg/m^2), mean (SD)	32.3 ± 5.3
Waist circumference (cm), mean (SD)	104.74 ± 11.7
Hip circumference (cm), mean (SD)	109.34 ± 9.94
Waist–hip ratio, mean (SD)	0.96 ± 0.074

SD—standard deviation; BMI—body mass index.

Table 2 outlines the patient's risk factors. The prevalence of obesity stands at 64.6%, HBP at 83%, with a mean systolic blood pressure (BP) of 132 ± 16.2 mm Hg and a mean diastolic BP of 80 ± 9.6 mm Hg; 13.9% of patients present with CVD. The lipid profile includes total-C at 185.1 ± 43.3 mg/dL, triglycerides at 142 (93) mg/dL, low-density lipoprotein cholesterol at 107.73 ± 36.01 mg/dL, and high-density lipoprotein cholesterol at 44.9 ± 11.8 mg/dL; the mean eGFR is 87.5 ± 20.6 mL/min/1.73 m^2. The distribution of patients across CV risk categories is as follows: 6.1% are classified in the moderate-risk category, 1.13% in the high-risk category, and 92.77% in the very-high-risk category.

Table 2. Patients' risk factors.

Characteristic	*n* = 885
Risk factors	
Mean SAF level, mean (SD)	2.6 ± 0.5
Mean SAF level in the moderate CV risk category, mean (SD)	2.44 ± 0.55
Mean SAF level in the high CV risk category, mean (SD)	2.22 ± 0.29
Mean SAF level in the very high CV risk category, mean (SD)	2.61 ± 0.5
Obesity, %, (*n*)	64.6% (572)
HBP, %, (*n*)	83% (737)
SBP (mm Hg), mean (SD)	132 ± 16.2
DBP (mm Hg), mean (SD)	80 ± 9.6
CVD, %, (*n*)	13.9% (123)
Total-C (mg/dL), mean (SD)	185.1 ± 43.3
HDL-C (mg/dL), mean (SD)	44.9 ± 11.8
TGs (mg/dL), median (interquartile range)	142 (104 to 197)
LDL-C (mg/dL), mean (SD)	107.7 ± 36.0
Creatinin (mg/dL), median (interquartile range)	0.83 (0.68 to 0.9)

Table 2. *Cont.*

Characteristic	n = 885
eGFR (mL/min/1.73 m^2)	87.5 ± 20.6
Moderate CV risk category, %, n	6.1% (54)
High CV risk category, %, n	1.13% (10)
Very high CV risk category, %, n	92.77% (821)

SAF—skin auto fluorescence; SD—standard deviation; HBP—high blood pressure; SBP—systolic blood pressure; DBP—diastolic blood pressure; CVD—cardiovascular disease; total-C—total cholesterol; HDL-C—high-density lipoprotein cholesterol; TGs—triglycerides; LDL-C—low-density lipoprotein cholesterol; eGFR—estimated glomerular filtration rate.

Table 3 provides a summary of the DM characteristics among the patients, indicating an average DM duration of 9.0 ± 4.4 years, with no significant differences observed across genders; the mean HbA1c level is 7.1% ± 1.3. Current DM complications include DSPN at 67.9%, DR at 4.29%, CKD at 8.7%, albuminuria at 6.21%, and ASCVD at 13.9%.

Table 3. Diabetes mellitus characteristics.

DM Characteristics	n = 885
DM mean duration, mean (SD)	9.0 ± 4.4
HbA1c (%), mean (SD)	7.1 ± 1.3
DSPN, %, n	67.9% (601)
DR, %, n	4.29% (38)
CKD, %, n	8.7% (76)
Stage V CKD, %, n	0
Stage IV CKD, %, n	3 (0.34%)
Stage IIIa CKD, %, n	13 (1.47%)
Stage IIIb CKD, %, n	60 (6.78%)
Albuminuria, %, n	6.21% (55)

DM—diabetes mellitus; DSPN—diabetic sensory peripheral neuropathy; DR—diabetic retinopathy; CKD—chronic kidney disease.

The use of pharmacological therapies in the analyzed cohort is presented in Table 4.

In univariate analysis, SAF was significantly correlated with age (Pearson coefficient = 0.294, $p < 0.001$), CAN score (Pearson coefficient = 0.136, $p < 0.001$), and NEPHRO score (Pearson coefficient = −0.230, $p < 0.001$). There was a small significant correlation between SAF and HbA1c (Spearman's rho = 0.091, $p = 0.007$, Figure 2).

Additionally, after adjusting for age and eGFR in the multivariate regression model, HbA1c values were found to correlate with SAF levels. Specifically, for each increase of 1 standard deviation (SD) in HbA1c value, an observed increase of 0.105 SDs in SAF levels was noted (Nagelkerke $R^2 = 0.110$; $p < 0.001$).

We compared the mean SAF levels in T2DM patients across different HbA1c targets. The mean SAF level was significantly higher in subjects with an HbA1c > 7% (2.65 ± 0.53) compared to those with an HbA1c < 7% (2.56 ± 0.53), $p = 0.018$. Additionally, SAF levels and DM duration were found to be independent, r = −0.028, $p = 0.401$. Regarding CV risk categories, patients in the very-high-risk CV group exhibited significantly higher SAF levels (2.61) compared to those in the high-risk (2.22) and moderate-risk (2.44) groups, respectively, $p = 0.003$, as illustrated in Figure 3.

Table 4. Therapies used in the studied cohort.

Characteristic	*n* = 885
Glucose-lowering medication usage	
Insulin, %, (*n*)	25.2% (223)
Metformin, %, (*n*)	87.0% (687)
DPP-4i, %, (*n*)	13.0% (115)
GLP-1 RAs, %, (*n*)	8.1% (71)
SGLT2i, %, (*n*)	3.9% (34)
Sulfonylurea, %, (*n*)	13.1% (116)
Thiazolidinediones, %, (*n*)	1.35% (12)
Other therapies	
ACEi/ARBs, %, (*n*)	61.5% (544)
Calcium channel blockers, %, (*n*)	33.1% (293)
Beta-blockers, %, (*n*)	54.57% (483)
Antiagregant, %, (*n*)	44.85% (397)
Statin, %, (*n*)	67.0% (593)
Ezetimibe, %, (*n*)	4.5% (40)
Fibrate, %, (*n*)	8.7% (77)
Non-vitamin K antagonist oral anticoagulants, %, (*n*)	1.8% (16)
Vitamin K antagonists, %, (*n*)	1.12% (10)

DPP-4i—dipeptidyl peptidase 4 inhibitors; GLP-1 RAs—glucagon-like peptide 1 receptor agonist; SGLT2i—sodium-glucose cotransporter-2 inhibitors; ACEI/ARBs—angiotensin-converting enzyme inhibitors/angiotensin receptor blockers.

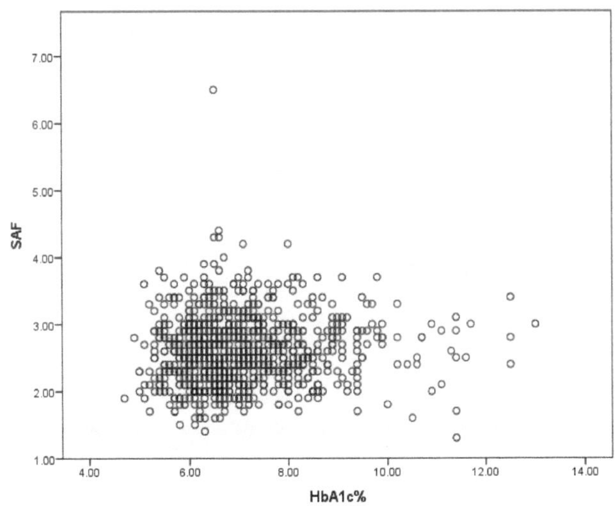

Figure 2. Correlation between SAF and HbA1c.

When we compared SAF levels in different DM complications groups—DSPN, DR, CKD, and albuminuria—we did not observe any significant differences, except for CKD, as seen in Table 5.

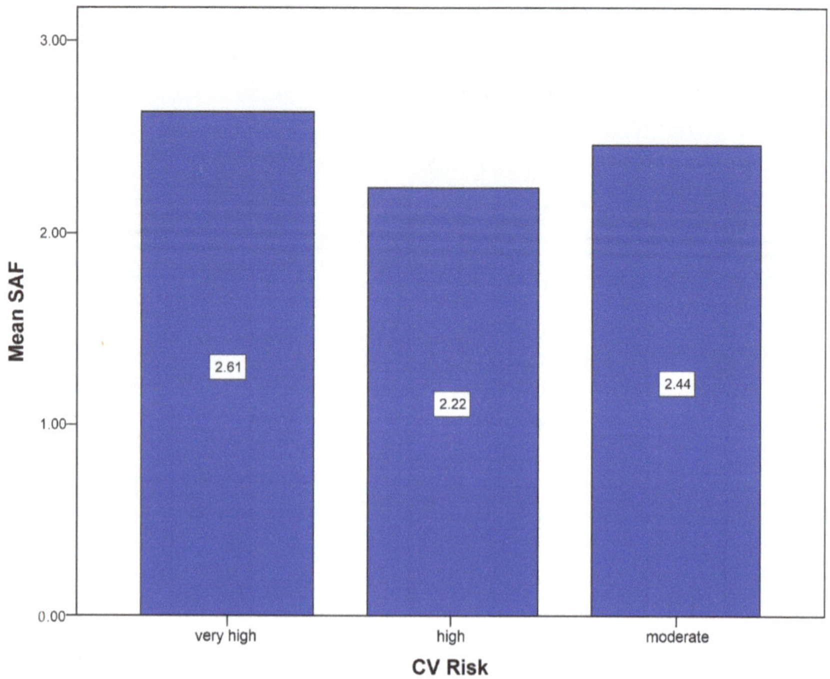

Figure 3. Mean SAF level depending on CV risk category.

Table 5. Mean SAF level depending on DM complications.

DM Complications	*n*	Present	Absent	Student's *t*-Test
DSPN	601	2.61 ± 0.51	2.60 ± 0.49	$p = 0.792$
DR	38	2.58 ± 0.44	2.61 ± 0.51	$p = 0.746$
CKD	76	2.76 ± 0.49	2.58 ± 0.51	$p = 0.003$
Albuminuria	55	2.56 ± 0.50	2.61 ± 0.51	$p = 0.497$

DM—diabetes mellitus; DSPN—diabetic sensory peripheral neuropathy; DR—diabetic retinopathy; CKD—chronic kidney disease.

When evaluating the cut-off value for SAF levels in predicting very high CV risk (2.35), an Sn of 67.7% and an Spe of 56.2% were achieved. The AUC value was 0.634 (95% CI: 0.560–0.709), with a *p*-value of 0.001, as depicted in Figure 4.

In the subgroup analysis, only age and HbA1c level showed significant differences in comparison to the 2.35 SAF cut-off level, with a *p*-value of 0.001. No significant differences were observed in terms of gender, DM duration, and DM complications (Table 6).

Logistic regression models (Table 7) for predictors of SAF > 2.35 have demonstrated that elevated SAF levels are significantly associated with very high CV risk, especially after adjusting for age, gender, and HbA1c level. Age was notably associated with an increased risk (OR: 1.072; 95% CI: 1.048–1.096; $p = 0.001$), indicating that older patients have a greater probability of being included in the very high CV risk group, with elevated SAF levels. Additionally, gender was identified as a significant factor (OR: 1.426; CI: 1.057–1.923; $p = 0.02$), as well as higher HbA1c levels, which further amplified this risk (OR: 1.171; CI: 1.039–1.321; $p = 0.01$).

Table 6. Subgroup analysis based on SAF cut-off.

Parameters	SAF ≤ 2.35 (*n* = 301)	SAF > 2.35 (*n* = 584)	Statistic Tests	*p*
Age, mean ± SD	60.06 ± 8.42	64.30 ± 6.86	Student's *t*-test	<0.001
Female, *n*, (%)	163 (54.2%)	312 (53.4%)	Chi2 test	0.837
Age of DM, mean ± SD	8.78 ± 4.68	9.06 ± 4.25	Student's *t*-test	0.369
HbA1c mean ± SD	6.86 ± 1.12	7.17 ± 1.30	Student's *t*-test	<0.001
DSPN, *n*, (%)	200 (66.44%)	401 (68.66%)	Chi2 test	0.733
DR, *n*, (%)	13 (4.31%)	25 (4.28%)	Chi2 test	0.937
CKD, *n*, (%)	5 (1.66%)	12 (2.05%)	Chi2 test	0.711
Albuminuria, *n*, (%)	19 (6.31%)	36 (6.16%)	Chi2 test	0.880

SAF—skin autofluorescence; SD—standard deviation; DM—diabetes mellitus; DSPN—diabetic sensory peripheral neuropathy; DR—diabetic retinopathy; CKD—chronic kidney disease.

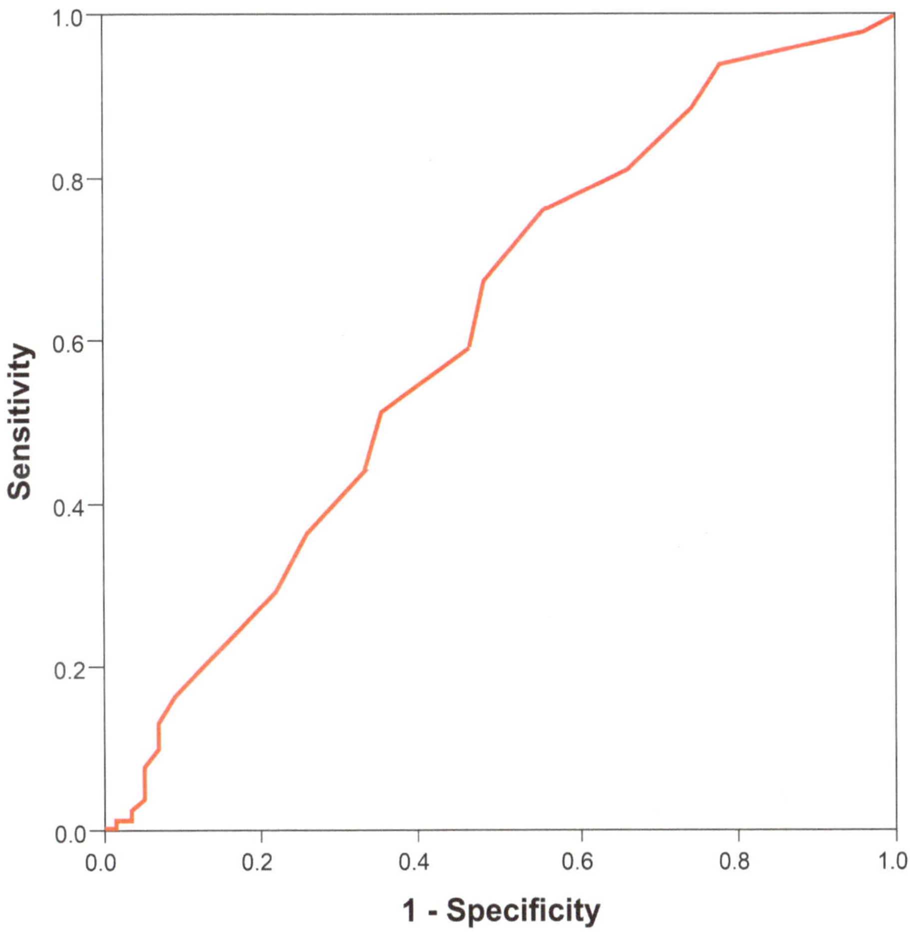

Figure 4. ROC curve for SAF levels in predicting very high CV risk.

Table 7. Logistic regression models for predictors of SAF > 2.35. Independent variables: age, female gender, HbA1c, DSPN, DR, CKD, albuminuria.

Logistic Regression Models (Very High Risk—Yes/No)	Predictors	OR (95% CI)	*p*
Model adjust 1	Age	1.071 (1.049–1.094)	0.001
Model adjust 2	Age	1.067 (1.044–1.091)	0.001
	Female gender	1.408 (1.046–1.895)	0.024
Model adjust 3	Age	1.072 (1.048–1.096)	0.001
	Female gender	1.430 (1.061–1.928)	0.019
	HbA1c	1.322 (1.149–1.522)	0.010

OR—odds ratio; CI—confidence interval.

4. Discussion

Our research assessed SAF in a large cohort of T2DM patients and examined the cross-sectional relationships between SAF and microvascular complications of DM. We also sought to identify associations between AGEs and CV risk categories and propose a cut-off value for SAF levels in predicting very high CV risk.

The findings indicate that, within this study's population, the SAF level potentially acts as a non-invasive indicator for detecting high CV risk among individuals with T2DM. In our cohort, significant associations were found between SAF levels and various parameters: such as age, CAN score reflecting CAN, NEPHRO score signaling kidney dysfunction risk, and glycemic control as measured by HbA1c levels. However, it is important to note that when we compared SAF levels across various DM complication groups—including DSPN, DR, CKD, and albuminuria—no significant differences were noted, with the exception of CKD. These findings underscore the importance of a multifactorial and individualized approach to risk assessment in patients with T2DM, integrating measures of glycemic control, as well as indicators of metabolic stress, such as SAF. In a multivariate regression analysis, accounting for factors, such as age and eGFR, HbA1c values were identified as independent predictors of SAF levels. Specifically, for each SD increase in HbA1c values, there was an observed rise of 0.105 SD in SAF levels. This indicates that as HbA1c levels (a measure of glycaemic control over the past 2 to 3 months) increase, so do SAF levels, suggesting a greater accumulation of AGEs, as assessed by SAF. This underscores the connection between inadequate glycaemic control and AGE accumulation, which contributes to the risk of complications in managing DM. Additionally, we assessed SAF's effectiveness as a screening tool, determining the optimal SAF threshold that balances Sn and Spe, along with the area under the ROC curve (AUC) as a measure of SAF's diagnostic accuracy.

4.1. *SAF and Microvascular Complications of DM (Neuropathy, Retinopathy, Nephropathy)*

Although no significant differences in SAF levels were observed among patients with various DM complications (DSPN, DR, albuminuria) within our study's population, an exception was noted for CKD. Patients with CKD (eGFR below 60 mL/min/1.73 m^2) exhibited higher SAF levels ($p < 0.003$), suggesting a potential association between elevated skin AGEs and CKD presence. This could imply that SAF as a marker of AGE accumulation might have a direct correlation with CKD in our cohort, in contrast to the other chronic complications included in the analysis. While our study indicates that SAF's effectiveness as a screening tool for microvascular complications, including DSPN, DR, and albuminuria, may be constrained within our study's population, it also underscores the potential for SAF's broader applicability and predictive value for other complications. This highlights the necessity for further research involving a more varied group of individuals and the importance of longitudinally assessing our study's population to fully explore SAF's utility.

Contrary to our findings, Hosseini et al., in their recent systematic review and meta-analysis of 29 studies—25 of which were cross-sectional, with 13 conducted within European cohorts—identified a pronounced link between SAF levels and suboptimal DM management, as reflected by the last HbA1c (0.21 with a 95% CI of 0.13–0.28), and a heightened risk for DR, DSPN, nephropathy, and macrovascular complications. This analysis highlighted heterogeneity among the reviewed studies, emphasizing the necessity for cautious interpretation of these findings due to the variations in study designs, methodologies, and the populations studied. Despite these variations, a consistent statistical significance was observed ($p < 0.05$), confirming SAF's relevance as a non-invasive marker for both micro- and macrovascular complications associated with DM. This insight into SAF's predictive capacity for the early detection of irreversible DM complications reflects its potential utility in further research with larger and more diverse cohorts alongside extended follow-up. So, it is essential to fully ascertain SAF's role in the clinical evaluation of DM complications, as underscored by Hosseini's analysis [34].

The connection between SAF and the complexity and severity of DM complications was also explored in a cross-sectional analysis of 825 patients with T2DM, revealing a significant correlation between skin AGEs and DR, diabetic kidney disease, CVD, and DSPN, noting that SAF levels increased with the progression of complications. Furthermore, the study identified distinct associations between SAF and demographic factors, including age, gender, and stimulated C-peptide, as well as clinical indicators like creatinine and fatty liver. The authors proposed an AGE-based risk score for diabetic complications, capable of predicting the likelihood of T2DM complications, thereby underscoring the significant predictive role of SAF [35].

As DSPN and its subtypes represent a prevalent complication of DM, SAF has been explored as a potential screening method for various forms of neuropathy in T2DM patients. In a study including 132 participants, the authors concluded that skin AGEs could serve as a screening tool for DSPN and CAN in T2DM patients, noting their potential utility in clinical settings. However, the study acknowledged the method's moderate to low Spe, suggesting that additional diagnostic procedures might be necessary for confirmation if screening results are positive, to enhance its accuracy. Compared to our cohort, T2DM patients in the Papachristou et al. study were slightly older (64.57 ± 8.21 years versus 62.9 ± 7.7 years) and had a longer history of hyperglycemia, as indicated by their higher DM duration (14.5 years [range 7.00–20.00] versus 9.0 ± 4.4 years). The optimal cut-off for overall DSPN was determined to be SAF ≥ 2.95, which is higher than our mean SAF score across the entire cohort (2.6 ± 0.5) [36]. In a recent longitudinal study, SAF levels were shown to predict foot ulcers in a cohort of 517 patients with a mean HbA1c of $8.7 \pm 1.8\%$, an average age of 62 ± 9 years, and an average DM duration of 14 ± 10 years [37]. The population studied in our research was notably different: our study included a cohort of T2DM patients who had a lower average HbA1c level and fewer patients diagnosed with chronic DM complications (DR, 4.29% versus 26.7%; CKD, 8.7% versus 44.9%; and macroangiopathies, 13.9% versus 33.8%). In a multicenter study involving 497 patients where neuropathy was assessed using three validated tools, the Toronto Clinical Neuropathy Score, the Neuropathy Disability Score, and the Neuropathy Symptoms Score, SAF levels were found to be not only elevated but also progressively increased with the severity of DSPN. These levels were associated with both the presence of symptoms and nerve deficits and demonstrated correlations with DM duration, glycemic control, and serum creatinine levels [38]. The various pathways leading to AGE formation—glycolytic dysfunction, lipid peroxidation, and glucotoxicity—play significant roles in the development of diabetic neuropathy, with varied impacts on its pathogenesis, including alterations in heart rate variability and changes in vibration perception thresholds, as underlined by Al-Saoudi et al. [39].

Increased levels of AGEs have been correlated with DR, as evidenced by multiple studies [40–45]. Yasuda and colleagues noted a correlation between skin AGE levels and the severity of DR in T2DM patients. In their examination of 67 T2DM patients and age-matched controls, significant increases in skin AGE levels were observed alongside DR progression. Furthermore, logistic regression analysis identified skin AGE levels as an independent predictor of proliferative DR (PDR) with an odds ratio (OR) of 17.2 ($p < 0.05$). The authors proposed measuring skin AGE levels as a useful tool for assessing DR risk and as a surrogate marker for non-invasively evaluating DR progression and severity in T2DM patients [42]. Hirano et al. reported similar findings in a population of 132 T2DM patients and non-diabetic controls, with a mean age of 63.7 ± 12.2 years, a HbA1c level of 7.5 ± 1.7%, and a DM duration of 13.2 ± 9.9 years. They found SAF to be correlated with DR severity but not with the prevalence or severity of diabetic macular edema (DME). This study also identified SAF as an independent factor indicating the occurrence of PDR and suggested that SAF can predict the risk of severe DR. The proportion of patients with DM complications was also higher than in their cohort, with only 26.1% showing no apparent DR [43]. A strong association between the presence of PDR and SAF, as well as its potential as a predictor for DR severity, was also reported by Takayanagi et al. Their study included T2DM patients with a mean age of 68.4 ± 13.7 years and a higher mean HbA1c [44].

A recent systematic review and meta-analysis, which included the majority of studies conducted in Caucasian populations, assessed the accuracy of SAF in the early detection of DR. This is particularly important because devices used for SAF readings have lower performance in highly pigmented skin. The review concluded that SAF demonstrates adequate accuracy for use in clinical settings for DR screening in patients with T2DM. The diagnostic OR (dOR) was 5.11 ($p < 0.001$), indicating a statistically significant association. However, there was notable heterogeneity in Spe but not in Sn. The authors concluded that while SAF is unlikely to replace ocular fundus inspection, it may serve as a convenient screening technique, especially in situations with limited resources. Further research is necessary to determine the reference values and the consistent usefulness of this tool across various populations [45]. Błaszkiewicz et al. describe the pathophysiological impact of AGEs on DR, noting their role in compromising neurovascular integrity through oxidative stress and inflammation The binding of AGEs to RAGE on retinal cells initiates harmful signaling pathways, especially in Müller cells. Once activated, these cells increase VEGF production and inflammatory responses, leading to neovascularization and retinal damage. Moreover, AGEs cause pathological alterations, including protein cross-linking and endoplasmic reticulum (ER) stress, worsening retinal cell dysfunction and blood–retinal barrier degradation. The authors underscore SAF's role in early DR detection and agree with previous studies on its efficacy as a non-invasive marker, suggesting that non-invasively measuring AGE accumulation offers a more accurate reflection of AGE levels than analyzing serum concentrations, which may not accurately represent tissue AGE levels and varies with the molecules' half-life [46].

In discussing renal complications in DM, AGEs are linked to the onset and progression of CKD. This process is initiated through AGEs' interaction with their receptor, triggering reactions that activate the inflammatory oxidative stress axis previously described, thereby influencing the progression of complications. Additionally, the AGE-RAGE interaction accelerates atherosclerosis, supports alterations in myocardial structure with cardiomyocyte impairment, leading to HF progression, and raises the risk for CVE, the main cause of death in patients living with both conditions, CKD and DM. Higher AGE levels in this population also contribute to muscle wasting and nutritional imbalances, accelerating the decline in the overall health status of this vulnerable patient group [47]. Fraser et al. investigated whether SAF could be a marker of all-cause mortality risk in stage 3 CKD. Initially, those in the highest quartile for skin AGEs faced a greater risk of all-cause mortality compared to patients with lower SAF levels. However, this association diminished after adjusting for confounding factors, such as CVD, glucose levels, BMI, albuminuria, and renal function. SAF in CKD reflects the accumulation of AGEs and cumulative metabolic stress, indicating the need for further research to understand SAF's predictive role in these patients [48]. On the contrary, Rigalleau et al. and Jin et al. reported that SAF is independently associated with renal outcomes in T2DM patients. Both studies affirm the significance of SAF; the former underscores its value as an independent prognostic marker for kidney dysfunction, separate from traditional renal markers, while the latter confirms its importance concerning macroangiopathy [49,50]. From these data, it seems evident that there exists a robust and firmly established correlation between SAF and DR, likely attributable to the direct influence of AGEs on retinal microvessels [45,51]. However, the connection between SAF and neuropathy or nephropathy is less pronounced, and SAF's role and predictive values may differ across various complications, potentially due to other pathophysiological pathways beyond AGE accumulation. The increased levels of AGEs in patients with DM also indicate the metabolic burden represented by elevated glucose levels, as well as atherogenic dyslipidemia, high levels of pro-inflammatory cytokines, and disruptions in reactive oxygen species homeostasis [34,52,53]. A systematic review and meta-analysis from 2023, evaluating 33 case–control studies mostly conducted in Europe, compared AGE levels in individuals with and without DM to assess SAF's validity in DM populations. Despite significant associations between SAF, a higher BMI, AGE accumulation, and DM complications, the authors conclude that more research is necessary to fully grasp SAF's usefulness as a surrogate biomarker in DM due to study variability. Factors like BMI, gender, age, and metabolic load significantly impact SAF, suggesting its potential utility as an indicator for MetS or DM [54].

Skin levels of AGEs, as determined by biopsy specimens, were linked to the onset and advancement of DM complications in both type 1 and T2DM in the DCCT-EDIC and UKPDS cohorts, even after adjusting for HbA1c levels [55,56]. High concentrations of AGEs in skin collagen, also determined through skin biopsies, were found to forecast the quick progression of microvascular complications in DM over a span of six years in T2DM patients, indicating the need for stricter DM control [57]. Still, evaluating AGEs through skin biopsy is impractical for routine clinical use [56].

Our study findings might be attributed to differences in demographic characteristics such as age, ethnicity, and DM durations. For instance, the rate of AGE accumulation and tissue impact is known to vary with age and a patient's metabolic control over time, which are specific to each study's population characteristics [58]. Additionally, DM duration is also likely to explain the differences in our findings, as a longer DM duration is generally linked to higher AGE accumulation, influencing relationships between SAF levels and complications in the microvasculature [56]. The relatively low prevalence of microvascular complications in our study population compared to other cohorts may have played a critical role in the observed lack of significant association between SAF levels and these complications. In populations where DM-related complications are less common, the statistical power to detect associations between SAF levels and such complications is diminished, which could result in a failure to observe significant associations, even if the pathophysiological link exists.

4.2. SAF and CV Risk in DM

AGEs are known to play a significant role in promoting CVD and mortality by influencing the structure and function of vascular and myocardial tissues. They cause stiffness in the vascular walls by altering the physical properties of extracellular matrix proteins and contribute to impaired vasodilation and reduced vascular flexibility by affecting endothelin-1 production and reducing nitric oxide levels. Additionally, AGEs, through their interaction with RAGEs, lead to a range of detrimental changes, including atherosclerosis, thrombosis, and further vascular constriction. RAGEs also facilitate fibrosis by increasing TGF-β levels [59] and altering calcium metabolism within the myocardium [60,61].

The vast majority of our patients (821, 92.77%) were in the very-high-risk category as per the ESC/EAS 2019 risk stratification applicable at the time of the cross-sectional evaluation of this cohort. Patients included in the very-high-risk category showed the highest average mean SAF levels. This finding, especially with a statistically significant difference (2.61AU; 2.22AU vs. 2.44AU; p = 0.003) suggests that higher SAF levels are associated with increased CV risk. This is consistent with previous research and suggests that AGEs, assessed by SAF, contribute to CVD progression in T2DM individuals. For the very high CV risk category, an SAF value greater than 2.35 was found to be a good predictor, having an Sn of 67.7% and an Spe of 56.2% (AUC = 0.634; 95% CI 0.560–0.709; $p = 0.001$). In a retrospective cohort study of 504 individuals followed for an average duration of 54 months, it was revealed that 14% of them experienced a CVE. The levels of SAF were significantly higher in patients who experienced a CVE compared to those who did not (2.89 vs. 2.63 AU, $p = 0.002$). This association between elevated SAF levels and the occurrence of CVEs remained statistically significant even after adjusting for other variables, including glycemic control, HBP, lipid profile abnormalities, other vascular complications, and DM treatments. Patients with higher SAF levels had a lower rate of CVE-free survival. Furthermore, even among those without macroangiopathy at baseline, individuals with higher SAF levels had a lower CVE-free survival rate compared to those with lower SAF levels. This suggests that SAF may serve as a useful marker for identifying patients at increased risk of CV complications [62].

A study incorporating 2349 participants from the Lifelines Cohort Study—all with T2DM, either newly diagnosed or already diagnosed, but without clinical CVD—investigated the predictive value of elevated SAF levels for the onset of CVD and mortality. Over an average follow-up of 3.7 years, the findings revealed that individuals with higher SAF levels had a substantially greater risk of experiencing CVEs or mortality. This correlation persisted even after adjusting for traditional risk factors, such as BP and cholesterol levels, indicating that SAF is a significant and independent predictor of new CVEs and mortality among T2DM patients (OR 2.59, 95% CI 2.10–3.20, $p < 0.001$). Notably, "new" T2DM cases exhibited lower SAF values than those with an established diagnosis, suggesting that prolonged hyperglycemia contributes to higher SAF levels in the latter group. In this study, SAF was a more potent predictor of CV complications and mortality in people with T2DM compared to traditional indicators [63].

Refining the identification in the T2DM populations of those patients with a higher risk for CVEs, and, hence, suitable candidates for revascularization procedures, has been investigated in a recent study by Alkhami et al. SAF levels were assessed at baseline in 477 T2DM patients, and new revascularizations (coronary and lower-limb arteries) were documented throughout a 54-month period. The study reported that patients with SAF levels greater than 2.6 AU, which was the median value for the study population, had a significantly higher incidence of revascularizations compared to individuals with lower SAF values, even after accounting for confounding factors. SAF offers the potential to improve the screening process for advanced investigations and possible revascularizations, resulting in better risk classification in T2DM patients [64]. The relationship between SAF and subclinical CVD was particularly strong in DM patients, emphasizing the importance of better understanding CV risk in T2DM populations, as shown in the Rotterdam study [65]. In our cohort, the optimum cut-off point for SAF levels in predicting a very high CV risk was 2.35 AU, which was associated with augmented HbA1c levels and older age.

4.3. Strengths and Limitations

Our results are restricted to a single-center cohort, necessitating caution when generalizing the results. Additionally, our group had a small sample size of participants with chronic DM microvascular complications, and the statistical power of our inference is restricted. As per the cross-sectional design of our study, this analysis provides a snapshot of a single point in time, describing the current state of a representative sample in Romanian T2DM patients, and it has the limitation that it cannot infer temporal sequences or causality from these associations described above. Nevertheless, the cohort examined was large and well characterized, and it remains a feasible scenario in our resource-constrained setting, as it provides a comprehensive overview of DM complications and the distribution of CV risk categories in correlation with SAF levels. One notable strength of our study is the cohort's well-controlled BP values and nearly optimal metabolic control. However, reports indicate that despite meeting glycemic control targets, most patients do not achieve a comprehensive treatment goal encompassing HbA1c, BP, and lipid levels, and there is an underutilization of newer antidiabetic medications with potential cardio- and reno-protective benefits [66–69]. Our study may constitute the baseline for a longitudinal follow-up of the same cohort. This approach will offer an opportunity for the development of targeted interventions to address the unmet needs of the Romanian T2DM population and enhance our understanding of disease progression and management over time.

In our research, we assessed the accumulation of AGEs using SAF. There are both advantages and drawbacks to this method. In a previous study, Lutgers et al. highlighted that autofluorescence readers have limitations, as non-fluorescent AGEs will not be measured by the method, and substances with auto-fluorescent properties may introduce confounding factors [70]. Several limitations have been identified in SAF measurements, including the fact that they are not suitable for individuals with high levels of skin pigmentation, and that various substances, such as creams, sunscreens, or extreme blood flow variations, may affect the accuracy of the measurements [71]. Despite this, evaluating skin AGEs by SAF has shown a stronger and more robust correlation with CVD risk than circulating AGE levels. Serum AGEs have had an unclear correlation with CVD due to their short activity duration and alterations influenced by various dietary habits and metabolic factors [72].

A noted constraint of our study is the incomplete available data regarding specific risk factors, such as smoking. Future studies should include smoking as a variable to provide a more comprehensive understanding of its effects on SAF's predictive accuracy for CV complications.

As far as we are aware, this research marks the first exploration into establishing a threshold for SAF's correlation with CV risk and DM complications among Romanian individuals with T2DM. We aim to extend this research longitudinally and expand the range of participants. This initiative is directed toward validating the reliability of SAF measurement as a tool for estimating CVD risk specifically within our demographics. By offering a non-invasive and accessible method, SAF holds the potential to significantly assist clinicians in identifying patients at high CV risk and making informed decisions about treatment adjustments, highlighting its critical clinical significance. Further research could explore how SAF, in conjunction with other biomarkers and clinical assessments, can be integrated into comprehensive risk assessment models to improve patient care and outcomes in DM management.

5. Conclusions

Increased CV risk and higher HbA1c values are associated with increased SAF levels in patients with T2DM. The optimal threshold of SAF levels in predicting very high CV risk in patients with T2DM is 2.35. The presence of CKD in patients with T2DM was associated with higher SAF levels, while the presence of other T2DM degenerative chronic complications was not significantly associated with SAF levels. SAF levels above the obtained threshold were associated with higher HbA1c and higher patient age.

Author Contributions: Conceptualization, D.R.-P. and B.T.; methodology, D.R.-P.; software, B.T. and R.-A.S.; validation, A.P.S. and B.T.; formal analysis, B.T. and R.-A.S.; investigation, D.R.-P.; resources, D.R.-P.; data curation, T.S.; writing—original draft preparation, D.R.-P., T.S. and L.M.-T.; writing—review and editing, R.-A.S., B.T. and S.L; visualization, A.P.S. and S.L.; supervision, A.P.S. and B.T.; project administration, D.R.-P.; funding acquisition, D.R.-P. All authors have read and agreed to the published version of the manuscript.

Funding: This research received no external funding.

Institutional Review Board Statement: The study was conducted in accordance with the Declaration of Helsinki and was approved by the Ethics Committee of Consultmed Iasi (protocol number CMD102018006, dated 18 October 2018).

Informed Consent Statement: Informed consent was obtained from all subjects involved in the study.

Data Availability Statement: The archived dataset analyzed during the study is available upon request.

Acknowledgments: We would like to thank the patients who agreed to participate in this study.

Conflicts of Interest: The authors declare no conflicts of interest related to the subject of this article.

References

1. Moldogazieva, N.T.; Mokhosoev, I.M.; Mel'nikova, T.I.; Porozov, Y.B.; Terentiev, A.A. Oxidative Stress and Advanced Lipoxidation and Glycation End Products (ALEs and AGEs) in Aging and Age-Related Diseases. *Oxid. Med. Cell. Longev.* **2019**, *2019*, 3085756. [CrossRef] [PubMed]
2. Sultana, R.; Parveen, A.; Kang, M.C.; Hong, S.M.; Kim, S.Y. Glyoxal-derived advanced glycation end products (GO-AGEs) with UVB critically induce skin inflammaging: In vitro and in silico approaches. *Sci. Rep.* **2024**, *14*, 1843. [CrossRef] [PubMed]
3. Rukadikar, C.; Rukadikar, A.; Kishore, S. A Review on Autonomic Functional Assessment in Diabetic Patients. *Cureus* **2023**, *15*, e34598. [CrossRef] [PubMed]
4. Mosquera-Sulbarán, J.A.; Hernández-Fonseca, J.P. Advanced Glycation End Products in Diabetes. In *Biomarkers in Diabetes*; Springer International Publishing: Cham, Switzerland, 2022; pp. 171–194.
5. Chaudhuri, J.; Bains, Y.; Guha, S.; Kahn, A.; Hall, D.; Bose, N.; Gugliucci, A.; Kapahi, P. The Role of Advanced Glycation End Products in Aging and Metabolic Diseases: Bridging Association and Causality. *Cell Metab.* **2018**, *28*, 337–352. [CrossRef] [PubMed]
6. Shardlow, A.; McIntyre, N.J.; Kolhe, N.V.; Nellums, L.B.; Fluck, R.J.; McIntyre, C.W.; Taal, M.W. The association of skin autofluorescence with cardiovascular events and all-cause mortality in persons with chronic kidney disease stage 3: A prospective cohort study. *PLoS Med.* **2020**, *17*, e1003163. [CrossRef]
7. Majchrzak, C.; Cougnard-Gregoire, A.; Le-Goff, M.; Féart, C.; Delcourt, C.; Reydit, M.; Helmer, C.; Rigalleau, V. Skin autofluorescence of Advanced Glycation End-products and mortality in older adults: The roles of chronic kidney disease and diabetes. *Nutr. Metab. Cardiovasc. Dis.* **2022**, *32*, 2526–2533. [CrossRef]
8. Kawamoto, H.; Hanatani, S.; Tsujita, K.; Ruparelia, N.; Chou, S.; Kono, Y.; Nakamura, S. Skin Autofluorescence and Clinical Outcomes in Patients with Coronary Artery Disease. *J. Atheroscler. Thromb.* **2024**, *31*, 316–325. [CrossRef]
9. Aoki, E.; Hirashima, T.; Kumamoto, Y.; Yamamoto, Y.; Suzuki, N.; Oshima, T.; Saito, D.; Hirano, T. Clinical significance of skin autofluorescence for diabetic macroangiopathy and comparison with conventional markers of atherosclerosis: A cross-sectional and prospective study. *Diabetol. Int.* **2022**, *14*, 145–154. [CrossRef]
10. Jujic, A.; Engström, G.; Nilsson, P.M.; Johansson, M. Accumulation of advanced glycation end products in skin and increased vascular ageing in the general population: The Malmö Offspring Study. *J. Hypertens.* **2024**, *42*, 530–537. [CrossRef]
11. Mengstie, M.A.; Chekol Abebe, E.; Behaile Teklemariam, A.; Tilahun Mulu, A.; Agidew, M.M.; Teshome Azezew, M.; Zewde, E.A.; Agegnehu Teshome, A. Endogenous advanced glycation end products in the pathogenesis of chronic diabetic complications. *Front. Mol. Biosci.* **2022**, *9*, 1002710. [CrossRef]
12. Twarda-Clapa, A.; Olczak, A.; Białkowska, A.M.; Koziołkiewicz, M. Advanced Glycation End-Products (AGEs): Formation, Chemistry, Classification, Receptors, and Diseases Related to AGEs. *Cells* **2022**, *11*, 1312. [CrossRef] [PubMed]
13. Prasad, A.; Bekker, P.; Tsimikas, S. Advanced Glycation End Products and Diabetic Cardiovascular Disease. *Cardiol. Rev.* **2012**, *20*, 177–183. [CrossRef] [PubMed]
14. Klonoff, D.C.; Aaron, R.E.; Tian, T.; DuNova, A.Y.; Pandey, A.; Rhee, C.; Fleming, G.A.; Sacks, D.B.; Pop-Busui, R.; Kerr, D. Advanced Glycation Endproducts: A Marker of Long-term Exposure to Glycemia. *J. Diabetes Sci. Technol.* 2024, *advance online publication*. [CrossRef]
15. Perrone, A.; Giovino, A.; Benny, J.; Martinelli, F. Advanced Glycation End Products (AGEs): Biochemistry, Signaling, Analytical Methods, and Epigenetic Effects. *Oxid. Med. Cell. Longev.* **2020**, *2020*, 3818196. [CrossRef]
16. van Waateringe, R.P. *Skin Autofluorescence in the General Population: Associations and Prediction*; Rijksuniversiteit Groningen: Groningen, The Netherland, 2019.
17. Da Moura Semedo, C.; Webb, M.; Waller, H.; Khunti, K.; Davies, M. Skin autofluorescence, a non-invasive marker of advanced glycation end products: Clinical relevance and limitations. *Postgrad. Med. J.* **2017**, *93*, 289–294. [CrossRef]
18. Stirban, A. Noninvasive Skin Fluorescence Spectroscopy for Diabetes Screening. *J. Diabetes Sci. Technol.* **2013**, *7*, 1001–1004. [CrossRef]
19. Fokkens, B.T.; van Waateringe, R.P.; Mulder, D.J.; Wolffenbuttel, B.H.R.; Smit, A.J. Skin autofluorescence improves the Finnish Diabetes Risk Score in the detection of diabetes in a large population-based cohort: The LifeLines Cohort Study. *Diabetes Metab.* **2018**, *44*, 424–430. [CrossRef]
20. Boersma, H.E.; van der Klauw, M.M.; Smit, A.J.; Wolffenbuttel, B.H.R. A non-invasive risk score including skin autofluorescence predicts diabetes risk in the general population. *Sci. Rep.* **2022**, *12*, 21794. [CrossRef]
21. van Waateringe, R.P.; Fokkens, B.T.; Slagter, S.N.; van der Klauw, M.M.; van Vliet-Ostaptchouk, J.V.; Graaff, R.; Paterson, A.D.; Smit, A.J.; Lutgers, H.L.; Wolffenbuttel, B.H.R. Skin autofluorescence predicts incident type 2 diabetes, cardiovascular disease and mortality in the general population. *Diabetologia* **2019**, *62*, 269–280. [CrossRef] [PubMed]
22. Reurean-Pintilei, D.; Pantea Stoian, A.; Potcovaru, C.-G.; Salmen, T.; Cinteză, D.; Stoica, R.-A.; Lazăr, S.; Timar, B. Skin Autofluorescence as a Potential Adjunctive Marker for Cardiovascular Risk Assessment in Type 2 Diabetes: A Systematic Review. *Int. J. Mol. Sci.* **2024**, *25*, 3889. [CrossRef]

23. Cavero-Redondo, I.; Soriano-Cano, A.; Álvarez-Bueno, C.; Cunha, P.G.; Martínez-Hortelano, J.A.; Garrido-Miguel, M.; Berlanga-Marcias, C.; Martinez-Vizcaino, V. Skin Autofluorescence–Indicated Advanced Glycation End Products as Predictors of Cardiovascular and All-Cause Mortality in High-Risk Subjects: A Systematic Review and Meta-analysis. *J. Am. Heart Assoc.* **2018**, *7*, e009833. [CrossRef]
24. Du, T.; Brandl, B.; Hauner, H.; Skurk, T. Skin Autofluorescence Mirrors Surrogate Parameters of Vascular Aging: An Enable Study. *Nutrients* **2023**, *15*, 1597. [CrossRef]
25. Pan, J.; Bao, X.; Gonçalves, I.; Jujić, A.; Engström, G. Skin autofluorescence, a measure of tissue accumulation of advanced glycation end products, is associated with subclinical atherosclerosis in coronary and carotid arteries. *Atherosclerosis* **2022**, *345*, 26–32. [CrossRef]
26. Kunimoto, M.; Yokoyama, M.; Shimada, K.; Matsubara, T.; Aikawa, T.; Ouchi, S.; Fukao, K.; Miyazaki, T.; Fujiwara, K.; Abulimiti, A.; et al. Relationship between skin autofluorescence levels and clinical events in patients with heart failure undergoing cardiac rehabilitation. *Cardiovasc. Diabetol.* **2021**, *20*, 208. [CrossRef]
27. Ahmad, M.S.; Kimhofer, T.; Ahmad, S.; AlAma, M.N.; Mosli, H.H.; Hindawi, S.I.; Mook-Kanamori, D.O.; Šebeková, K.; Damanhouri, Z.A.; Holmes, E. Ethnicity and skin autofluorescence-based risk-engines for cardiovascular disease and diabetes mellitus. *PLoS ONE* **2017**, *12*, e0185175. [CrossRef]
28. Stirban, A.; Heinemann, L. Skin Autofluorescence—A Non-invasive Measurement for Assessing Cardiovascular Risk and Risk of Diabetes. *Eur. Endocrinol.* **2014**, *10*, 106–110. [CrossRef]
29. Mach, F.; Baigent, C.; Catapano, A.L.; Koskinas, K.C.; Casula, M.; Badimon, L.; Chapman, M.J.; De Backer, G.G.; Delgado, V.; Ference, B.A. 2019 ESC/EAS Guidelines for the management of dyslipidaemias: Lipid modification to reduce cardiovascular risk. *Eur. Heart J.* **2020**, *41*, 111–188. [CrossRef]
30. Casellini, C.M.; Parson, H.K.; Richardson, M.S.; Nevoret, M.L.; Vinik, A.I. Sudoscan, a noninvasive tool for detecting diabetic small fiber neuropathy and autonomic dysfunction. *Diabetes Technol. Ther.* **2013**, *15*, 948–953. [CrossRef]
31. Freedman, B.I.; Bowden, D.W.; Smith, S.C.; Xu, J.; Divers, J. Relationships between electrochemical skin conductance and kidney disease in Type 2 diabetes. *J. Diabetes Complicat.* **2014**, *28*, 56–60. [CrossRef]
32. Yajnik, C.S.; Kantikar, V.; Pande, A.; Deslypere, J.P.; Dupin, J.; Calvet, J.H.; Bauduceau, B. Screening of cardiovascular autonomic neuropathy in patients with diabetes using non-invasive quick and simple assessment of sudomotor function. *Diabetes Metab.* **2013**, *39*, 126–131. [CrossRef]
33. Stevens, P.E.; Ahmed, S.B.; Carrero, J.J.; Foster, B.; Francis, A.; Hall, R.K.; Herrington, W.G.; Hill, G.; Inker, L.A.; Kazancıoğlu, R.; et al. KDIGO 2024 Clinical Practice Guideline for the Evaluation and Management of Chronic Kidney Disease. *Kidney Int.* **2024**, *105*, S117–S314.
34. Hosseini, M.S.; Razavi, Z.; Ehsani, A.H.; Firooz, A.; Afazeli, S. Clinical Significance of Non-invasive Skin Autofluorescence Measurement in Patients with Diabetes: A Systematic Review and Meta-analysis. *EClinicalMedicine* **2021**, *42*, 101194. [CrossRef]
35. Wang, X.; Zhao, X.; Lian, T.; Wei, J.; Yue, W.; Zhang, S.; Chen, Q. Skin autofluorescence and the complexity of complications in patients with type 2 diabetes mellitus: A cross-sectional study. *BMC Endocr. Disord.* **2021**, *21*, 58. [CrossRef]
36. Papachristou, S.; Pafili, K.; Trypsianis, G.; Papazoglou, D.; Vadikolias, K.; Papanas, N. Skin advanced glycation end products as a screening tool of neuropathy in type 2 diabetes mellitus. *J. Diabetes Complicat.* **2022**, *36*, 108356. [CrossRef]
37. Borderie, G.; Foussard, N.; Larroumet, A.; Blanco, L.; Barbet-Massin, M.A.; Ducos, C.; Rami-Arab, L.; Domenge, F.; Mohammedi, K.; Ducasse, E.; et al. The skin autofluorescence of advanced glycation end-products relates to the development of foot ulcers in type 2 diabetes: A longitudinal observational study. *J. Diabetes Complicat.* **2023**, *37*, 108595. [CrossRef]
38. Stirban, A.O.; Bondor, C.I.; Florea, B.; Veresiu, I.A.; Gavan, N.A. Skin autofluorescence: Correlation with measures of diabetic sensorimotor neuropathy. *J. Diabetes Complicat.* **2018**, *32*, 851–856. [CrossRef]
39. Al-Saoudi, E.; Christensen, M.M.B.; Nawroth, P.; Fleming, T.; Hommel, E.E.; Jørgensen, M.E.; Fleischer, J.; Hansen, C.S. Advanced glycation end-products are associated with diabetic neuropathy in young adults with type 1 diabetes. *Front. Endocrinol.* **2022**, *13*, 891442. [CrossRef]
40. Guo, H.; Xu, Y. Role of advanced glycation end products in the progression of diabetes mellitus. *Glob. J. Obes. Diabetes Metab. Syndr.* **2017**, *4*, 024–035.
41. Al-Mesallamy, H.O.; Hammad, L.N.; El-Mamoun, T.A.; Khalil, B.M. Role of advanced glycation end product receptors in the pathogenesis of diabetic retinopathy. *J. Diabetes Complicat.* **2011**, *25*, 168–174. [CrossRef]
42. Yasuda, M.; Shimura, M.; Kunikata, H.; Kanazawa, H.; Yasuda, K.; Tanaka, Y.; Konno, H.; Takahashi, M.; Kokubun, T.; Maruyama, K. Relationship of Skin Autofluorescence to Severity of Retinopathy in Type 2 Diabetes. *Curr. Eye Res.* **2015**, *40*, 338–345. [CrossRef]
43. Hirano, T.; Iesato, Y.; Toriyama, Y.; Imai, A.; Chiba, D.; Murata, T. Correlation between diabetic retinopathy severity and elevated skin autofluorescence as a marker of advanced glycation end-product accumulation in type 2 diabetic patients. *J. Diabetes Complicat.* **2014**, *28*, 729–734. [CrossRef]
44. Takayanagi, Y.; Yamanaka, M.; Fujihara, J.; Matsuoka, Y.; Gohto, Y.; Obana, A.; Tanito, M. Evaluation of Relevance between Advanced Glycation End Products and Diabetic Retinopathy Stages Using Skin Autofluorescence. *Antioxidants* **2020**, *9*, 1100. [CrossRef]

45. Martínez-García, I.; Cavero-Redondo, I.; Álvarez-Bueno, C.; Pascual-Morena, C.; Gómez-Guijarro, M.D.; Saz-Lara, A. Non-invasive skin autofluorescence as a screening method for diabetic retinopathy. *Diabetes Metab. Res. Rev.* **2024**, *40*, e3721. [CrossRef]
46. Błaszkiewicz, M.; Walulik, A.; Florek, K.; Górecki, I.; Sławatyniec, O.; Gomułka, K. Advances and Perspectives in Relation to the Molecular Basis of Diabetic Retinopathy—A Review. *Biomedicines* **2023**, *11*, 2951. [CrossRef]
47. Dozio, E.; Caldiroli, L.; Molinari, P.; Castellano, G.; Delfrate, N.W.; Romanelli, M.M.C.; Vettoretti, S. Accelerated AGEing: The Impact of Advanced Glycation End Products on the Prognosis of Chronic Kidney Disease. *Antioxidants* **2023**, *12*, 584. [CrossRef]
48. Fraser, S.D.; Roderick, P.J.; McIntyre, N.J.; Harris, S.; McIntyre, C.W.; Fluck, R.J.; Taal, M.W. Skin autofluorescence and all-cause mortality in stage 3 CKD. *Clin. J. Am. Soc. Nephrol.* **2014**, *9*, 1361–1368. [CrossRef]
49. Rigalleau, V.; Cougnard-Gregoire, A.; Nov, S.; Gonzalez, C.; Maury, E.; Lorrain, S.; Gin, H.; Barberger-Gateau, P. Association of advanced glycation end products and chronic kidney disease with macroangiopathy in type 2 diabetes. *J. Diabetes Complicat.* **2015**, *29*, 270–274. [CrossRef]
50. Jin, Q.; Lau, E.S.; Luk, A.O.; Ozaki, R.; Chow, E.Y.; So, T.; Yeung, T.; Loo, K.M.; Lim, C.K.; Kong, A.P.; et al. Skin autofluorescence is associated with progression of kidney disease in type 2 diabetes: A prospective cohort study from the Hong Kong diabetes biobank. *Nutr. Metab. Cardiovasc. Dis.* **2022**, *32*, 436–446. [CrossRef]
51. Kulkarni, A.; Thool, A.R.; Daigavane, S.; Aditi, S. Understanding the Clinical Relationship Between Diabetic Retinopathy, Nephropathy, and Neuropathy: A Comprehensive Review. *Cureus* **2024**, *16*, e56674. [CrossRef]
52. Bhatti, J.S.; Sehrawat, A.; Mishra, J.; Sidhu, I.S.; Navik, U.; Khullar, N.; Kumar, S.; Bhatti, G.K.; Reddy, P.H. Oxidative stress in the pathophysiology of type 2 diabetes and related complications: Current therapeutics strategies and future perspectives. *Free Radic. Biol. Med.* **2022**, *184*, 114–134. [CrossRef]
53. Welsh, K.J.; Kirkman, M.S.; Sacks, D.B. Role of Glycated Proteins in the Diagnosis and Management of Diabetes: Research Gaps and Future Directions. *Diabetes Care* **2016**, *39*, 1299–1306. [CrossRef]
54. Hosseini, M.S.; Razavi, Z.; Bahri, R.A.; Ehsani, A.H.; Firooz, A.; Aryanian, Z.; Ehsani, A.; Sadeghi, Y. Is skin autofluorescence a novel non-invasive marker in diabetes? A systematic review and meta-analysis of case-control studies. *J. Res. Med. Sci.* **2023**, *28*, 68.
55. Genuth, S.; Sun, W.; Cleary, P.; Sell, D.R.; Dahms, W.; Malone, J.; Sivitz, W.; Monnier, V.M.; DCCT Skin Collagen Ancillary Study Group. Glycation and carboxymethyllysine levels in skin collagen predict the risk of future 10-year progression of diabetic retinopathy and nephropathy in the diabetes control and complications trial and epidemiology of diabetes interventions and complications participants with type 1 diabetes. *Diabetes* **2005**, *54*, 3103–3111.
56. Planas, A.; Simó-Servat, O.; Hernández, C.; Simó, R. Advanced Glycations End Products in the Skin as Biomarkers of Cardiovascular Risk in Type 2 Diabetes. *Int. J. Mol. Sci.* **2022**, *23*, 6234. [CrossRef]
57. Sternberg, M.; M'bemba, J.; Urios, P.; Borsos, A.M.; Selam, J.L.; Peyroux, J.; Slama, G. Skin collagen pentosidine and fluorescence in diabetes were predictors of retinopathy progression and creatininemia increase already 6years after punch-biopsy. *Clin. Biochem.* **2016**, *49*, 225–231. [CrossRef]
58. Koetsier, M.; Lutgers, H.L.; de Jonge, C.; Links, T.P.; Smit, A.J.; Graaff, R. Reference values of skin autofluorescence. *Diabetes Technol. Ther.* **2010**, *12*, 399–403. [CrossRef]
59. Zgutka, K.; Tkacz, M.; Tomasiak, P.; Tarnowski, M. A Role for Advanced Glycation End Products in Molecular Ageing. *Int. J. Mol. Sci.* **2023**, *24*, 9881. [CrossRef]
60. Kohn, J.C.; Lampi, M.C.; Reinhart-King, C.A. Age-related vascular stiffening: Causes and consequences. *Front. Genet.* **2015**, *6*, 112. [CrossRef]
61. de la Cruz-Ares, S.; Cardelo, M.P.; Gutiérrez-Mariscal, F.M.; Torres-Peña, J.D.; García-Rios, A.; Katsiki, N.; Malagón, M.M.; López-Miranda, J.; Pérez-Martínez, P.; Yubero-Serrano, E.M. Endothelial Dysfunction and Advanced Glycation End Products in Patients with Newly Diagnosed Versus Established Diabetes: From the CORDIOPREV Study. *Nutrients* **2020**, *12*, 238. [CrossRef]
62. Alkhami, F.; Borderie, G.; Foussard, N.; Larroumet, A.; Blanco, L.; Barbet-Massin, M.A.; Ferriere, A.; Ducos, C.; Mohammedi, K.; Fawaz, S.; et al. Skin autofluorescence of advanced glycation end-products relates to new cardiovascular events in type 2 diabetes: A longitudinal observational study. *Diabetes Metab.* **2024**, *50*, 101524. [CrossRef]
63. Boersma, H.E.; van Waateringe, R.P.; van der Klauw, M.M.; Graaff, R.; Paterson, A.D.; Smit, A.J.; Wolffenbuttel, B.H.R. Skin autofluorescence predicts new cardiovascular disease and mortality in people with type 2 diabetes. *BMC Endocr. Disord.* **2021**, *21*, 14. [CrossRef] [PubMed]
64. Alkhami, F.; Borderie, G.; Foussard, N.; Larroumet, A.; Blanco, L.; Barbet-Massin, M.A.; Ferriere, A.; Ducos, C.; Mohammedi, K.; Fawaz, S.; et al. The skin autofluorescence may help to select patients with Type 2 diabetes candidates for screening to revascularization procedures. *Cardiovasc. Diabetol.* **2024**, *23*, 32. [CrossRef] [PubMed]
65. Chen, J.; Arshi, B.; Waqas, K.; Lu, T.; Bos, D.; Ikram, M.A.; Uitterlinden, A.G.; Kavousi, M.; Zillikens, M.C. Advanced glycation end products measured by skin autofluorescence and subclinical cardiovascular disease: The Rotterdam Study. *Cardiovasc. Diabetol.* **2023**, *22*, 326. [CrossRef] [PubMed]
66. Reurean-Pintilei, D.; Potcovaru, C.-G.; Salmen, T.; Mititelu-Tartau, L.; Cinteză, D.; Lazăr, S.; Pantea Stoian, A.; Timar, R.; Timar, B. Assessment of Cardiovascular Risk Categories and Achievement of Therapeutic Targets in European Patients with Type 2 Diabetes. *J. Clin. Med.* **2024**, *13*, 2196. [CrossRef]

67. McGurnaghan, S.; Blackbourn, L.A.K.; Mocevic, E.; Haagen Panton, U.; McCrimmon, R.J.; Sattar, N.; Wild, S.; Colhoun, H.M. Cardiovascular disease prevalence and risk factor prevalence in Type 2 diabetes: A contemporary analysis. *Diabet. Med.* **2019**, *36*, 718–725. [CrossRef] [PubMed]
68. Rungby, J.; Schou, M.; Warrer, P.; Ytte, L.; Andersen, G.S. Prevalence of cardiovascular disease and evaluation of standard of care in type 2 diabetes: A nationwide study in primary care. *Cardiovasc. Endocrinol.* **2017**, *6*, 145–151. [CrossRef] [PubMed]
69. Ray, K.K.; Haq, I.; Bilitou, A.; Manu, M.C.; Burden, A.; Aguiar, C.; Arca, M.; Connolly, D.L.; Eriksson, M.; Ferrieres, J.; et al. Treatment gaps in the implementation of LDL cholesterol control among high- and very high-risk patients in Europe between 2020 and 2021: The multinational observational SANTORINI study. *Lancet Reg. Health Eur.* **2023**, *29*, 100624. [CrossRef] [PubMed]
70. Lutgers, H.L.; Gerrits, E.G.; Graaff, R.; Links, T.P.; Sluiter, W.J.; Gans, R.O.; Bilo, H.J.; Smit, A.J. Skin autofluorescence provides additional information to the UK Prospective Diabetes Study (UKPDS) risk score for the estimation of cardiovascular prognosis in type 2 diabetes mellitus. *Diabetologia* **2009**, *5*, 789–797. [CrossRef] [PubMed]
71. Noordzij, M.J.; Mulder, D.J.; Oomen, P.H.; Brouwer, T.; Jager, J.; Castro Cabezas, M.; Lefrandt, J.D.; Smit, A.J. Skin autofluorescence and risk of micro- and macrovascular complications in patients with Type 2 diabetes mellitus-a multi-centre study. *Diabet. Med.* **2012**, *29*, 1556–1561. [CrossRef]
72. Chen, J.; Waqas, K.; Tan, R.C.; Voortman, T.; Ikram, M.A.; Nijsten, T.E.C.; de Groot, L.C.P.G.M.; Uitterlinden, A.G.; Zillikens, M.C. The association between dietary and skin advanced glycation end products: The Rotterdam Study. *Am. J. Clin. Nutr.* **2020**, *112*, 129–137. [CrossRef]

biomedicines

Article

Cross-Sectional and Longitudinal Associations between Skin Autofluorescence and Tubular Injury Defined by Urinary Excretion of Liver-Type Fatty Acid-Binding Protein in People with Type 2 Diabetes

Hiroki Yamagami [1,2], Tomoyo Hara [1], Saya Yasui [2], Minae Hosoki [2], Taiki Hori [1,2], Yousuke Kaneko [2], Yukari Mitsui [1], Kiyoe Kurahashi [3], Takeshi Harada [1], Sumiko Yoshida [1], Shingen Nakamura [4], Toshiki Otoda [4], Tomoyuki Yuasa [4], Akio Kuroda [5], Itsuro Endo [6], Munehide Matsuhisa [5], Masahiro Abe [7] and Ken-ichi Aihara [2,4,*]

1 Department of Hematology, Endocrinology and Metabolism, Graduate School of Biomedical Sciences, Tokushima University, 3-18-15 Kuramoto-cho, Tokushima 770-8503, Japan; yamagami.hiroki@tokushima-u.ac.jp (H.Y.); hara.tomoyo@tokushima-u.ac.jp (T.H.); takeshi_harada@tokushima-u.ac.jp (T.H.); yoshida.sumiko@tokushima-u.ac.jp (S.Y.)
2 Department of Internal Medicine, Anan Medical Center, 6-1 Kawahara Takarada-cho, Tokushima 774-0045, Japan; minae.energy.flow@gmail.com (M.H.)
3 Department of Community Medicine for Respirology, Hematology and Metabolism, Graduate School of Biomedical Sciences, Tokushima University, 3-18-15 Kuramoto-cho, Tokushima 770-8503, Japan; kurahashi.kiyoe@tokushima-u.ac.jp
4 Department of Community Medicine and Medical Science, Graduate School of Biomedical Sciences, Tokushima University, 3-18-15 Kuramoto-cho, Tokushima 770-8503, Japan; shingen@tokushima-u.ac.jp (S.N.); otoda.toshiki@tokushima-u.ac.jp (T.O.); yuasa.tomoyuki@tokushima-u.ac.jp (T.Y.)
5 Diabetes Therapeutics and Research Center, Institute of Advanced Medical Sciences, Tokushima University, 3-18-15 Kuramoto-cho, Tokushima 770-8503, Japan; kurodaakio@tokushima-u.ac.jp (A.K.); matuhisa@tokushima-u.ac.jp (M.M.)
6 Department of Bioregulatory Sciences, Graduate School of Biomedical Sciences, Tokushima University, 3-18-15 Kuramoto-cho, Tokushima 770-8503, Japan; endoits@tokushima-u.ac.jp
7 Department of Hematology, Kawashima Hospital, 6-1 Kitasakoichiban-cho, Tokushima 770-8548, Japan; masabe@tokushima-u.ac.jp
* Correspondence: aihara@tokushima-u.ac.jp; Tel.: +81-88-633-9267; Fax: +81-88-633-7121

Citation: Yamagami, H.; Hara, T.; Yasui, S.; Hosoki, M.; Hori, T.; Kaneko, Y.; Mitsui, Y.; Kurahashi, K.; Harada, T.; Yoshida, S.; et al. Cross-Sectional and Longitudinal Associations between Skin Autofluorescence and Tubular Injury Defined by Urinary Excretion of Liver-Type Fatty Acid-Binding Protein in People with Type 2 Diabetes. *Biomedicines* **2023**, *11*, 3020. https://doi.org/10.3390/biomedicines11113020

Academic Editors: Ana Dascalu and Dragos Serban

Received: 7 October 2023
Revised: 6 November 2023
Accepted: 7 November 2023
Published: 10 November 2023

Abstract: It has previously been unclear whether the accumulation of advanced glycation end products, which can be measured using skin autofluorescence (SAF), has a significant role in diabetic kidney disease (DKD), including glomerular injury and tubular injury. This study was therefore carried out to determine whether SAF correlates with the progression of DKD in people with type 2 diabetes (T2D). In 350 Japanese people with T2D, SAF values were measured using an AGE Reader®, and both urine albumin-to-creatinine ratio (uACR), as a biomarker of glomerular injury, and urine liver-type fatty acid-binding protein (uLFABP)-to-creatinine ratio (uL-FABPCR), as a biomarker of tubular injury, were estimated as indices of the severity of DKD. Significant associations of SAF with uACR ($p < 0.01$), log-transformed uACR ($p < 0.001$), uL-FABPCR ($p < 0.001$), and log-transformed uL-FABPCR ($p < 0.001$) were found through a simple linear regression analysis. Although SAF was positively associated with increasing uL-FABPCR ($p < 0.05$) and increasing log-transformed uL-FABPCR ($p < 0.05$), SAF had no association with increasing uACR or log-transformed uACR after adjusting for clinical confounding factors. In addition, the annual change in SAF showed a significant positive correlation with annual change in uL-FABPCR regardless of confounding factors ($p = 0.026$). In conclusion, SAF is positively correlated with uL-FABP but not with uACR in people with T2D. Thus, there is a possibility that SAF can serve as a novel predictor for the development of diabetic tubular injury.

Keywords: skin autofluorescence; type 2 diabetes; diabetic kidney disease; albuminuria; tubular injury; L-FABP

1. Introduction

Type 2 diabetes mellitus (T2D) is a metabolic disorder that promotes the development of macroangiopathy and microangiopathy, including cardiovascular disease (CVD), diabetic nephropathy, retinopathy, and neuropathy [1]. Since urinary albumin excretion reflects glomerular disorders that are due to hyperglycemia, albuminuria is used as a typical biomarker of diabetic nephropathy. In addition, recent studies have shown that renal insufficiency without overt albuminuria often occurs in people with T2D [2–6], and the involvement of a tubular disorder and glomerulosclerosis has been suggested to be involved. Therefore, those diabetic renal disorders have been broadly defined as diabetic kidney disease (DKD), including renal dysfunction, regardless of albuminuria. Since DKD, the incidence of which has been increasing, is a major cause of morbidity and mortality in people with T2D, the assessment of the risk factors for the development of DKD is a crucial clinical issue.

Advanced glycation end products (AGEs) are heterogeneous molecules derived from the nonenzymatic products that result from reactions between glucose or other saccharide derivatives and proteins or lipids [7]. The accumulation of AGEs is thought to be an independent predictor and risk factor for CVD and renal failure in people with T2D. Although the level of AGEs in the human body can be measured in serum or plasma, the serum or plasma level of AGEs does not accurately reflect the level of AGEs in tissue [8]. Therefore, the correct evaluation of the accumulation of AGEs in tissue requires invasive skin biopsies.

The AGE Reader® (Diagnoptics Technologies BV, Groningen, The Netherlands) is a non-invasive monitoring device that uses ultraviolet light to excite autofluorescence in human skin tissue; skin autofluorescence (SAF) is indicative of the amount of AGEs present. The measurement of SAF using this device has been extensively validated and levels of AGEs determined using SAF have been shown to strongly correlate with the accumulation of AGEs found in the dermal tissue skin biopsies taken from the same site as that of the SAF measurement in individuals [9]. Because of the non-invasive character, portability, and ease in performing measurements of the AGE Reader® device, SAF is more suitable for evaluating the accumulation of AGEs in daily medical care than tissue skin biopsies are.

In studies using the AGE Reader®, it was shown that subjects with T1D and T2D who had micro- or macrovascular complications had higher SAF levels than control subjects [10–12]. Although two studies showed that SAF was positively associated with albuminuria [13,14], there have been no studies on the association between SAF and diabetic tubular injury.

Liver-type fatty acid-binding protein (L-FABP) is a low-molecular-weight protein expressed in the proximal tubule that has been recognized as a biomarker specific to tubular injury. Since an experimental model demonstrated that urinary L-FABP is correlated with tubular injury such as that resulting from stress from protein overload [15], much attention has been paid to not only albuminuria but also L-FABP for assessing the severity of DKD.

Taken together, there is a possibility that the quantification of the skin accumulation of AGEs using SAF provides predictive value for assessing the severity of DKD, including glomerular injury and tubular injury. However, it has not been fully determined whether AGEs have clinical significance in DKD. This study was therefore carried out to determine whether the accumulation of AGEs, as measured using SAF, is correlated with the progression of DKD, represented by the urine albumin-to-creatinine ratio (uACR), as a biomarker of glomerular injury, and the urine L-FABP-to-creatinine ratio (uL-FABPCR), as a biomarker of tubular injury, in individuals with T2D.

2. Materials and Methods

2.1. Subjects

We consecutively recruited 350 Japanese people (198 men and 152 women) with T2D who were outpatients or inpatients at the Department of Internal Medicine, Anan Medical Center, Tokushima, Japan, during the period from May 2020 to March 2022. T2D was diagnosed in accordance with the criteria proposed by the Expert Committee on the

Diagnosis and Classification of Diabetes Mellitus [16]. We collected clinical data, blood samples, and urine samples and measured SAF levels at baseline and 1 year later to perform cross-sectional and longitudinal analyses. The study design is shown in Figure 1. Physical examinations including anthropometry were performed on all of the participants in this study. Current smokers were defined as individuals who had smoked in the last two years. Body mass index was calculated as obesity index. Blood pressure was measured two times and averaged. Subjects with hypertension were those who had systolic blood pressure (SBP) $\geq$ 140 mmHg and/or diastolic blood pressure (DBP) $\geq$ 90 mmHg, or those receiving antihypertensive drugs. Subjects with dyslipidemia were those who had low-density lipoprotein cholesterol (LDL-C) level $\geq$ 140 mg/dL (3.6204 mmol/L) or triglycerides (TG) $\geq$ 150 mg/dL (1.6935 mmol/L) or a high-density lipoprotein cholesterol (HDL-C) level of less than 40 mg/dL (1.0344 mmol/L) or those receiving hypolipidemic drugs. Exclusion criteria included known malignancy, liver cirrhosis, malnutrition, and if the patient was undergoing hemodialysis.

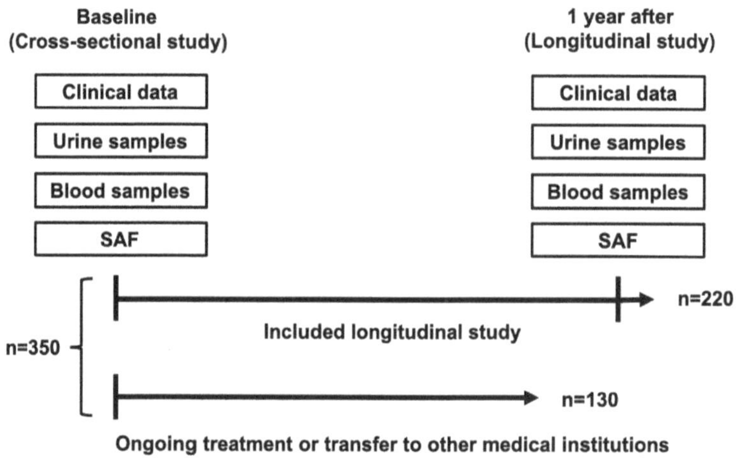

Figure 1. Schematic representation of the study protocol.

2.2. SAF Measurement

The AGE Reader® contains a UV-A light emitter with a peak wavelength of 360 to 370 nm. The light reflected and emitted from the skin in the 300 to 600 nm range is measured with a built-in spectrometer using a UV glass fiber. SAF is measured on the volar side of the forearm. To correct for light absorption differences, SAF is calculated as the ratio of emitted fluorescence (420 to 600 nm) to the reflected excitation light (300 to 420 nm). Consequently, SAF is expressed in arbitrary units (AU). Intra-observer variation in repeated autofluorescence measurements is 5% to 6% throughout the day [17]. In 350 Japanese people with T2D, SAF values were measured using the AGE Reader®, and both urine albumin-to-creatinine ratio (uACR), as a biomarker of glomerular injury, and urine liver-type fatty acid-binding protein (uL-FABP)-to-creatinine ratio (uL-FABPCR), as a biomarker of tubular injury, were estimated as indices of the severity of DKD.

2.3. Biochemical Analyses

Blood and spot urine samples were collected and used for the determination of blood cell counts, plasma glucose (PG), HbA1c, and serum biochemical parameters including LDL-C, TG, HDL-C, ALB, uric acid (UA), and Cr. PG and serum levels of LDL-C, TG, HDL-C, ALB, UA, and Cr were measured through enzymatic methods using an automatic analyzing apparatus (LABOSPECT 008, Hitachi High-Tech Co., Tokyo, Japan). HbA1c was assayed via high-performance liquid chromatography using an analyzing apparatus (HLC-723 G11, Tosoh Co., Tokyo, Japan).

Urine albumin was measured using a turbidimetric immunoassay, and uL-FABP was measured using a chemiluminescent enzyme immunoassay. eGFR was calculated in accordance with the following formula from the Japanese Society of Nephrology: eGFR (mL/min/1.73 m^2) = 194 × serum and creatinine $level^{-1.094}$ × $age^{-0.287}$ (×0.739 if female).

2.4. Statistical Analyses

The Shapiro–Wilk test was used to evaluate the normality of continuous variables. Continuous variables with a normal distribution were expressed as means ± standard deviation (SD) and those with a non-normal distribution were expressed as medians (Q1, Q3). Categorical parameters were expressed as percentages and numbers. Males, presence of hypertension, dyslipidemia, and current smokers were coded as dummy variables. Multiple regression analyses were used to determine the independent associations of urinary DKD biomarkers (uACR, uACR with logarithmic transformation, uL-FABPCR, and uL-FABPCR with logarithmic transformation) with each variable, including sex, age, BMI, SBP, serum lipid parameters, UA, Cr, HbA1c, SAF, current smoker status, hypertension, dyslipidemia, and duration of T2D. These analyses were performed using Excel (Microsoft Office Excel 16.78.3; Microsoft, Richmond, CA, USA) and GraphPad Prism 9.5.1 (GraphPad Software, San Diego, CA, USA). Statistical significance was considered to be $p < 0.05$.

3. Results

3.1. Baseline Characteristics of the Subjects

The physical and laboratory characteristics of the study participants are presented in Table 1. On average, the levels of HDL-C, uL-FABP, and log-transformed uL-FABP were higher in females than in males. Casual PG levels, serum levels of UA and Cr, and SAF levels were higher in males. There were no significant gender differences in age, BMI, SBP, LDL-C, TG, HbA1c, eGFR, uACR, or log-transformed uACR. The percentage of individuals who were current smokers was much higher among males than among females. The percentage of females who used statins was higher than that of males. Greater percentages of males used antiplatelets and sulfonyl urea.

Table 1. Clinical characteristics of the subjects in the cross-sectional study.

	Total	Males	Females	*p* Value (Males vs. Females)
Number of subjects	350	198	152	
Age (years)	70 (61, 75)	71 (61, 76)	69 (61, 75)	0.706
BMI (kg/m^2)	24.2 (22.0, 26.9)	24.2 (22.1, 26.4)	24.2(21.6, 28.0)	0.943
SBP (mmHg)	132.0 (120.3, 143.0)	131.0 (120.0, 141.8)	133.0 (121.8, 144.0)	0.257
TG (mmol/L)	1.3 (0.9, 1.8)	1.4 (0.9, 1.9)	1.1 (0.8, 1.7)	0.050
HDL-C (mmol/L)	1.3 (1.1, 1.6)	1.3 (1.1, 1.5)	1.5 (1.3, 1.7)	<0.001
LDL-C (mmol/L)	2.5 (2.1, 3.1)	2.5 (2.0, 3.1)	2.6 (2.1, 3.1)	0.435
Casual PG (mmol/L)	7.5 (6.3, 9.7)	7.7 (6.8, 10.4)	6.7 (5.8, 8.7)	<0.001
HbA1c (%)	6.8 (6.4, 7.4)	6.7 (6.3, 7.3)	6.9 (6.5, 7.5)	0.854
HbA1c (mmol/mol)	51 (46, 57)	50 (45, 56)	52 (48, 58)	0.854

Table 1. *Cont.*

	Total	Males	Females	*p* Value (Males vs. Females)
UA (umol/L)	297.4 (249.8, 355.4)	321.2 (273.6, 368.8)	258.7 (218.6, 304.8)	<0.001
Cr (umol/L)	66.7 (55.7, 82.2)	75.1 (55.6, 83.8)	55.7 (46.9, 63.9)	<0.001
eGFR (mL/min)	71.5 ± 20.1	70.2 ± 19.4	73.2 ± 20.9	0.171
uACR (mg/gCr)	16.5 (7.7, 53.0)	15.6 (7.1, 72.4)	16.7 (8.9, 37.8)	0.880
Log-transformed uACR	1.22 (0.89, 1.72)	1.19 (0.85, 1.86)	1.22 (0.95, 1.58)	0.880
uL-FABPCR (μg/gCr)	2.76 (1.74, 5.01)	2.55 (1.47, 5.02)	3.04 (2.06, 4.94)	0.045
Log-transformed uL-FABPCR	0.44 (0.24, 0.70)	0.41 (0.17, 0.70)	0.48 (0.31, 0.69)	0.045
SAF (AU)	2.4 (2.1, 2.7)	2.4 (2.1, 2.8)	2.3 (2.0, 2.6)	0.025
Current smoker (n, (%))	59 (16.9)	53 (26.8)	6 (3.9)	<0.001
Hypertension (n, (%))	227 (64.9)	121 (61.1)	106 (69.7)	0.114
Dyslipidemia (n, (%))	260 (74.3)	144 (72.7)	116 (76.3)	0.462
Duration of T2D (years)	10 (3, 19)	10 (4, 18)	10 (2, 19)	0.632
ARB or ACEi (n, (%))	142 (40.6)	76 (38.4)	66 (43.4)	0.380
CCB (n, (%))	130 (37.1)	74 (37.4)	56 (36.8)	0.999
β blocker (n, (%))	15 (4.3)	9 (4.5)	6 (3.9)	0.999
MR blocker (n, (%))	4 (1.1)	3 (1.5)	1 (0.7)	0.636
Statin (n, (%))	174 (49.7)	87 (43.9)	87 (57.2)	0.018
Ezetimibe (n, (%))	27 (7.7)	14 (7.1)	13 (8.6)	0.596
Other hypolipidemic drugs (n, (%))	21 (6.0)	13 (6.6)	8 (5.3)	0.657
Antiplatelets (n, (%))	35 (10.0)	28 (14.1)	7 (4.6)	0.004
SU or Glinide (n, (%))	66 (18.9)	46 (23.2)	20 (13.2)	0.019
Metformin (n, (%))	184 (52.6)	106 (53.5)	78 (51.3)	0.746
DPP-4i (n, (%))	209 (59.7)	120 (60.6)	89 (58.6)	0.742
SGLT2i (n, (%))	149 (42.6)	87 (43.9)	62 (40.8)	0.587
αGI (n, (%))	46 (13.1)	25 (12.6)	21 (13.8)	0.752
Pioglitazone (n, (%))	11 (3.1)	5 (2.5)	6 (3.9)	0.542
Insulin (n, (%))	73 (20.9)	39 (19.7)	34 (22.4)	0.596
GLP-1RA (n, (%))	35 (10.0)	18 (9.1)	17 (11.2)	0.591

The values are presented as means ± SD or medians (Q1, Q3). Abbreviations: BMI: body mass index, SBP: systolic blood pressure, TG: triglycerides, HDL-C: high-density lipoprotein cholesterol, LDL-C: low-density lipoprotein cholesterol, PG: plasma glucose, HbA1c: hemoglobin A1c, UA: uric acid, Cr: creatinine, eGFR: estimated glomerular filtration rate, uACR: urinary albumin-to-creatinine ratio, uL-FABPCR: urinary liver-type fatty acid-binding protein-to-creatinine ratio, SAF: skin autofluorescence, ARB: angiotensin II receptor blocker, ACEi: angiotensin-converting enzyme inhibitor, CCB: calcium channel blocker, MR: mineral corticoid receptor, SU: sulfonylurea, DPP-4i: dipeptidyl peptidase-4 inhibitor, SGLT2i: sodium glucose cotransporter 2 inhibitor, αGI: alpha-glucosidase inhibitor, GLP-1RA: glucagon-like peptide-1 receptor agonist.

3.2. Associations of SAF with uACR and Log-Transformed uACR without Adjusting for Confounding Factors

In the simple linear regression analysis, uACR level showed a significant positive correlation with SAF ($R^2 = 0.0214$, $p < 0.01$, as shown in Figure 2a). Log-transformed uACR also showed a significant positive correlation with SAF ($R^2 = 0.0341$, $p < 0.001$, as shown in Figure 2b).

3.3. Associations of SAF with uL-FABPCR and Log-Transformed uL-FABPCR without Adjusting for Confounding Factors

In the simple linear regression analysis, uL-FABPCR level showed a significant positive correlation with SAF ($R^2 = 0.0443$, $p < 0.001$, as shown in Figure 2c). Log-transformed uL-FABPCR also showed a significant positive correlation with SAF ($R^2 = 0.0516$, $p < 0.001$, as shown in Figure 2d).

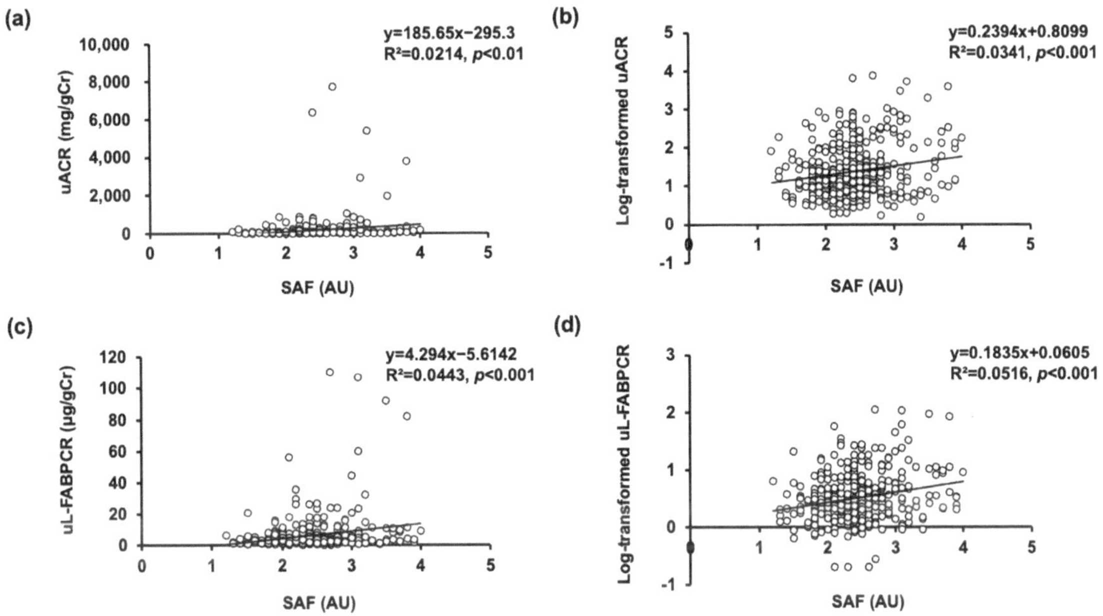

Figure 2. Scatter plots between SAF and urinary DKD biomarkers. (**a**) Scatter plot between SAF and uACR. (**b**) Scatter plot between SAF and log-transformed uACR. (**c**) Scatter plot between SAF and uL-FABPCR. (**d**) Scatter plot between SAF and log-transformed uL-FABPCR.

3.4. Associations of SAF with uACR and Log-Transformed uACR after Adjusting for Confounding Factors

A multiple linear regression analysis was carried out using the univariate baseline parameters including Cr but not eGFR (Table 2). The common risk factors for greater uACR and log-transformed uACR were Cr and duration of T2D. No significant correlation of SAF with uACR or log-transformed uACR was found. In addition, no association of SAF with uACR or log-transformed uACR was found in the multiple linear regression analysis including eGFR but not Cr (Table S1).

Table 2. Multiple linear regression analysis for determinants of DKD biomarkers.

Variables	uACR			Log-Transformed uACR			uL-FABPCR			Log-Transformed uL-FABPCR		
	t Value	VIF	*p* Value	*t* Value	VIF	*p* Value	*t* Value	VIF	*p* Value	*t* Value	VIF	*p* Value
Age	−2.378	1.620	0.018	0.628	1.620	0.531	−0.223	1.620	0.823	2.055	1.620	0.041
Male	−2.960	1.539	0.003	−1.913	1.539	0.057	−3.164	1.539	0.002	−3.250	1.539	0.001
BMI	1.557	1.556	0.121	3.21	1.556	0.002	2.003	1.556	0.046	2.105	1.556	0.036
Current smoker	1.311	1.228	0.191	2.16	1.228	0.032	1.899	1.228	0.058	2.074	1.228	0.039
SBP	2.210	1.309	0.028	1.765	1.309	0.079	1.023	1.309	0.307	0.469	1.309	0.639
TG	2.963	1.473	0.003	1.272	1.473	0.204	0.713	1.473	0.476	−0.502	1.473	0.616
HDL-C	0.099	1.388	0.921	−0.708	1.388	0.479	0.183	1.388	0.855	−0.290	1.388	0.772
LDL-C	3.259	1.252	0.001	1.700	1.252	0.090	2.503	1.252	0.013	0.858	1.252	0.391
Casual PG	2.494	1.651	0.013	0.805	1.651	0.422	0.857	1.651	0.392	0.687	1.651	0.493
HbA1c	−0.604	1.687	0.547	0.529	1.687	0.598	0.180	1.687	0.858	1.742	1.687	0.082
UA	−2.772	1.506	0.006	−2.144	1.506	0.033	−2.409	1.506	0.017	−2.189	1.506	0.029
Cr	7.308	1.660	<0.001	4.779	1.660	<0.001	6.406	1.660	<0.001	4.472	1.660	<0.001
Hypertension	0.506	1.329	0.613	2.982	1.329	0.003	1.023	1.329	0.307	2.017	1.329	0.045
Duration of T2D	3.539	1.260	<0.001	3.443	1.260	0.001	3.680	1.260	<0.001	2.857	1.260	0.005
Dyslipidemia	−0.347	1.150	0.729	−0.010	1.150	0.992	−0.770	1.150	0.442	0.127	1.150	0.899
SAF	1.549	1.294	0.122	1.644	1.294	0.101	2.255	1.294	0.025	2.022	1.294	0.044

VIF: variance inflation factor.

3.5. Associations of SAF with uL-FABPCR and Log-Transformed uL-FABPCR after Adjusting for Confounding Factors

As in the case of uACR analysis, univariate baseline parameters including Cr but not eGFR were entered into a multiple linear regression analysis (Table 2). The common risk factors for higher levels of these urinary markers, uL-FABPCR and log-transformed uL-FABPCR, were female gender, BMI, Cr, duration of T2D, and SAF. Significant positive associations of SAF with uL-FABPCR and log-transformed uL-FABPCR were also found in a multiple linear regression analysis including eGFR but not Cr (Table S1).

3.6. Associations of SAF with uL-FABPCR and Log-Transformed uL-FABPCR after Adjusting for Identified Confounding Factors and Medications Used

Because pharmacological interventions (including treatments with hypoglycemic agents, antihypertensive agents, and statins) can change the development of DKD, we next performed multiple linear regression analysis with the confirmed independent variables shown in Table 2 and Table S1 as well as certain medications used. Model 1 included the addition of cardiovascular drugs and Model 2 included the addition of hypoglycemic agents. Multiple linear regression analysis showed that female gender, Cr (Table 3) or eGFR (Table S2), and SAF were positive contributors to increases in uL-FABPCR and log-transformed uL-FABPCR. Furthermore, SAF levels remained positively associated with the severity of tubular injury regardless of subgroup analyses regarding sex, age, BMI, duration of T2D, and use of SGLT2i (Figures S1–S5).

Table 3. Multiple linear regression analysis including identified confounding factors and medications used for determinants of uL-FABPCR.

	Model 1						Model 2					
Variables	**uL-FABPCR**			**Log-Transformed uL-FABPCR**			**uL-FABPCR**			**Log-Transformed uL-FABPCR**		
	***t* Value**	**VIF**	***p* Value**	***T* Value**	**VIF**	***p* Value**	***t* Value**	**VIF**	***p* Value**	***t* Value**	**VIF**	***p* Value**
Age	-	-	-	1.570	1.591	0.117	-	-	-	1.901	1.663	0.058
Male	−2.842	1.370	0.005	−3.570	1.511	<0.001	−2.670	1.347	0.008	−3.398	1.532	<0.001
BMI	2.669	1.257	0.008	2.445	1.483	0.015	2.424	1.340	0.016	1.988	1.577	0.048
Current smoker	-	-	-	2.008	1.221	0.045	-	-	-	2.224	1.213	0.027
LDL-C	3.135	1.279	0.002	-	-	-	2.877	1.089	0.004	-	-	-
UA	−2.005	1.502	0.046	−1.927	1.511	0.055	−2.132	1.538	0.034	−1.511	1.536	0.132
Cr	6.224	1.597	<0.001	4.672	1.708	<0.001	6.217	1.553	<0.001	4.019	1.702	<0.001
Hypertension	-	-	-	1.621	1.871	0.106	-	-	-	2.271	1.175	0.024
Duration of T2D	3.515	1.205	<0.001	2.979	1.250	0.003	3.465	1.550	<0.001	1.591	1.673	0.113
SAF	3.063	1.193	0.002	2.606	1.258	0.010	3.115	1.229	0.002	2.268	1.322	0.024
ARB or ACEi	−0.004	1.344	0.997	−0.862	1.695	0.390	-	-	-	-	-	-
CCB	3.107	1.385	0.002	1.842	1.525	0.066	-	-	-	-	-	-
β blocker	−2.267	1.085	0.024	−2.227	1.094	0.027	-	-	-	-	-	-
MR blocker	−2.091	1.078	0.037	−1.897	1.085	0.059	-	-	-	-	-	-
Statin	−0.480	1.272	0.632	−0.905	1.141	0.366	-	-	-	-	-	-
Ezetimibe	0.066	1.080	0.948	0.950	1.070	0.343	-	-	-	-	-	-
Other hypolipidemic drugs	−0.372	1.049	0.710	0.253	1.064	0.800	-	-	-	-	-	-
Antiplatelets	0.149	1.189	0.882	0.276	1.200	0.783	-	-	-	-	-	-
SU or Glinide	-	-	-	-	-	-	−0.569	1.297	0.570	0.897	1.300	0.370
Metformin	-	-	-	-	-	-	0.089	1.282	0.929	−1.151	1.287	0.251
DPP-4i	-	-	-	-	-	-	−1.121	1.505	0.263	0.028	1.531	0.978
SGLT2i	-	-	-	-	-	-	0.134	1.256	0.894	2.746	1.262	0.006
αGI	-	-	-	-	-	-	−0.470	1.226	0.639	0.015	1.225	0.988
Pioglitazone	-	-	-	-	-	-	1.122	1.083	0.263	−0.065	1.085	0.948
Insulin	-	-	-	-	-	-	−0.221	1.241	0.825	1.011	1.342	0.313
GLP-1RA	-	-	-	-	-	-	0.705	1.496	0.482	0.201	1.482	0.841

3.7. Association between Annual Changes in uL-FABPCR and Those in SAF

For our longitudinal analysis, we evaluated 220 individuals (128 males and 92 females; Table S3)of the original 350 individuals in our cross-sectional study for annual changes in their ⊿uL-FABPCR and ⊿SAF values. The ⊿SAF was positively correlated with ⊿uL-FABPCR (R^2 = 0.0206, p = 0.033; Figure 3). This association between ⊿SAF and ⊿uL-FABCR remained significant even after adjusting for confounding factors (p = 0.026) (Table 4).

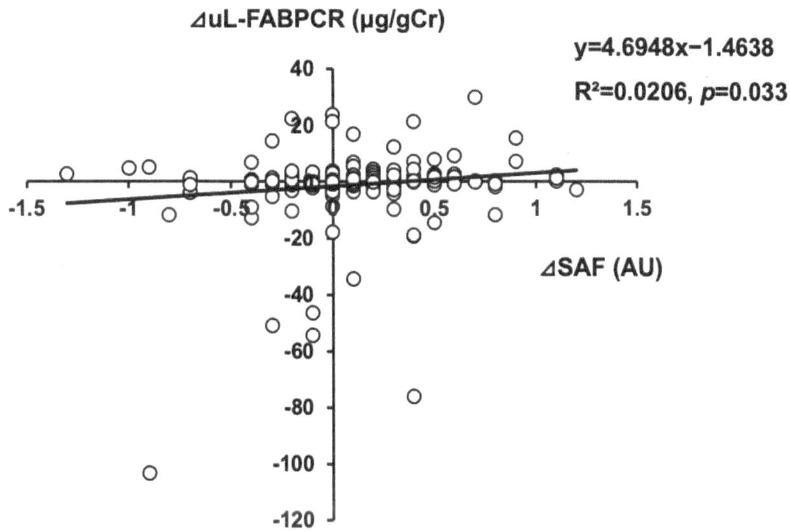

Figure 3. Scatter plot between ⊿SAF and ⊿uL-FABPCR in our longitudinal study.

Table 4. Association between annual changes in SAF and those in uL-FABPCR, adjusting for confounding factors.

Variables	⊿uL-FABPCR		
	t Value	VIF	*p* Value
Age	−0.183	1.329	0.854
Male	0.282	1.159	0.778
BMI	−0.558	1.349	0.577
Hypertension	−0.249	1.131	0.804
Duration of T2D	−1.820	1.173	0.070
Dyslipidemia	2.020	1.086	0.045
Current smoker	−0.490	1.173	0.624
⊿SAF	2.240	1.023	0.026

4. Discussion

Although widespread medical knowledge regarding risk management for diabetic microvascular complications and macroangiopathy has extended the lifespan of people with T2D, the early detection of DKD without excessive urinary albumin excretion has remained a pivotal clinical problem. Given the large number of people with T2D who are expected to be screened, a simple and non-invasive method for evaluating the severity of DKD is desirable.

SAF showed a significant positive association with urinary excretion of L-FABP in the subjects with T2D in our cross-sectional study, and annual changes in SAF also showed a significant positive association with annual changes in uL-FABPCR in our longitudinal study. As an increase in urine L-FABP is typically associated with tubular injury due to hypoxia [18], urinary excretion of L-FABP accurately reflects the severity and progression of DKD [15,19,20]. In addition, increased urine L-FABP might be found in the early stages of DKD, even before the development of albuminuria. Thus, the early-stage detection of tubular injury and stratification of subjects who have a considerable risk of tubular injury using a non-invasive and simple examination method such as SAF measurement could be useful for the prevention of DKD.

Simple linear regression analysis in previous studies demonstrated a positive correlation between SAF and uACR in people with diabetes, including T1D and T2D [13,14,21], and Gerrits E.G. and colleagues showed, using multiple regression analysis, that SAF had a significant correlation with uACR in subjects with T2D [22]. However, we did not find a significant correlation between SAF and albuminuria after adjusting for clinical confounders. The results of this study may differ from the results of those previous studies due to differences in characteristics of subjects and the inclusion of clinical confounding factors in our analysis.

AGEs have been identified in renal structures such as the glomerular basement membrane, mesangium, and tubules [23]. The involvement of AGEs in the development of glomerular lesions in people with T1D has been suggested through several lines of evidence [24,25]. Moreover, previous in vitro studies showed that AGEs could induce tubular cell apoptosis and dysfunction, contributing to glomerular hyperfiltration, an early manifestation of renal dysfunction in diabetes [26,27]. AGEs include various protein adducts, such as pentosidine, nepsilon-(carboxymethyl) lysine (CML), and pyrraline, the accumulation of which alters the structure and function of tissue proteins and stimulates cellular responses. They have been shown to be involved in tissue damage associated with diabetic microvascular complications including DKD. Horie et al. showed, using immunohistochemistry, that CML and pentosidine accumulate in the expanded mesangial matrix and thickened glomerular capillary walls of early-stage DKD and in nodular lesions and arterial walls of advanced DKD [28].

The validity of SAF as a marker for the accumulation of AGEs in the human body has been established [29]. Meerwaldt et al. demonstrated a significant correlation, validated with skin biopsies, between SAF and the fluorescent AGE pentosidine in people with diabetes, including T1D and T2D, and in those who were undergoing hemodialysis [17,30,31].

Miyata et al. reported that pentosidine accumulation in the proximal renal tubules of healthy rats was detected using immunohistochemistry one hour after the intravenous administration of synthetic pentosidine and that no further immunostaining was detected after twenty-four hours [32]. In addition, following administration of radiolabeled pentosidine, levels of radioactivity peaked in the kidney within an hour and then declined as they gradually increased in the urine [32]. Thus, the authors concluded that most of the pentosidine is catabolized during the process of tubular reabsorption. Waanders et al. demonstrated that pentosidine accumulates in damaged renal tubules in a rat model with adriamycin-induced nephropathy (AN) and that renoprotective treatment with an angiotensin-converting enzyme inhibitor reduces the accumulation of pentosidine in injured tubules in the rat model with AN [33].

Taken together, these observations indicate that intact tubules play pivotal roles in the disposal of AGEs including pentosidine into urine and that injured tubules fail in the in vivo clearance of pentosidine, leading to increased skin accumulation of pentosidine, the amount of which is detectable through SAF measurement.

Horie et al. reported that pentosidine was immunohistochemically detected in both glomeruli and tubules in individuals with T2D who had nephropathy and that pentosidine was detected in tubules but not glomeruli in individuals with T2D who did not have nephropathy [28]. Since almost all of our study subjects did not have macroalbuminuria, the lack of association between SAF values and uACR in our study might be due to the presence of only minor glomerular injury in the majority of the subjects.

These observations are in agreement with the results of our study showing that SAF is a simple and non-invasive tubular injury-specific biomarker in people with T2D.

Limitations

Nevertheless, our study is not without limitations, including a relatively small sample size and the inclusion of only people with T2D. Since the results are based on cross-sectional data from single measurements of SAF, albuminuria, and urinary L-FABP, a causal relationship between SAF and the development of tubular injury could not be shown in this study. Therefore, larger-scale and longitudinal studies are needed to elucidate this clinical question.

5. Conclusions

In conclusion, SAF was positively correlated with uL-FABP but not with uACR in subjects with T2D in our study. These results suggest that SAF can be used as a novel predictor for the development of tubular injury in people with T2D.

Supplementary Materials: The following supporting information can be downloaded at: https://www.mdpi.com/article/10.3390/biomedicines11113020/s1, Table S1: Multiple linear regression analysis for determinants (including eGFR but not Cr) of DKD biomarkers; Table S2: Multiple linear regression analysis including identified confounding factors (including eGFR but not Cr) and medications used for determinants of uL-FABPCR; Table S3: Clinical characteristics of subjects at baseline in the longitudinal study; Figure S1: Associations of SAF value with uL-FABPCR and log-transformed uL-FABPCR in sex subgroup; Figure S2: Associations of SAF value with uL-FABPCR and log-transformed uL-FABPCR in age subgroup; Figure S3: Associations of SAF value with uL-FABPCR and log-transformed uL-FABPCR in BMI subgroup; Figure S4: Associations of SAF value with uL-FABPCR and log-transformed uL-FABPCR in SGLT2i subgroup; Figure S5: Associations of SAF value with uL-FABPCR and log-transformed uL-FABPCR in duration of T2D subgroup.

Author Contributions: Conceptualization, K.-i.A.; methodology, H.Y., S.Y. (Saya Yasui), Y.K. and K.-i.A.; formal analysis, T.H. (Taiki Hori), S.Y. (Sumiko Yoshida), S.N. and K.-i.A.; data curation, M.H., T.H. (Tomoyo Hara), Y.M. and K.K.; writing—original draft preparation, H.Y. and K.-i.A.; writing—review and editing, T.H. (Takeshi Harada) and K.-i.A.; visualization, A.K.; supervision, M.A. and M.M.; project administration, T.O., T.Y. and I.E. All authors have read and agreed to the published version of the manuscript.

Funding: This work was supported by the Japan Society for the Promotion of Science (JSPS) KAKENHI Grant Number 22K08033.

Institutional Review Board Statement: This study was conducted in accordance with the Declaration of Helsinki and was approved by the Institutional Review Board of Anan Medical Center (approval date: 21 April 2020, approval number: 202019).

Informed Consent Statement: Prior informed consent was obtained from all of the participants in this study.

Data Availability Statement: The datasets generated in the present study are available from the corresponding author upon reasonable request.

Acknowledgments: We would like to thank S.E.S Translation and Proofreading Services for English language editing.

Conflicts of Interest: The authors declare no conflict of interest.

References

1. Zheng, Y.; Ley, S.H.; Hu, F.B. Global aetiology and epidemiology of type 2 diabetes mellitus and its complications. *Nat. Rev. Endocrinol.* **2018**, *14*, 88–98. [CrossRef] [PubMed]
2. Dwyer, J.P.; Parving, H.H.; Hunsicker, L.G.; Ravid, M.; Remuzzi, G.; Lewis, J.B. Renal Dysfunction in the Presence of Normoalbuminuria in Type 2 Diabetes: Results from the DEMAND Study. *Cardiorenal Med.* **2012**, *2*, 1–10. [CrossRef]
3. Yoshida, Y.; Kashiwabara, K.; Hirakawa, Y.; Tanaka, T.; Noso, S.; Ikegami, H.; Ohsugi, M.; Ueki, K.; Mita, T.; Watada, H.; et al. Conditions, pathogenesis, and progression of diabetic kidney disease and early decliner in Japan. *BMJ Open Diabetes Res. Care* **2020**, *8*, e000902. [CrossRef]
4. Kramer, H.J.; Nguyen, Q.D.; Curhan, G.; Hsu, C.Y. Renal insufficiency in the absence of albuminuria and retinopathy among adults with type 2 diabetes mellitus. *JAMA* **2003**, *289*, 3273–3277. [CrossRef] [PubMed]
5. Penno, G.; Solini, A.; Bonora, E.; Fondelli, C.; Orsi, E.; Zerbini, G.; Trevisan, R.; Vedovato, M.; Gruden, G.; Cavalot, F.; et al. Clinical significance of nonalbuminuric renal impairment in type 2 diabetes. *J. Hypertens.* **2011**, *29*, 1802–1809. [CrossRef] [PubMed]
6. Thomas, M.C.; Macisaac, R.J.; Jerums, G.; Weekes, A.; Moran, J.; Shaw, J.E.; Atkins, R.C. Nonalbuminuric renal impairment in type 2 diabetic patients and in the general population (national evaluation of the frequency of renal impairment cO-existing with NIDDM [NEFRON] 11). *Diabetes Care* **2009**, *32*, 1497–1502. [CrossRef] [PubMed]
7. Semba, R.D.; Nicklett, E.J.; Ferrucci, L. Does accumulation of advanced glycation end products contribute to the aging phenotype? *J. Gerontol. A Biol. Sci. Med. Sci.* **2010**, *65*, 963–975. [CrossRef]
8. Meertens, J.H.; Nienhuis, H.L.; Lefrandt, J.D.; Schalkwijk, C.G.; Nyyssönen, K.; Ligtenberg, J.J.; Smit, A.J.; Zijlstra, J.G.; Mulder, D.J. The Course of Skin and Serum Biomarkers of Advanced Glycation Endproducts and Its Association with Oxidative Stress,

Inflammation, Disease Severity, and Mortality during ICU Admission in Critically Ill Patients: Results from a Prospective Pilot Study. *PLoS ONE* **2016**, *11*, e0160893. [CrossRef]
9. Mulder, D.J.; Water, T.V.; Lutgers, H.L.; Graaff, R.; Gans, R.O.; Zijlstra, F.; Smit, A.J. Skin autofluorescence, a novel marker for glycemic and oxidative stress-derived advanced glycation endproducts: An overview of current clinical studies, evidence, and limitations. *Diabetes Technol. Ther.* **2006**, *8*, 523–535. [CrossRef]
10. Lutgers, H.L.; Graaff, R.; Links, T.P.; Ubink-Veltmaat, L.J.; Bilo, H.J.; Gans, R.O.; Smit, A.J. Skin autofluorescence as a noninvasive marker of vascular damage in patients with type 2 diabetes. *Diabetes Care* **2006**, *29*, 2654–2659. [CrossRef]
11. Lutgers, H.L.; Gerrits, E.G.; Graaff, R.; Links, T.P.; Sluiter, W.J.; Gans, R.O.; Bilo, H.J.; Smit, A.J. Skin autofluorescence provides additional information to the UK Prospective Diabetes Study (UKPDS) risk score for the estimation of cardiovascular prognosis in type 2 diabetes mellitus. *Diabetologia* **2009**, *52*, 789–797. [CrossRef]
12. Januszewski, A.S.; Xu, D.; Cho, Y.H.; Benitez-Aguirre, P.Z.; O'Neal, D.N.; Craig, M.E.; Donaghue, K.C.; Jenkins, A.J. Skin autofluorescence in people with type 1 diabetes and people without diabetes: An eight-decade cross-sectional study with evidence of accelerated aging and associations with complications. *Diabet. Med.* **2021**, *38*, e14432. [CrossRef] [PubMed]
13. McIntyre, N.J.; Fluck, R.J.; McIntyre, C.W.; Taal, M.W. Skin autofluorescence and the association with renal and cardiovascular risk factors in chronic kidney disease stage 3. *Clin. J. Am. Soc. Nephrol.* **2011**, *6*, 2356–2363. [CrossRef]
14. Skrha, J., Jr.; Soupal, J.; Loni Ekali, G.; Prázný, M.; Kalousová, M.; Kvasnička, J.; Landová, L.; Zima, T.; Skrha, J. Skin autofluorescence relates to soluble receptor for advanced glycation end-products and albuminuria in diabetes mellitus. *J. Diabetes Res.* **2013**, *2013*, 650694. [CrossRef] [PubMed]
15. Fiseha, T.; Tamir, Z. Urinary Markers of Tubular Injury in Early Diabetic Nephropathy. *Int. J. Nephrol.* **2016**, *2016*, 4647685. [CrossRef] [PubMed]
16. American Diabetes Association. 2. Classification and Diagnosis of Diabetes: Standards of Medical Care in Diabetes-2019. *Diabetes Care* **2019**, *42*, S13–S28. [CrossRef]
17. Meerwaldt, R.; Graaff, R.; Oomen, P.H.N.; Links, T.P.; Jager, J.J.; Alderson, N.L.; Thorpe, S.R.; Baynes, J.W.; Gans, R.O.B.; Smit, A.J. Simple non-invasive assessment of advanced glycation endproduct accumulation. *Diabetologia* **2004**, *47*, 1324–1330. [CrossRef]
18. Kamijo-Ikemori, A.; Sugaya, T.; Kimura, K. Novel urinary biomarkers in early diabetic kidney disease. *Curr. Diab Rep.* **2014**, *14*, 513. [CrossRef]
19. Suzuki, K.; Babazono, T.; Murata, H.; Iwamoto, Y. Clinical significance of urinary liver-type fatty acid-binding protein in patients with diabetic nephropathy. *Diabetes Care* **2005**, *28*, 2038–2039. [CrossRef]
20. Kamijo-Ikemori, A.; Sugaya, T.; Yasuda, T.; Kawata, T.; Ota, A.; Tatsunami, S.; Kaise, R.; Ishimitsu, T.; Tanaka, Y.; Kimura, K. Clinical significance of urinary liver-type fatty acid-binding protein in diabetic nephropathy of type 2 diabetic patients. *Diabetes Care* **2011**, *34*, 691–696. [CrossRef]
21. Genevieve, M.; Vivot, A.; Gonzalez, C.; Raffaitin, C.; Barberger-Gateau, P.; Gin, H.; Rigalleau, V. Skin autofluorescence is associated with past glycaemic control and complications in type 1 diabetes mellitus. *Diabetes Metab.* **2013**, *39*, 349–354. [CrossRef]
22. Gerrits, E.G.; Lutgers, H.L.; Kleefstra, N.; Graaff, R.; Groenier, K.H.; Smit, A.J.; Gans, R.O.; Bilo, H.J. Skin autofluorescence: A tool to identify type 2 diabetic patients at risk for developing microvascular complications. *Diabetes Care* **2008**, *31*, 517–521. [CrossRef] [PubMed]
23. Tanji, N.; Markowitz, G.S.; Fu, C.; Kislinger, T.; Taguchi, A.; Pischetsrieder, M.; Stern, D.; Schmidt, A.M.; D'Agati, V.D. Expression of advanced glycation end products and their cellular receptor RAGE in diabetic nephropathy and nondiabetic renal disease. *J. Am. Soc. Nephrol.* **2000**, *11*, 1656–1666. [CrossRef] [PubMed]
24. McCance, D.R.; Dyer, D.G.; Dunn, J.A.; Bailie, K.E.; Thorpe, S.R.; Baynes, J.W.; Lyons, T.J. Maillard reaction products and their relation to complications in insulin-dependent diabetes mellitus. *J. Clin. Investig.* **1993**, *91*, 2470–2478. [CrossRef]
25. Beisswenger, P.J.; Moore, L.L.; Brinck-Johnsen, T.; Curphey, T.J. Increased collagen-linked pentosidine levels and advanced glycosylation end products in early diabetic nephropathy. *J. Clin. Investig.* **1993**, *92*, 212–217. [CrossRef] [PubMed]
26. Maeda, S.; Matsui, T.; Takeuchi, M.; Yamagishi, S. Sodium-glucose cotransporter 2-mediated oxidative stress augments advanced glycation end products-induced tubular cell apoptosis. *Diabetes Metab. Res. Rev.* **2013**, *29*, 406–412. [CrossRef]
27. Qi, W.; Niu, J.; Qin, Q.; Qiao, Z.; Gu, Y. Glycated albumin triggers fibrosis and apoptosis via an NADPH oxidase/Nox4-MAPK pathway-dependent mechanism in renal proximal tubular cells. *Mol. Cell Endocrinol.* **2015**, *405*, 74–83. [CrossRef]
28. Horie, K.; Miyata, T.; Maeda, K.; Miyata, S.; Sugiyama, S.; Sakai, H.; van Ypersole de Strihou, C.; Monnier, V.M.; Witztum, J.L.; Kurokawa, K. Immunohistochemical colocalization of glycoxidation products and lipid peroxidation products in diabetic renal glomerular lesions. Implication for glycoxidative stress in the pathogenesis of diabetic nephropathy. *J. Clin. Investig.* **1997**, *100*, 2995–3004. [CrossRef] [PubMed]
29. Atzeni, I.M.; van de Zande, S.C.; Westra, J.; Zwerver, J.; Smit, A.J.; Mulder, D.J. The AGE Reader: A non-invasive method to assess long-term tissue damage. *Methods* **2022**, *203*, 533–541. [CrossRef]
30. Meerwaldt, R.; Hartog, J.W.; Graaff, R.; Huisman, R.J.; Links, T.P.; den Hollander, N.C.; Thorpe, S.R.; Baynes, J.W.; Navis, G.; Gans, R.O.; et al. Skin autofluorescence, a measure of cumulative metabolic stress and advanced glycation end products, predicts mortality in hemodialysis patients. *J. Am. Soc. Nephrol.* **2005**, *16*, 3687–3693. [CrossRef]
31. Meerwaldt, R.; Links, T.; Graaff, R.; Thorpe, S.R.; Baynes, J.W.; Hartog, J.; Gans, R.; Smit, A. Simple noninvasive measurement of skin autofluorescence. *Ann. N. Y. Acad. Sci.* **2005**, *1043*, 290–298. [CrossRef] [PubMed]

32. Miyata, T.; Ueda, Y.; Horie, K.; Nangaku, M.; Tanaka, S.; van Ypersele de Strihou, C.; Kurokawa, K. Renal catabolism of advanced glycation end products: The fate of pentosidine. *Kidney Int.* **1998**, *53*, 416–422. [CrossRef] [PubMed]
33. Waanders, F.; Greven, W.L.; Baynes, J.W.; Thorpe, S.R.; Kramer, A.B.; Nagai, R.; Sakata, N.; van Goor, H.; Navis, G. Renal accumulation of pentosidine in non-diabetic proteinuria-induced renal damage in rats. *Nephrol. Dial. Transplant.* **2005**, *20*, 2060–2070. [CrossRef] [PubMed]

biomedicines

Article

Angiopoietin-2 and Angiopoietin-like Proteins with a Prospective Role in Predicting Diabetic Nephropathy

Eman Alshawaf [1], Mohamed Abu-Farha [1,2], Anwar Mohammad [1], Sriraman Devarajan [3], Irina Al-Khairi [1], Preethi Cherian [1], Hamad Ali [4,5], Hawra Al-Matrouk [6], Fahd Al-Mulla [4,7], Abdulnabi Al Attar [2,*] and Jehad Abubaker [3,*]

1 Department of Biochemistry and Molecular Biology, Dasman Diabetes Institute, Dasman 15462, Kuwait; eman.alshawaf@dasmaninstitute.org (E.A.); mohamed.abufarha@dasmaninstitute.org (M.A.-F.); anwar.mohammad@dasmaninstitute.org (A.M.); irina.alkhairi@dasmaninstitute.org (I.A.-K.); preethi.cherian@dasmaninstitute.org (P.C.)
2 Diabetology Unit, Amiri Hospital, Dasman Diabetes Institute, Dasman 15462, Kuwait
3 National Dasman Diabetes Biobank, Dasman Diabetes Institute, Dasman 15462, Kuwait; sriraman.devarajan@dasmaninstitute.org
4 Functional Genomic Unit, Dasman Diabetes Institute, Dasman 15462, Kuwait; hamad.ali@ku.edu.kw (H.A.); fahd.almulla@dasmaninstitute.org (F.A.-M.)
5 Department of Medical Laboratory Sciences, Faculty of Allied Health Sciences, Health Sciences Center, Kuwait University, Kuwait 15462, Kuwait
6 Medical Department, Amiri Hospital, Ministry of Health, Kuwait 15462, Kuwait; hawra.h.almatrouk@gmail.com
7 Department of Translational Research, Dasman Diabetes Institute, Dasman 15462, Kuwait
* Correspondence: abdulnabi.alattar@dasmaninstitute.org (A.A.A.); jehad.abubakr@dasmaninstitute.org (J.A.)

Abstract: Angiopoietins are crucial growth factors for maintaining a healthy, functional endothelium. Patients with type 2 diabetes (T2D) exhibit significant levels of angiogenic markers, particularly Angiopoietin-2, which compromises endothelial integrity and is connected to symptoms of endothelial injury and failure. This report examines the levels of circulating angiopoietins in people with T2D and diabetic nephropathy (DN) and explores its link with ANGPTL proteins. We quantified circulating ANGPTL3, ANGPTL4, ANGPTL8, Ang1, and Ang2 in the fasting plasma of 117 Kuwaiti participants, of which 50 had T2D and 67 participants had DN. The Ang2 levels increased with DN (4.34 ± 0.32 ng/mL) compared with T2D (3.42 ± 0.29 ng/mL). This increase correlated with clinical parameters including the albumin-to-creatinine ratio (ACR) ($r = 0.244$, $p = 0.047$), eGFR ($r = -0.282$, $p = 0.021$), and SBP ($r = -0.28$, $p = 0.024$). Furthermore, Ang2 correlated positively to both ANGPTL4 ($r = 0.541$, $p < 0.001$) and ANGPTL8 ($r = 0.41$, $p = 0.001$). Multiple regression analysis presented elevated ANGPTL8 and ACRs as predictors for Ang2's increase in people with DN. In people with T2D, ANGPTL4 positively predicted an Ang2 increase. The area under the curve (AUC) in receiver operating characteristic (ROC) analysis of the combination of Ang2 and ANGPTL8 was 0.77 with 80.7% specificity. In conclusion, significantly elevated Ang2 in people with DN correlated with clinical markers such as the ACR, eGFR, and SBP, ANGPTL4, and ANGPTL8 levels. Collectively, this study highlights a close association between Ang2 and ANGPTL8 in a population with DN, suggesting them as DN risk predictors.

Keywords: Ang1; Ang2; ANGPTL8; ANGPTL4; DN

Citation: Alshawaf, E.; Abu-Farha, M.; Mohammad, A.; Devarajan, S.; Al-Khairi, I.; Cherian, P.; Ali, H.; Al-Matrouk, H.; Al-Mulla, F.; Al Attar, A.; et al. Angiopoietin-2 and Angiopoietin-like Proteins with a Prospective Role in Predicting Diabetic Nephropathy. *Biomedicines* **2024**, *12*, 949. https://doi.org/10.3390/biomedicines12050949

Academic Editors: Ana Dascalu and Dragos Serban

Received: 26 March 2024
Revised: 14 April 2024
Accepted: 20 April 2024
Published: 24 April 2024

1. Introduction

Diabetes mellitus is a significant global public health challenge [1]. Epidemiological studies indicate that the proportion of people affected by diabetes was approximately 8.8% in 2015, and this percentage is projected to increase to 10.4% by 2024. An analysis released in 2020, utilizing data from 2014, determined that the prevalence of diabetes in Kuwait was projected to reach 21.8% [2]. Diabetic nephropathy (DN) is a significant microvascular complication and the primary cause of mortality among individuals with diabetes [3]. According to the tenth report of the International Diabetes Federation (IDF), approximately 7.8% of people in Kuwait are diagnosed with DN as a consequence of T2D [4]. Thus, DN imposes a substantial cost on individuals affected by the condition and on healthcare systems.

The angiopoietin (Ang)/Tie ligand-receptor system is of paramount importance in the maintenance of endothelial integrity and vascular development [5]. Alterations in the Ang/Tie system have been suggested to contribute to the advancement of kidney injuries in individuals with DN [6]. Dysregulation of the angiopoietin balance, specifically Ang1 and Ang2, and their interaction with Tie receptors (Tie1 and Tie2) can have detrimental effects on endothelial function and vascular stability in the renal microvasculature [6,7]. Disruption in Tie receptor signaling and an imbalance between Ang1 and Ang2 collectively contribute to heightened vascular permeability, inflammation, and oxidative stress. These various factors have the potential to induce glomerular dysfunction and structural alterations in the kidneys, thereby playing a role in the advancement of diabetic nephropathy [6,7]. In general, the pathogenesis of diabetic nephropathy is characterized by an intricate interplay of metabolic and hemodynamic factors. The role of angiogenesis in the development of DN is well acknowledged [8], and consequently, growth factors associated with angiogenesis, such as Ang1 and Ang2, will have a substantial impact on the development of DN. Patients with T2D were found to have elevated levels of circulating Ang2 [9], which led to microalbuminuria [10]. The rise in Ang2 levels acted as an independent predictor of microalbuminuria in patients with T2D [10] and was reported in connection with a rapid decrease in kidney function [11]. Significantly elevated Ang2 levels were also positively linked with the progression of albuminuria in people with DN [8,12,13].

Angiopoietin-like proteins (ANGPTLs) are a family of proteins structurally similar to angiopoietins. To date, eight ANGPTLs have been discovered, namely ANGPTL1–ANGPTL8. ANGPTLs, particularly ANGPTL3, ANGPTL4, and ANGPTL8, play pivotal roles in regulating lipid metabolism and energy homeostasis, with emerging implications for diabetic nephropathy. ANGPTL3 and ANGPTL4 are known inhibitors of lipoprotein lipase (LPL), a key enzyme involved in triglyceride metabolism. By inhibiting LPL activity, these proteins modulate lipid clearance from the circulation, resulting in alterations in plasma lipid levels. ANGPTL8, often acting in conjunction with ANGPTL3, also regulates LPL activity and lipid metabolism, albeit with some distinct roles in glucose homeostasis and insulin sensitivity [14]. Dysregulation of these ANGPTLs has been implicated in the pathogenesis of metabolic disorders, including obesity, dyslipidemia, and type 2 diabetes mellitus (T2DM), all of which are risk factors for diabetic nephropathy [15]. ANGPTL8 is a hepatic protein that is produced in the liver and adipose tissue. Despite its lack of direct involvement in angiogenesis, ANGPTL8 has been identified as a significant contributor to regulating metabolism processes by adjusting glucose and lipid levels, in addition to its crucial role in maintaining lipid balance [14]. In addition to its association with T2D [16,17], ANGPTL8 is linked to diabetes complications and other concomitant disorders such as DN [15]. Chen et al. reported that ANGPTL8 is considerably higher in T2D patients with various stages of albuminuria [18]. Having said that, ANGPTL8 is thought to be involved in DN, and further research into its mechanism in T2D and DN is needed. The collective involvement of Ang proteins and ANGPTL8 in vascular biology and metabolic equilibrium renders them highly promising targets for diagnostic and therapeutic interventions in conditions such as metabolic syndromes and cardiovascular disorders. This study examines the correlation between Ang2, ANGPTL8, and DN to assess the potential predictive significance of ANGPTL8 and Ang2 in DN.

2. Materials and Methods

2.1. Study Population

A total of 117 participants joined this study, and the participants were segregated into two groups. Throughout the report, people diagnosed with diabetes are referred to as T2D, and those diagnosed with diabetes and nephropathy are designated as DN. The participants were body mass index (BMI) and age matched among the individuals with T2D (n = 50) and people with T2D and nephropathy (DN group, n = 67). All participants provided written consent at the Dasman Diabetes Institute (Dasman, Kuwait) to be enrolled in the study. Ethical approval for this study was granted by the Ethical Review Board of the Dasman Diabetes Institute (DDI), abiding by the ethical guidelines outlined in the Declaration of Helsinki.

2.2. *Anthropometric and Biochemical Measurements*

Blood pressure was measured with an Omron HEM-907XL digital sphygmomanometer, and the presented values are the average of three consecutive readings. Plasma was extracted from fasting blood samples collected in vacutainer-EDTA tubes (centrifugation for 10 min at 400× *g*). The plasma samples were aliquoted and stored at −80 °C for further testing. The fasting blood glucose (FBG), serum total cholesterol (TC), low-density lipoprotein (LDL-C), high-density lipoprotein (HDL-C) and triglycerides (TG) were measured with a Siemens Dimension RXL chemical analyzer (Diamond Diagnostics, Holliston, MA, USA). Quantification of albumin and creatinine in the urine samples was performed with a CLINITEK Novus Automated Urine Chemical Analyzer (Siemens Healthineers, Erlangen, Germany).

2.3. *Quantification of Creatinine and Urinary Protein*

The urinary and serum creatinine levels were measured with a VITROS 250 automatic analyzer (New York City, NY, USA). Urinary proteins were quantified using a Coomassie Plus protein assay kit according to the manufacturer's protocol (pierce, Rockford, IL, USA). The estimated glomerular filtration rate (eGFR) was calculated using a modification of the Diet in Renal Disease study equation.

2.4. *ANGPTL3, 4, and 8 Enzyme-Linked Immunosorbent Assays (ELISAs)*

The levels of plasma in the ANGPTL3, ANGPTL4 and ANGPTL8 proteins were quantified using a Magnetic Luminex Assay kit (R&D Systems Europe, Ltd., Abingdon, UK), following the manufacturer's protocol. Repeated freeze-thaw cycles of the plasma samples were avoided, and all samples were thawed on ice before assaying. Cross-reactivity with other proteins was not significant.

2.5. *Quantification of Ang1 and Ang2*

The levels of Ang1 and Ang2 were determined by the Magnetic Luminex Assay kit (R&D Systems Europe, Ltd., Abingdon, UK), following the manufacturer's protocol.

2.6. *Statistical Analysis*

An unpaired student's *t*-test was used to compare the two study groups (i.e., people with T2D and people with DN) to determine statistical significance. The correlation analysis between Ang1, Ang2, and various parameters was estimated by Pearson's correlation coefficient. A stepwise multiple linear regression model was performed to identify the parameters independently associated with Ang1 and Ang2. The diagnostic strength of Ang2 as well as ANGPTL8 as biomarkers for DN was calculated through area under the receiver operating characteristic (ROC) curve analysis. Similarly, the diagnostic strength of Ang2 with Ang1 as biomarkers for DN was calculated by ROC curve analysis. All data are presented as the mean ± SEM, with a p value < 0.05 indicating significance. All statistical analysis was performed using Graphpad Prism Software version 9 (La Jolla, CA, USA) and SPSS for Windows version 25.0 (IBM SPSS Inc., New York City, NY, USA).

3. Results

Our study involved a total of 117 Arab participants recruited from the population of the state of Kuwait. Descriptions of the population demographic and various clinical parameters are detailed in Table 1. The participants were segregated into two main study groups: people diagnosed with T2D (n = 50) and people diagnosed with diabetes and nephropathy (DN = 67). Diagnosis of type 2 diabetes was determined by having an elevated fasting blood sugar (i.e., ≥7 mmol/L) and glycated hemoglobin (A1C) (i.e., ≥6.5%) test results. Diagnosis of diabetic nephropathy was determined with the presence of a severely increased albumin-to-creatinine ratio (ACR) (i.e., >30 mg/g).

Table 1. Anthropometric and clinical parameters of people with T2D and diabetic nephropathy.

Parameter	T2D N = 50	DN N = 67	*p* Value
Gender (M/F)	18/32	45/22	
Age (years)	58.96 ± 1.02	60.09 ± 1.38	0.512
BMI (kg/m^2)	33.94 ± 0.88	34.23 ± 0.85	0.816
SBP (mmHg)	132.98 ± 3.88	132.03 ± 3.41	0.911
DBP (mmHg)	69.72 ± 2.26	68.78 ± 1.98	0.909
Height (cm)	161.90 ± 1.30	162.04 ± 3.64	0.004
Weight (kg)	88.79 ± 2.35	92.03 ± 2.79	0.055
Fasting Glucose (mmol/L)	8.27 ± 0.36	9.61 ± 0.48	0.028
HbA1C (%)	9.53 ± 1.73	8.09 ± 0.22	0.415
TChol (mmol/L)	4.15 ± 0.13	4.02 ± 0.12	0.472
TG (mmol/L)	1.41 ± 0.16	1.77 ± 0.11	0.066
HDL-C (mmol/L)	1.25 ± 0.05	1.13 ± 0.03	0.067
LDL-C (mmol/L)	2.28 ± 0.11	2.10 ± 0.10	0.203
VLDL (mmol/L)	0.56 ± 0.06	0.71 ± 0.04	0.067
C Peptide (pg/mL)	0.65 ± 0.05	0.77 ± 0.06	0.136
Serum Creatinine (mg/L)	79.42 ± 3.54	118.36 ± 6.57	0.001
eGFR (ml/min/1.73 m^2)	79.22 ± 3.19	59.70 ± 3.00	0.001
BUN	5.10 ± 0.29	7.53 ± 0.52	0.001
Albumin (mcg/L)	37.94 ± 0.50	37.28 ± 0.42	0.316
Insulin (mU/L)	22.08 ± 3.13	22.22 ± 1.93	0.969
ACR	137.27 ± 69.22	569.94 ± 174.39	0.005
Urine Creatinine (mg/day)	11.38 ± 0.84	10.17 ± 0.91	0.015
Microalbumin (mg/day)	157.58 ± 85.09	460.34 ± 169.37	0.001

Data are mean ± standard error mean. DN = diabetes with nephropathy; SBP = systolic blood pressure; DBP = diastolic blood pressure; ACR = albumin/creatinine ratio.

3.1. Circulating Angiopoietins 1 and 2 Are Elevated in People with DN

In this study, we detected an increase in the circulating levels of Ang1 and Ang2 in people with DN compared to people with T2D. People with DN showed elevated levels of Ang1, but the increase in circulating Ang1 was statistically insignificant compared with the people with T2D (Figure 1A). On the other hand, there was a significant increase in the levels of Ang2 in the people with DN ($p = 0.034$, Figure 1B) compared with those with T2D (Table 1). In addition to the angiopoietins, our data showed an increase in some angiopoietin-like proteins in the people with DN. Both ANGPTL4 ($p = 0.029$, Figure 1C) and ANGPTL8 ($p \leq 0.001$, Figure 1D) showed a significant increase in their circulating levels compared with people with T2D (Table 1).

3.2. Increased Ang2 Correlated with Clinical Parameters of DN

To further examine the significance of the angiopoietins' elevation, for both Ang1 and Ang2, under the conditions of diabetes and kidney disease, we employed Pearson's correlation analysis to explore the link between the clinical parameters indicative of DN and angiopoietin proteins. When performing the correlation analysis, we found a significant negative correlation between the elevation of Ang1 in the plasma and albumin ($r = -0.24$, $p = 0.045$) and a positive correlation with both microalbumin ($r = 0.35$, $p = 0.003$) and ANGPTL3 ($r = 0.25$, $p = 0.03$) in people with DN (Table 2). By implementing Pearson's analysis, we identified a correlation between Ang2 and several clinical parameters in people with DN and T2D (Figure 2). Here, we present a negative correlation between increased Ang2 levels and eGFR in people with DN ($r = -0.28$, $p = 0.021$; Figure 2A) and people with

T2D (r = −0.307, p = 0.034), while elevated Ang2 levels showed a negative correlation with systolic blood pressure (r = −0.28, p = 0.024) only in people with DN (Table 3). Furthermore, increased levels of Ang2 showed positive correlations with elevated levels of ACR (r = 0.24, p 0.047; Figure 2B), ANGPTL4 (r = 0.54, $p \leq 0.001$; Figure 2C), and ANGPTL8 (r = 0.41, p = 0.001; Figure 2D) in people with DN (Table 3). Our analysis also showed that increased Ang2 levels in people with T2D was positively correlated with the levels of serum creatinine (r = 0.3, p = 0.036), BUN (r = 0.32, p = 0.025), and ANGPTL4 (r = 0.55, $p \leq 0.001$) (Table 3).

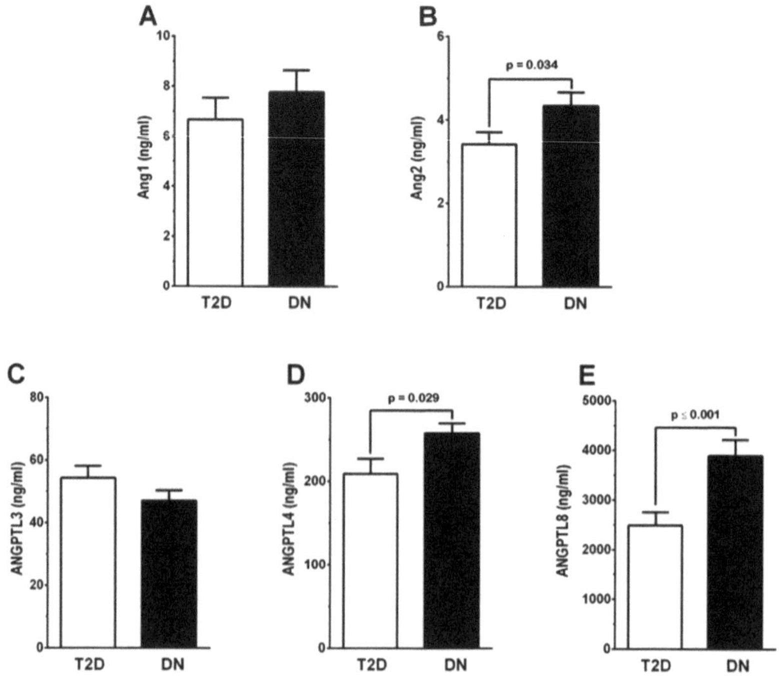

Figure 1. A comparison of circulating levels of Ang1, Ang2, ANGPTL3, ANGPTL4, and ANGPTL8 in people with T2D (white bar, n = 50) and DN (black bar, n = 67). (**A**) Levels of circulating Ang1 were higher in people with DN compared with people with T2D. Difference showed no statistical significance. (**B**) People with DN showed a significant increase in levels of Ang2 (p = 0.034) compared with people with T2D. (**C**) ANGPTL3 levels did not show a significant difference between people with T2D and those with DN. (**D**) Circulating levels of ANGPTL4 were significantly higher (p = 0.029) in people with DN compared with those with T2D. (**E**) Levels of circulating ANGPTL8 were significantly increased ($p \leq 0.001$) in people with DN compared with those with T2D.

Table 2. Pearson's correlation analysis for Ang1 in study groups T2D and DN.

Parameters	Ang1			
	T2D		DN	
	r	p	r	p
Age (years)	−0.036	0.808	−0.045	0.715
BMI (kg/m^2)	0.151	0.301	−0.001	0.994
SBP (mmHg)	0.177	0.228	0.024	0.852
DBP (mmHg)	−0.121	0.414	0.007	0.957
Fasting Glucose (mmol/L)	0.079	0.591	−0.042	0.735
HbA1C (%)	−0.003	0.985	−0.140	0.257
T. Chol (mmol/L)	−0.083	0.570	0.047	0.703
TGL (mmol/L)	0.061	0.675	0.009	0.941

Table 2. *Cont.*

Parameters	Ang1			
	T2D		DN	
	r	*p*	r	*p*
HDL (mmol/L)	−0.175	0.228	0.073	0.556
LDL (mmol/L)	−0.078	0.599	0.020	0.872
VLDL (mmol/L)	0.063	0.667	0.008	0.952
C peptide (pg/mL)	−0.014	0.927	−0.049	0.696
Serum Creatinine (mg/L)	0.291	0.043	−0.049	0.696
eGFR (mL/min/1.73 m^2)	−0.225	0.123	0.018	0.883
BUN	0.198	0.174	−0.069	0.581
Albumin (mcg/L)	−0.274	0.057	−0.246	0.045
Insulin (mU/L)	−0.013	0.931	0.131	0.291
ACR (mg/g)	0.065	0.659	0.177	0.152
Urine Creatinine (mg/day)	−0.002	0.991	−0.014	0.909
Microalbumin (mg/day)	0.057	0.697	0.353	0.003
Ang2 (ng/mL)	0.197	0.175	0.093	0.455
ANGPTL3 (ng/mL)	0.235	0.104	0.257	0.036
ANGPTL4 (ng/mL)	0.138	0.343	−0.002	0.986
ANGPTL8 (ng/mL)	0.214	0.139	0.089	0.472

r, Pearson's correlation coefficient with significance at $p < 0.05$.

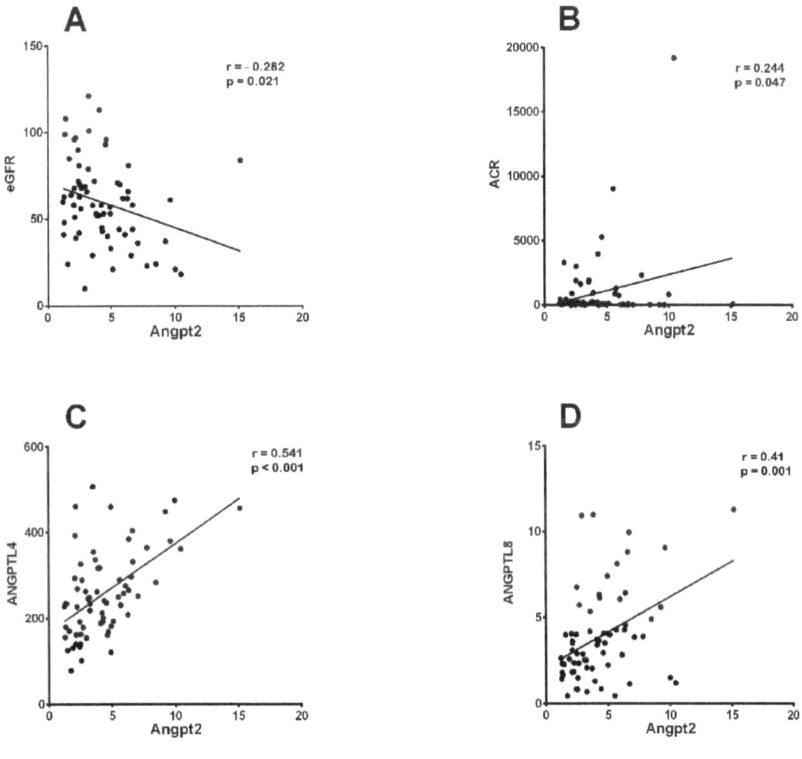

Figure 2. Correlation analysis between Ang2 and parameters associated with DN. Pearson's correlation coefficient showed a significant (**A**) negative correlation between eGFR and Ang2 (r = −0.282, $p = 0.021$), while elevated Ang2 showed positive correlations with (**B**) ACR (r = 0.244, $p = 0.047$), (**C**) ANGPTL4 (r = 0.541, $p = 0.001$), and (**D**) ANGPTL8 (r = 0.410, $p = 0.001$).

Table 3. Pearson's correlation analysis for Ang2 in study groups T2D and DN.

Parameters	Ang2			
	T2D		DN	
	r	*p*	r	*p*
Age (years)	0.277	0.054	0.135	0.277
BMI (kg/m^2)	0.240	0.096	0.070	0.580
SBP (mmHg)	0.013	0.932	−0.280	0.024
DBP (mmHg)	−0.211	0.150	−0.209	0.095
Fasting Glucose (mmol/L)	0.203	0.162	0.100	0.419
HbA1C (%)	−0.024	0.871	0.028	0.820
T. Chol (mmol/L)	−0.204	0.159	−0.007	0.955
TGL (mmol/L)	0.063	0.669	0.196	0.113
HDL (mmol/L)	−0.176	0.227	−0.049	0.692
LDL (mmol/L)	−0.221	0.132	−0.099	0.428
VLDL (mmol/L)	0.063	0.666	0.195	0.114
C peptide (pg/mL)	0.032	0.827	−0.047	0.706
Serum Creatinine (mg/L)	0.300	0.036	0.215	0.081
eGFR (mL/min/1.73 m^2)	−0.307	0.034	−0.282	0.021
BUN	0.320	0.025	0.236	0.054
Albumin(mcg/L)	−0.220	0.129	−0.189	0.126
Insulin (mU/L)	−0.060	0.683	−0.017	0.889
ACR (mg/g)	0.157	0.281	0.244	0.047
Urine Creatinine (mg/day)	−0.160	0.272	−0.157	0.204
Microalbumin (mg/day)	−0.040	0.783	0.092	0.461
Ang1 (ng/mL)	0.197	0.175	0.093	0.455
ANGPTL3 (ng/mL)	−0.123	0.401	0.190	0.123
ANGPTL4 (ng/mL)	0.555	0.001	0.541	0.001
ANGPTL8 (ng/mL)	0.020	0.889	0.410	0.001

r, Pearson's correlation coefficient with significance at $p < 0.05$.

3.3. *Predictive Analysis Suggests a Potential Ang2-ANGPTL8 Link with DN*

We implemented multiple stepwise regression analysis with a set of predictors to gain further insight into the potential connection between Ang2 and the biochemical diagnostic parameters (Table 4). According to our model, elevated ANGPTL8 levels and ACRs are both predictive markers with a significant positive regression weight for the elevation of Ang2 in people with DN (Table 4). These markers showed an independent correlation with Ang2 and ANGPTL8 levels ($F_{1,58} = 3.171$, $p < 0.001$, and $r^2 = 34\%$) and ACRs ($F_{1,58} = 2.611$, $p = 0.031$, and $r^2 = 30\%$), while SBP acted as a negative predictor of increased Ang2 levels in people with DN ($F_{1,58} = 4.102$, $p = 0.039$, and $r^2 = 32\%$) (Table 4). Collectively, these markers independently correlated with the increase in the Ang2 level. On the other hand, the people with T2D presented ANGPTL4 ($\beta = 0.552$, $p < 0.001$; Table 4) as a positive independent predictor for increased Ang2 levels ($F_{1,47} = 11.008$, $p < 0.001$, and $r^2 = 30\%$). To sum up, our data revealed an independent correlation between SBP, the ACR and ANGPTL8 with Ang2 in people with DN, thus highlighting these markers as significant predictors, whereby Ang2 is a dependent variable in people with diabetic nephropathy.

Table 4. Multiple regression analysis to identify parameters associated with Ang2. Adjusted for age, gender, and BMI.

Parameters	T2D		DN	
	β Coefficient	*p* Value	β Coefficient	*p* Value
SBP	−0.047	0.706	−0.273	0.012
ACR	0.078	0.535	0.345	0.002
ANGPTL4	0.552	<0.001	0.082	0.614
ANGPTL8	−0.099	0.438	0.424	<0.001

Additionally, we performed ROC curve analysis to identify the cut-off value of Ang2 and ANGPTL8 and to further evaluate the predictive accuracy of these biomarkers for people with DN (Figure 3). Our analysis showed that the area under the curve (AUC) (95% CI) was 0.74 (0.66–0.81, $p < 0.001$) for Ang2, 0.79 (0.70–0.89, $p < 0.001$) for ANGPTL8, and 0.77 (0.70–0.85, $p < 0.001$) for the combination of Ang2-ANGPTL8. The optimal cut-off value for predicting DN with Ang2 was higher than 1426.78 ng/mL with 93% sensitivity and a specificity of 86%. The optimal cut-off value for ANGPTL8 as a predictive marker for DN was higher than 1135.75 ng/mL, with 93.9% sensitivity and a specificity of 81.1%. Additionally, the optimal cut-off value for the combination of Ang2 and ANGPTL8 was higher than 2681.08 ng/mL, with 94.1% sensitivity and a specificity of 80.7%. This finding indicates that ANGPTL8 may possess superior diagnostic accuracy compared with Ang2 and the combined biomarkers in distinguishing diabetic nephropathy from other conditions or outcomes.

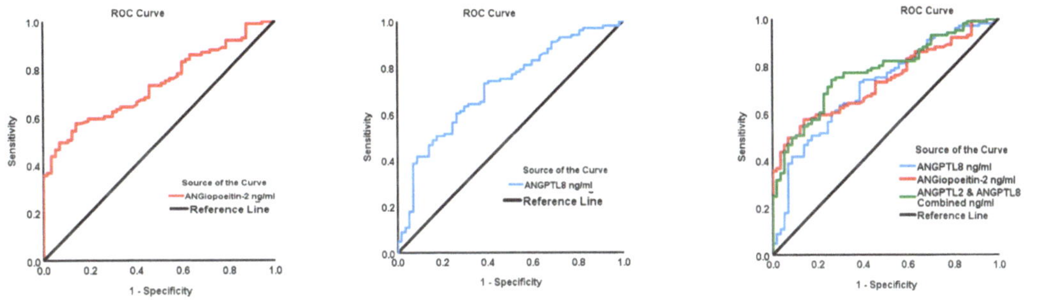

(A) (B) (C)

Figure 3. ROC curve analysis performed to identify the cut-off values of Ang2 and ANGPTL8 as biomarkers for DN. (**A**) AUC for Ang2 (0.74 (0.66–0.81) $p < 0.001$). (**B**) AUC for ANGPTL8 (0.79 (0.70–0.89) $p < 0.001$). (**C**) AUC of the combination of Ang2 and ANGPTL8 (0.77 (0.70–0.85) $p < 0.001$).

ROC curve analysis was also performed to identify the cut-off value of Ang2 and Ang1 as potential predictive biomarkers for people with DN in the study population (Figure 4). Our analysis showed that the AUC (%95 CI) was 0.72 (0.65–0.80, $p < 0.001$) for Ang1, 0.74 (0.66–0.81, $p < 0.001$) for Ang2, and 0.81 (0.74–0.88, $p < 0.001$) for the combination of Ang1-Ang2. The identified cut-off value for predicting DN with Ang1 was above 1590.92 ng/mL, with 94% sensitivity and a specificity of 77%. As for Ang2, the optimal cut-off value for predicting DN was above 1426.78 ng/mL, with 93% sensitivity and a specificity of 86%. The combination of Ang1 and Ang2 demonstrated an optimal cut-off value exceeding 1518.91 ng/mL, with a sensitivity of 92% and a specificity of 89%.

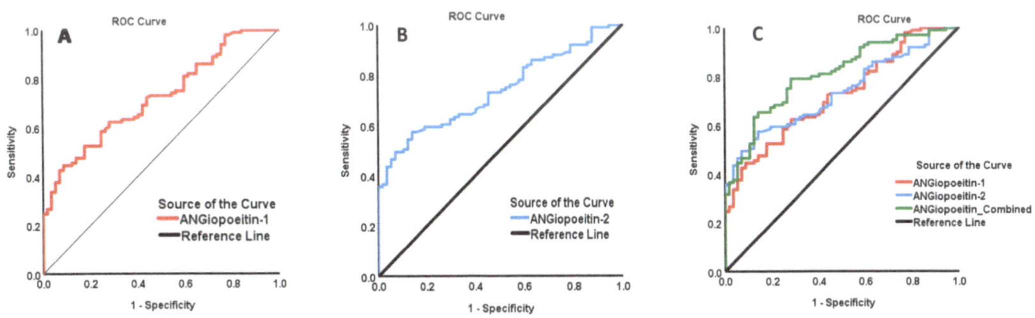

Figure 4. ROC curve analysis performed to identify the cut-off values of Ang1 and Ang2 as biomarkers for DN. (**A**) ANG1 (0.72 (0.65–0.80), $p < 0.001$). (**B**) AUC for Ang2 (0.74 (0.66–0.81) $p < 0.001$). (**C**) AUC of the combination of Ang1 and Ang2 (0.81 (0.74–0.88), $p < 0.001$).

4. Discussion

In this study, we verified that individuals with DN have elevated levels of circulating Ang2 and ANGPTL8 compared with those with T2D. Our analysis revealed for the first time a positive correlation between the elevation in Ang2 and rising ANGPTL8 and ANGPTL4 levels in patients with DN. Additionally, the ROC curve analysis indicated the sensitivity and specificity of using Ang2 in combination with ANGPTL8 as a diagnostic tool for people with DN, suggesting their potential as significant predictors for nephropathy in patients with T2D.

In agreement with previous studies, we report that blood levels of Ang2 [11,12] and ANGPTL8 [19] were higher in the people with DN. A noteworthy finding in the current report was that the rise in Ang2 exhibited a direct and strong link with ANGPTL8 and the clinical parameters of DN, such as the ACR and eGFR. Ang2 is a growth factor belonging to the angiopoietin/tyrosine kinase signaling pathway that is upregulated in animal models of kidney disease [20–23] and in diabetic nephropathy [24–26].

Angiopoietins are a group of vascular growth factors that have several physiological roles associated with vascular development and repair. The actions of Angs are facilitated by endothelial Tie receptor tyrosine kinases via the Ang-Tie signaling pathway, which helps with angiogenesis in health and disease (i.e., vascular diseases, systemic inflammation, and cancers). Angiopoietins are present in two isoforms, Ang1 and Ang2, as they contribute to regulating vascular homeostasis. Ang1 and Ang2 share significant amino acid similarities (≥65%) (Figure S1A), whereby the fibrinogen-like domains (FLDs) of Ang1 or Ang2 possess the same structural fold (Figure S1B) [27]. The FLD domains of Ang1 or Ang2 bind to the Ig2 domain of the Tie2 protein (Figure S1C). The Ang1/Tie2 interface is similar to the Ang2/Tie2 interface (Figure S1D). Studies have demonstrated that the binding affinity of Ang2 to Tie2 is less than Ang1 and less potent in activating Tie2 [8]. However, when Ang1 and Ang2 are added simultaneously, Ang1-Tie2 phosphorylation is inhibited, diminishing the protective effect of Ang1 [8]. Functionally, Ang1 plays a critical protective role in endothelial cells (ECs) by initiating an anti-inflammatory response and inducing vessel wall stabilization. The activity of Ang1 is attained via binding and inducing autophosphorylation, thus activating Tie2. On the contrary, Ang2 acts as a natural antagonist of a pro-inflammatory nature toward Ang1 and its Tie2 receptor [28]. The Ang-Tie signaling pathway has critical importance, particularly in the development and maturation of the kidney, as well as acute and chronic kidney diseases like DN, lupus nephropathy, hemolytic uremic syndrome, end-stage renal diseases, and renal cell carcinoma [6]. Ang2 is mainly produced by endothelial cilia [29] and stored in the Weibel–Palade bodies within ECs [30]. The occurrence of ischemia, hypoxia, or inflammation induces the release of Ang2 into circulation [31], where it competes with Ang1 to inhibit the phosphorylation and activation of the Tie2 receptor. Consequently, Ang2 facilitates permeability of the endothelium, which leads to destabilization of the blood vessel wall [32].

Our data presents a significant rise in the plasma levels of Ang2 in people with DN, which is in agreement with Tsai et al., who found a significant connection between elevated Ang2 levels and an increased risk of poor renal outcomes in patients with diabetic nephropathy [11]. However, the significance of Ang2 and its role in kidney pathophysiology is still obscure. A growing body of research indicates that the elevation of Ang2 has negative effects on renal physiology and function. Additionally, increased Ang2 expression impacts podocytes through paracrine signaling, leading to glomerular EC destabilization that consequently deteriorates the function of the glomerular filtration barrier [33]. Furthermore, normoalbuminuric patients with T2D presented elevated levels of Ang2 in their blood and urine [10,34]. Collectively, this compiling evidence highlights the importance of Ang2 as an early marker of tubular damage before the appearance of clinical symptoms like microalbuminuria.

ANGPTL8 moderates plasma triglyceride levels by blocking lipoprotein lipases. Multiple studies have indicated an increase in circulating ANGPTL8 in individuals with T2D [35–41]. Consistent with the findings showing elevated ANGPTL8 levels in individuals with T2D, we previously reported a threefold rise in circulating ANGPTL8 in people with T2D compared with healthy individuals. Additionally, we have shown that the circulating ANGPTL8 level is higher in people with DN compared with people with T2D. The increase in ANGPTL8 correlated with the clinical parameters of nephropathy in the same population with DN [42]. This rise aligns with the reported link between varying levels of ANGPTL8 and different phases of albuminuria [18,43]. ANGPTL8 was proposed as a new regulator in DN development and as a new predictive risk indicator for all-cause death in individuals with T2D [41], and it was highlighted as a predictive marker for diabetic complications, specifically DN and declining kidney function [19,42]. In this report, individuals diagnosed with DN exhibited microalbuminuria (460.34 $\pm$ 169.37 mg/day, $p = 0.001$), which is consistent with the documented association between elevated circulating ANGPTL8 levels and albuminuria [18]. The increase in ANGPTL8 synthesis was ascribed to albumin loss, a condition that leads to insulin resistance and heightened insulin requirements in individuals with T2D and albuminuria.

In individuals with diabetic nephropathy, our data presents a notable rise in the levels of circulating Ang2, ANGPTL8, and ANGPTL4. Among these, Ang2 showed significant associations with certain clinical markers of DN. The Ang2 levels increased in conjunction with higher levels of ANGPTL8 and ANGPTL4 and the ACR but showed a negative relationship with the eGFR and systolic blood pressure, suggesting possible involvement in diabetic nephropathy. The proven association between Ang2 and nephropathy [6,7,9,11], as well as the link between ANGPTL8 and renal dysregulation [18,41,42], and the presence of a positive correlation between Ang2 and ANGPTL8 all point to a possible interplay between these proteins contributing to DN's progression or severity. The detected rise in circulating ANGPTL4 is in agreement with a prior report [44], and it exhibited a significant positive correlation with Ang2 in both people with T2D and DN. This implicated a potential link between the two markers in T2D, which was emphasized by the significant multiple regression analysis for ANGPTL4 ($\beta = 0.552$, $p < 0.001$; Table 4). Although the Ang2-ANGPTL4 link is currently obscure, multiple regression data (Table 4) highlights the importance of changes in ANGPTL4 in T2D as a predictive marker for changes in Ang2. On the other hand, the larger AUC for ANGPTL8 underscores its potential as a robust biomarker for diabetic nephropathy, offering promise for improved diagnostic precision and patient stratification in clinical practice. Finally, the ROC analysis of the amalgamation of Ang2 and ANGPTL8 accentuates the significance of these proteins as prospective indicators for the progression into a state of nephropathy in people with T2D. Future research is needed to explain the relationship between Ang2 and ANGPTLs, particularly ANGPTL8 and ANGPTL4, as well as the mechanism by which they contribute to diabetic microvascular problems.

The main limitation of this study is the cross-sectional design of the study, which hindered the ability to determine the biological significance of these proteins and their potential involvement in the pathophysiology of DN and sequentially establish causality. Future research should include prognosis and mechanistic studies to confirm the correlation between Ang2 and ANGPTL(s) in DN. This would also clarify the cause-effect link between DN and these proteins.

5. Conclusions

In conclusion, we report a positive correlation between the elevation of Ang2 and increasing levels of ANGPTL8 in individuals with DN and ANGPTL4 in people with T2D, thus suggesting a potential interplay between Ang2 and ANGPTL(s) that influences the development and advancement of DN. This finding suggests that increased levels of Ang2, ANGPTL8, and ANGPTL4 in the bloodstream could be used as predictive indicators to help detect the advancement of early nephropathy in people with T2D. ANGPTL8's role in lipid regulation suggests that a continuous rise in its levels can cause dyslipidemia and heightened insulin resistance, exacerbating metabolic stress and contributing to the onset of metabolic disorders such as T2D and DN. ROC curve analysis also emphasized the sensitivity and specificity of Ang2 alone and in combination with ANGPTL8 and Ang1 as diagnostic tools for diabetic nephropathy.

Supplementary Materials: The following supporting information can be downloaded at https://www.mdpi.com/article/10.3390/biomedicines12050949/s1, Figure S1. Sequence alignment of Ang1 (PDB ID: 4K0V) and Ang2 (PDB ID: 2GY7) FLD domains. Figure S2. A comparison of circulating levels of Ang1, Ang2, ANGPTL3, ANGPTL4 and ANGPTL8 in 3 groups. Table S1. Circulatory levels of ANGPTL8 and Ang2 in the T2D and DN groups were assessed in normal and high serum creatine stratification.

Author Contributions: Study design, data interpretation, directing the laboratory investigation, and writing the manuscript, E.A. and M.A.-F.; data analysis, management, and statistical analysis, S.D.; structural analyses and writing the manuscript, A.M.; performing the ELISA assay, I.A.-K. and P.C.; blood processing, storage, and data analysis, H.A.-M.; data interpretation and critical revising of the manuscript, H.A. and F.A.-M.; study design, data interpretation, and management, A.A.A.; study design, data interpretation, and writing and critically revising the manuscript, J.A. All authors have read and agreed to the published version of the manuscript.

Funding: This work was supported by the Kuwait Foundation for the Advancement of Sciences (KFAS) under projects RA-2015-012, PR17-13MM-07, and RA HM 2019-008.

Institutional Review Board Statement: This study was conducted in accordance with the Declaration of Helsinki and approved by the Ethical Review Board of the Dasman Diabetes Institute (protocol code RA-2015-012 and approval date 29 March 2017).

Informed Consent Statement: Informed consent was obtained from all subjects involved in the study.

Data Availability Statement: The data presented in this study are available on request from the corresponding author.

Acknowledgments: We are grateful to the Clinical Laboratory and Special Services Facility at the DDI for their contribution in handling the samples. The corresponding authors had full access to all the data in the study and had final responsibility for the decision to submit for publication. The authors would like to thank the staff at the Tissue Bank and Clinical Laboratory for their assistance throughout this study. Eman Al Shawaf and Mohamed Abu-Farha are co-first authors.

Conflicts of Interest: The authors declare no conflicts of interest.

References

1. Koye, D.N.; Magliano, D.J.; Nelson, R.G.; Pavkov, M.E. The Global Epidemiology of Diabetes and Kidney Disease. *Adv. Chronic Kidney Dis.* **2018**, *25*, 121–132. [CrossRef] [PubMed]
2. Alkandari, A.; Alarouj, M.; Elkum, N.; Sharma, P.; Devarajan, S.; Abu-Farha, M.; Al-Mulla, F.; Tuomilehto, J.; Bennakhi, A. Adult Diabetes and Prediabetes Prevalence in Kuwait: Data from the Cross-Sectional Kuwait Diabetes Epidemiology Program. *J. Clin. Med.* **2020**, *9*, 3420. [CrossRef] [PubMed]
3. Qi, C.; Mao, X.; Zhang, Z.; Wu, H. Classification and Differential Diagnosis of Diabetic Nephropathy. *J. Diabetes Res.* **2017**, *2017*, 8637138. [CrossRef] [PubMed]
4. Federation, I.D. IDF Diabetes Atlas. Available online: https://diabetesatlas.org/data/en/country/106/kw.html (accessed on 15 February 2024).
5. Brindle, N.P.; Saharinen, P.; Alitalo, K. Signaling and functions of angiopoietin-1 in vascular protection. *Circ. Res.* **2006**, *98*, 1014–1023. [CrossRef] [PubMed]
6. He, F.F.; Zhang, D.; Chen, Q.; Zhao, Y.; Wu, L.; Li, Z.Q.; Zhang, C.; Jiang, Z.H.; Wang, Y.M. Angiopoietin-Tie signaling in kidney diseases: An updated review. *FEBS Lett.* **2019**, *593*, 2706–2715. [CrossRef] [PubMed]
7. Li, M.; Popovic, Z.; Chu, C.; Reichetzeder, C.; Pommer, W.; Krämer, B.K.; Hocher, B. Impact of Angiopoietin-2 on Kidney Diseases. *Kidney Dis.* **2023**, *9*, 143–156. [CrossRef] [PubMed]
8. Yuan, H.T.; Khankin, E.V.; Karumanchi, S.A.; Parikh, S.M. Angiopoietin 2 is a partial agonist/antagonist of Tie2 signaling in the endothelium. *Mol. Cell. Biol.* **2009**, *29*, 2011–2022. [CrossRef] [PubMed]
9. Lim, H.S.; Blann, A.D.; Chong, A.Y.; Freestone, B.; Lip, G.Y. Plasma vascular endothelial growth factor, angiopoietin-1, and angiopoietin-2 in diabetes: Implications for cardiovascular risk and effects of multifactorial intervention. *Diabetes Care* **2004**, *27*, 2918–2924. [CrossRef] [PubMed]
10. Bontekoe, J.; Lee, J.; Bansal, V.; Syed, M.; Hoppensteadt, D.; Maia, P.; Walborn, A.; Liles, J.; Brailovsky, E.; Fareed, J. Biomarker Profiling in Stage 5 Chronic Kidney Disease Identifies the Relationship between Angiopoietin-2 and Atrial Fibrillation. *Clin. Appl. Thromb./Hemost. Off. J. Int. Acad. Clin. Appl. Thromb./Hemost.* **2018**, *24*, 269s–276s. [CrossRef] [PubMed]

11. Tsai, Y.C.; Lee, C.S.; Chiu, Y.W.; Lee, J.J.; Lee, S.C.; Hsu, Y.L.; Kuo, M.C. Angiopoietin-2, Renal Deterioration, Major Adverse Cardiovascular Events and All-Cause Mortality in Patients with Diabetic Nephropathy. *Kidney Blood Press. Res.* **2018**, *43*, 545–554. [CrossRef] [PubMed]
12. Tsai, Y.C.; Lee, C.S.; Chiu, Y.W.; Kuo, H.T.; Lee, S.C.; Hwang, S.J.; Kuo, M.C.; Chen, H.C. Angiopoietin-2, Angiopoietin-1 and subclinical cardiovascular disease in Chronic Kidney Disease. *Sci. Rep.* **2016**, *6*, 39400. [CrossRef] [PubMed]
13. Shroff, R.C.; Price, K.L.; Kolatsi-Joannou, M.; Todd, A.F.; Wells, D.; Deanfield, J.; Johnson, R.J.; Rees, L.; Woolf, A.S.; Long, D.A. Circulating angiopoietin-2 is a marker for early cardiovascular disease in children on chronic dialysis. *PLoS ONE* **2013**, *8*, e56273. [CrossRef] [PubMed]
14. Ren, G.; Kim, J.Y.; Smas, C.M. Identification of RIFL, a novel adipocyte-enriched insulin target gene with a role in lipid metabolism. *Am. J. Physiol. Endocrinol. Metab.* **2012**, *303*, E334–E351. [CrossRef] [PubMed]
15. Luo, M.; Peng, D. ANGPTL8: An Important Regulator in Metabolic Disorders. *Front. Endocrinol.* **2018**, *9*, 169. [CrossRef]
16. Yin, Y.; Ding, X.; Peng, L.; Hou, Y.; Ling, Y.; Gu, M.; Wang, Y.; Peng, Y.; Sun, H. Increased Serum ANGPTL8 Concentrations in Patients with Prediabetes and Type 2 Diabetes. *J. Diabetes Res.* **2017**, *2017*, 8293207. [CrossRef] [PubMed]
17. Abu-Farha, M.; Abubaker, J.; Tuomilehto, J. ANGPTL8 (betatrophin) role in diabetes and metabolic diseases. *Diabetes/Metab. Res. Rev.* **2017**, *33*, e2919. [CrossRef] [PubMed]
18. Chen, C.-C.; Susanto, H.; Chuang, W.-H.; Liu, T.-Y.; Wang, C.-H. Higher serum betatrophin level in type 2 diabetes subjects is associated with urinary albumin excretion and renal function. *Cardiovasc. Diabetol.* **2016**, *15*, 3. [CrossRef] [PubMed]
19. Issa, Y.A.; Abd ElHafeez, S.S.; Amin, N.G. The potential role of angiopoietin-like protein-8 in type 2 diabetes mellitus: A possibility for predictive diagnosis and targeted preventive measures? *EPMA J.* **2019**, *10*, 239–248. [CrossRef] [PubMed]
20. Xue, L.; Shuyan, T.; Xiaoli, L.; Zilong, L.; Qiuling, F.; Lining, W.; Yanqiu, L.; Li, Y. Glomerular Proteomic Profiles in the NZB/W F1 Hybrid Mouse Model of Lupus Nephritis. *Med. Sci. Monit. Int. Med. J. Exp. Clin. Res.* **2019**, *25*, 2122–2131. [CrossRef] [PubMed]
21. Lu, Y.H.; Deng, A.G.; Li, N.; Song, M.N.; Yang, X.; Liu, J.S. Changes in angiopoietin expression in glomeruli involved in glomerulosclerosis in rats with daunorubicin-induced nephrosis. *Acta Pharmacol. Sin.* **2006**, *27*, 579–587. [CrossRef] [PubMed]
22. Yuan, H.T.; Tipping, P.G.; Li, X.Z.; Long, D.A.; Woolf, A.S. Angiopoietin correlates with glomerular capillary loss in anti-glomerular basement membrane glomerulonephritis. *Kidney Int.* **2002**, *61*, 2078–2089. [CrossRef]
23. Davis, B.; Dei Cas, A.; Long, D.A.; White, K.E.; Hayward, A.; Ku, C.-H.; Woolf, A.S.; Bilous, R.; Viberti, G.; Gnudi, L. Podocyte-Specific Expression of Angiopoietin-2 Causes Proteinuria and Apoptosis of Glomerular Endothelia. *J. Am. Soc. Nephrol.* **2007**, *18*, 2320–2329. [CrossRef] [PubMed]
24. Rizkalla, B.; Forbes, J.M.; Cao, Z.; Boner, G.; Cooper, M.E. Temporal renal expression of angiogenic growth factors and their receptors in experimental diabetes: Role of the renin–angiotensin system. *J. Hypertens.* **2005**, *23*, 153–164. [CrossRef] [PubMed]
25. Yamamoto, Y.; Maeshima, Y.; Kitayama, H.; Kitamura, S.; Takazawa, Y.; Sugiyama, H.; Yamasaki, Y.; Makino, H. Tumstatin peptide, an inhibitor of angiogenesis, prevents glomerular hypertrophy in the early stage of diabetic nephropathy. *Diabetes* **2004**, *53*, 1831–1840. [CrossRef] [PubMed]
26. Ichinose, K.; Maeshima, Y.; Yamamoto, Y.; Kitayama, H.; Takazawa, Y.; Hirokoshi, K.; Sugiyama, H.; Yamasaki, Y.; Eguchi, K.; Makino, H. Antiangiogenic Endostatin Peptide Ameliorates Renal Alterations in the Early Stage of a Type 1 Diabetic Nephropathy Model. *Diabetes* **2005**, *54*, 2891–2903. [CrossRef] [PubMed]
27. Yu, X.; Seegar, T.C.; Dalton, A.C.; Tzvetkova-Robev, D.; Goldgur, Y.; Rajashankar, K.R.; Nikolov, D.B.; Barton, W.A. Structural basis for angiopoietin-1-mediated signaling initiation. *Proc. Natl. Acad. Sci. USA* **2013**, *110*, 7205–7210. [CrossRef] [PubMed]
28. Maisonpierre, P.C.; Suri, C.; Jones, P.F.; Bartunkova, S.; Wiegand, S.J.; Radziejewski, C.; Compton, D.; McClain, J.; Aldrich, T.H.; Papadopoulos, N.; et al. Angiopoietin-2, a natural antagonist for Tie2 that disrupts in vivo angiogenesis. *Science* **1997**, *277*, 55–60. [CrossRef] [PubMed]
29. Gale, N.W.; Thurston, G.; Hackett, S.F.; Renard, R.; Wang, Q.; McClain, J.; Martin, C.; Witte, C.; Witte, M.H.; Jackson, D.; et al. Angiopoietin-2 is required for postnatal angiogenesis and lymphatic patterning, and only the latter role is rescued by Angiopoietin-1. *Dev. Cell* **2002**, *3*, 411–423. [CrossRef]
30. Fiedler, U.; Scharpfenecker, M.; Koidl, S.; Hegen, A.; Grunow, V.; Schmidt, J.M.; Kriz, W.; Thurston, G.; Augustin, H.G. The Tie-2 ligand angiopoietin-2 is stored in and rapidly released upon stimulation from endothelial cell Weibel-Palade bodies. *Blood* **2004**, *103*, 4150–4156. [CrossRef] [PubMed]
31. Krikun, G.; Schatz, F.; Finlay, T.; Kadner, S.; Mesia, A.; Gerrets, R.; Lockwood, C.J. Expression of Angiopoietin-2 by Human Endometrial Endothelial Cells: Regulation by Hypoxia and Inflammation. *Biochem. Biophys. Res. Commun.* **2000**, *275*, 159–163. [CrossRef]
32. Hakanpaa, L.; Sipila, T.; Leppanen, V.-M.; Gautam, P.; Nurmi, H.; Jacquemet, G.; Eklund, L.; Ivaska, J.; Alitalo, K.; Saharinen, P. Endothelial destabilization by angiopoietin-2 via integrin β1 activation. *Nat. Commun.* **2015**, *6*, 5962. [CrossRef] [PubMed]
33. Fu, J.; Lee, K.; Chuang, P.Y.; Liu, Z.; He, J.C. Glomerular endothelial cell injury and cross talk in diabetic kidney disease. *Am. J. Physiol. Ren. Physiol.* **2015**, *308*, F287–F297. [CrossRef] [PubMed]
34. Chen, S.; Li, H.; Zhang, C.; Li, Z.; Wang, Q.; Guo, J.; Luo, C.; Wang, Y. Urinary angiopoietin-2 is associated with albuminuria in patients with type 2 diabetes mellitus. *Int. J. Endocrinol.* **2015**, *2015*, 163120. [CrossRef] [PubMed]
35. Abu-Farha, M.; Abubaker, J.; Al-Khairi, I.; Cherian, P.; Noronha, F.; Hu, F.B.; Behbehani, K.; Elkum, N. Higher plasma betatrophin/ANGPTL8 level in Type 2 Diabetes subjects does not correlate with blood glucose or insulin resistance. *Sci. Rep.* **2015**, *5*, 10949. [CrossRef]

36. Al-Shawaf, E.; Al-Ozairi, E.; Al-Asfar, F.; Al-Beloushi, S.; Kumari, S.; Tuomilehto, J.; Arefanian, H. Biphasic changes in angiopoietin-like 8 level after laparoscopic sleeve gastrectomy and type 2 diabetes remission during a 1-year follow-up. *Surg. Obes. Relat. Dis. Off. J. Am. Soc. Bariatr. Surg.* **2018**, *14*, 1284–1294. [CrossRef] [PubMed]
37. Chen, X.; Lu, P.; He, W.; Zhang, J.; Liu, L.; Yang, Y.; Liu, Z.; Xie, J.; Shao, S.; Du, T.; et al. Circulating betatrophin levels are increased in patients with type 2 diabetes and associated with insulin resistance. *J. Clin. Endocrinol. Metab.* **2015**, *100*, E96–E100. [CrossRef]
38. Fu, Z.; Berhane, F.; Fite, A.; Seyoum, B.; Abou-Samra, A.B.; Zhang, R. Elevated circulating lipasin/betatrophin in human type 2 diabetes and obesity. *Sci. Rep.* **2014**, *4*, 5013. [CrossRef]
39. Gao, T.; Jin, K.; Chen, P.; Jin, H.; Yang, L.; Xie, X.; Yang, M.; Hu, C.; Yu, X. Circulating Betatrophin Correlates with Triglycerides and Postprandial Glucose among Different Glucose Tolerance Statuses—A Case-Control Study. *PLoS ONE* **2015**, *10*, e0133640. [CrossRef] [PubMed]
40. Yamada, H.; Saito, T.; Aoki, A.; Asano, T.; Yoshida, M.; Ikoma, A.; Kusaka, I.; Toyoshima, H.; Kakei, M.; Ishikawa, S.E. Circulating betatrophin is elevated in patients with type 1 and type 2 diabetes. *Endocr. J.* **2015**, *62*, 417–421. [CrossRef]
41. Zou, H.; Xu, Y.; Chen, X.; Yin, P.; Li, D.; Li, W.; Xie, J.; Shao, S.; Liu, L.; Yu, X. Predictive values of ANGPTL8 on risk of all-cause mortality in diabetic patients: Results from the REACTION Study. *Cardiovasc. Diabetol.* **2020**, *19*, 121. [CrossRef]
42. AlMajed, H.T.; Abu-Farha, M.; Alshawaf, E.; Devarajan, S.; Alsairafi, Z.; Elhelaly, A.; Cherian, P.; Al-Khairi, I.; Ali, H.; Jose, R.M.; et al. Increased Levels of Circulating IGFBP4 and ANGPTL8 with a Prospective Role in Diabetic Nephropathy. *Int. J. Mol. Sci.* **2023**, *24*, 14244. [CrossRef] [PubMed]
43. Yang, L.; Song, J.; Zhang, X.; Xiao, L.; Hu, X.; Pan, H.; Qin, L.; Liu, H.; Ge, B.; Zheng, T. Association of Serum Angiopoietin-like Protein 8 with Albuminuria in Type 2 Diabetic Patients: Results from the GDMD Study in China. *Front. Endocrinol.* **2018**, *9*, 414. [CrossRef] [PubMed]
44. Al Shawaf, E.; Abu-Farha, M.; Devarajan, S.; Alsairafi, Z.; Al-Khairi, I.; Cherian, P.; Ali, H.; Mathur, A.; Al-Mulla, F.; Al Attar, A.; et al. ANGPTL4: A Predictive Marker for Diabetic Nephropathy. *J. Diabetes Res.* **2019**, *2019*, 4943191. [CrossRef] [PubMed]

biomedicines

Article

The Value of White Cell Inflammatory Biomarkers as Potential Predictors for Diabetic Retinopathy in Type 2 Diabetes Mellitus (T2DM)

Ana Maria Dascalu [1,2], Dragos Serban [1,*], Denisa Tanasescu [3,†], Geta Vancea [1,†], Bogdan Mihai Cristea [1], Daniela Stana [2], Vanessa Andrada Nicolae [1,2], Crenguta Serboiu [1,*], Laura Carina Tribus [4], Corneliu Tudor [1], Adriana Georgescu [2], Mihail Silviu Tudosie [1], Daniel Ovidiu Costea [5,6,†] and Dan Georgian Bratu [7,8]

1 Faculty of Medicine, Carol Davila University of Medicine and Pharmacy Bucharest, 020021 Bucharest, Romania; geta.vancea@umfcd.ro (G.V.)
2 Ophthalmology Department, Emergency University Hospital Bucharest, 050098 Bucharest, Romania
3 Department of Nursing and Dentistry, Faculty of General Medicine, 'Lucian Blaga' University of Sibiu, 550169 Sibiu, Romania
4 Faculty of Dental Medicine, Carol Davila University of Medicine and Pharmacy Bucharest, 020021 Bucharest, Romania
5 Faculty of Medicine, Ovidius University Constanta, 900470 Constanta, Romania
6 General Surgery Department, Emergency County Hospital Constanta, 900591 Constanta, Romania
7 Faculty of Medicine, University "Lucian Blaga", 550169 Sibiu, Romania
8 Department of Surgery, Emergency County Hospital Sibiu, 550245 Sibiu, Romania
* Correspondence: dragos.serban@umfcd.ro (D.S.); crenguta.serboiu@umfcd.ro (C.S.)
† These authors contributed equally to this work.

Abstract: The pathogenesis of diabetic retinopathy is still challenging, with recent evidence proving the key role of inflammation in the damage of the retinal neurovascular unit. This study aims to investigate the predictive value of the neutrophil-to-lymphocyte ratio (NLR), platelet-to-lymphocyte (PLR), lymphocyte-to-monocyte ratio (LMR), and systemic inflammation index (SII) for diabetic retinopathy (DR) and its severity. We performed a retrospective study on 129 T2DM patients, divided into three groups: without retinopathy (NDR), non-proliferative DR (NPDR), and proliferative DR (PDR). NLR, MLR, and SII were significantly higher in the PDR group when compared to NDR and NPDR (3.2 ± 1.6 vs. 2.4 ± 0.9 and 2.4 ± 1.1; $p = 0.005$; 0.376 ± 0.216 vs. 0.269 ± 0.083 and 0.275 ± 0.111, $p = 0.001$; 754.4 ± 514.4 vs. 551.5 ± 215.1 and 560.3 ± 248.6, $p = 0.013$, respectively). PDR was correlated with serum creatinine (OR: 2.551), NLR (OR: 1.645), MPV (OR: 1.41), and duration of diabetes (OR: 1.301). Logistic regression analysis identified three predictive models with very good discrimination power for PDR (AUC ROC of 0.803, 0.809, and 0.830, respectively): combining duration of diabetes with NLR, MLR, and, respectively, PLR, MPV, and serum creatinine. NLR, MPV, SII, and LMR were associated with PDR and could be useful when integrated into comprehensive risk prediction models.

Keywords: diabetic retinopathy; biomarkers; neutrophil-to-lymphocyte ratio (NLR); platelet-to-lymphocyte ratio (PLR); monocyte-to-lymphocyte ratio (MLR); systemic inflammation index (SII)

Citation: Dascalu, A.M.; Serban, D.; Tanasescu, D.; Vancea, G.; Cristea, B.M.; Stana, D.; Nicolae, V.A.; Serboiu, C.; Tribus, L.C.; Tudor, C.; et al. The Value of White Cell Inflammatory Biomarkers as Potential Predictors for Diabetic Retinopathy in Type 2 Diabetes Mellitus (T2DM). *Biomedicines* **2023**, *11*, 2106. https://doi.org/10.3390/biomedicines11082106

Academic Editor: Victor Chong

Received: 3 July 2023
Revised: 19 July 2023
Accepted: 24 July 2023
Published: 26 July 2023

1. Introduction

Diabetes mellitus is a major health problem globally, accounting for 537 million patients worldwide and with an expected ascendant trend reaching 700 million by 2045 [1,2]. For type 2 diabetes mellitus (T2DM), which accounts for approximately 90% of the total, this rising trend can be attributed to aging, a rapid increase in urbanization, and obesogenic environments. Insulin resistance and low-grade systemic inflammation lead to multiple organ damage by microvascular and macrovascular complications. Despite significant achievements in early diagnosis and therapy, diabetic retinopathy is currently the major cause of blindness and visual impairment in working-age adults worldwide [2], causing an increased burden on national healthcare systems worldwide [3]. The prevalence of diabetic retinopathy is estimated to be 27.0% in diabetic patients, which leads to 0.4 million incidences of blindness in the world [4]. The World Health Organization (WHO) Universal Eye Health: A Global Action Plan 2014–2019 outlines the need to achieve a reduction in the prevalence of avoidable visual impairment and blindness, including that related to diabetes, which is currently among the five most common causes of both moderate or severe visual impairment and blindness [5]. DR is listed as a priority eye disease in the 2030 IN SIGHT strategy [6]. However, an effective screening may be challenging due to limited available retina specialists. Finding new biomarkers with potential predictive value may be a valuable tool in the management of diabetic patients by identifying early those at risk for the development of sight-threatening complications.

Chronic hyperglycemia induces both neurodegeneration and apoptosis of the retinal ganglion cells and photoreceptors and causes microvascular retinal damages by multiple pathways: activation of aldose-reductase and polyol pathway, activation of the protein kinase C (PKC) pathway, accumulation of reactive oxygen species (ROS), nitric oxide deficiency, and changes in the blood flow, with increased platelet adhesion and aggregation [7,8]. Nonenzymatic glycation and glycoxidation of proteins lead to the accumulation of advanced glycation end products (AGEs) in the extracellular matrix and subendothelial space, leading to basal membrane thickening, pericyte loss and monocyte migration, and activation of nuclear factor (NF)-κB along with activation of pro-inflammatory pathways [9,10]. Animal studies proved that monocytes use pigment epithelium (RPE) as a gateway for trafficking to the retina [11].

The role of ocular and systemic inflammation in the pathogenesis of diabetic retinopathy has been confirmed by many previous articles [12–14]. Impaired glucose and lipid metabolism lead to the accumulation of toxic metabolites and pro-inflammatory changes in the retina, leading to damage to the neurovascular unit [15,16]. Several inflammatory markers such as interleukin (IL)-1, IL6, IL8, transforming growth factor β1, and tumor necrosis factor-α have been linked to end organ damage in diabetes [17–19]. However, their practical use is limited due to the high cost and lack of availability in clinical practice.

Blood cell changes contribute to the pathophysiological changes observed in diabetic retinopathy through various mechanisms, including endothelial dysfunction, leukocyte adhesion and infiltration, cytokine and chemokine production, platelet activation, and neovascularization [20]. Increased levels of systemic inflammation were correlated with activated circulant platelets, presenting bigger and inequal volumes, resulting in higher MPV (mean platelet volume) and PDW (platelets distribution width) values [21,22]. Similarly, destruction and fragmentation of the erythrocytes in an inflammatory environment, with endothelial activation, sludge, and microcapillary occlusion, was associated with higher RDW (red cell distribution width) values [23].

White cell inflammatory biomarkers, such as neutrophil-to-lymphocyte ratio (NLR), platelet-to-lymphocyte (PLR), lymphocyte-to-monocyte ratio (LMR), and systemic inflammation index (SII), can be easily calculated based on the complete blood cells count (CBC). In recent studies, they showed good predictive value for several inflammatory, oncologic, and cardiovascular diseases [24,25]. Previous research showed higher values in diabetic patients when compared to normal subjects. Moreover, higher values of NLR and PLR were found to be predictive for diabetic microvascular and macrovascular complications, such as diabetic foot ulcer and the risk of amputation [2], diabetic nephropathy [26] and coronary artery disease (CAD), and end-organ damage in type2 diabetes mellitus (T2DM) [17].

However, there is still conflicting evidence regarding the significance of these biomarkers in clinical evaluation and individual management of diabetic patients. There are relatively wide differences reported for the "cut-off" values that could be predictive of diabetic retinopathy and its severity.

In this paper, we analyzed the correlations between diabetic retinopathy and NLR, PLR, MLR, and SII. Also, we analyzed the RDW and MPV values in T2DM patients with noDR comparatively, with non-proliferative DR, and with proliferative DR. The potential predictive value of these biomarkers was studied both for the onset of any changes of DR in T2DM patients and the presence of proliferative DR (PDR).

2. Materials and Methods

A 1-year retrospective study was performed on patients with type 2 diabetes mellitus (T2DM) admitted to the Ophthalmology Department of Surgery, Emergency University Hospital Bucharest, for cataract surgery, between January 2022 and December 2022. Data were collected from observation charts and electronic patient files.

Age, sex, duration of diabetes, and associated pathologies were documented for each patient. Each patient underwent a complete ophthalmological exam. The diagnosis and staging of diabetic retinopathy were performed according to the Guidelines of the International Council of Ophthalmology for Diabetic Eye Care [27,28]. The patients included in the study group were classified based on the fundoscopic findings into the noDR group (NDR), non-proliferative DR group (NPDR), and proliferative diabetic retinopathy group (PDR).

Blood routine tests and complete blood count with differential were performed after collecting fasting peripheral blood from each patient. Systemic inflammatory indices NLR, PLR, and MLR were calculated as the ratios of the neutrophils, platelets, and monocytes to lymphocytes, respectively. The systemic inflammatory index (SII) was calculated by the formula SII = $P \times N/L$, where P, N, and L are the count for platelets, neutrophils, and lymphocytes, respectively [29]. All counts were determined from the same automated blood sample measurement and expressed as a value in cells/L.

Patients with systemic or ocular inflammatory conditions and oncologic or previously diagnosed hematologic conditions were excluded. From the remaining patients, paired individuals in terms of age and sex were selected to be included in the 3 study groups (NDR, NPDR, and PDR), to limit the effect on the final statistical analysis of these two potential cofounders [29].

Statistical Analysis

MedCalc®Statistical Softwareversion 22.006 (MedCalcSoftwareLtd., Ostend, Belgium; https://www.medcalc.org; accessed on 5 June 2023) was used for statistical analysis. An ANOVA test was used for continuous variables. For the statistically significant results, a post hoc analysis was performed to establish the differences within the three groups by using the Tukey–Kramer test for all pairwise comparisons. A Pearson chi-square test, Fisher's exact test, and Linear by Linear test were used to evaluate the association between discrete variables.

The specificity and sensitivity of NLR, PLR, MLR, and SII in predicting DR and PDR were analyzed by ROC curves. A minimum value of 0.6 area under the curve (AUC) was considered as a criterion for an acceptable discrimination model [30]. Logistic regression analysis was used to study correlations between several biological parameters and DR. The best regression models were compared by an ROC AUC (area under the curve) score in terms of efficiency of prediction.

3. Results

A total of 129 Caucasian patients with T2DM were divided into three study groups according to the presence and the severity of DR: NDR (36 patients), NPDR (49 patients), and PDR (44 patients). The general data of the patients included in the study group are presented in Table 1.

There were no statistically significant differences in age, sex ratio, associated diseases, lipidic profile, fasting blood glucose (FBG), Hb, and HbA1C between the three study groups. The duration of diabetes was significantly different among the three groups, being well correlated with the severity of DR. The RDW distribution was different among groups. However, the differences were not statistically significant in the post hoc analysis.

Table 1. General data of the patients included in the study group.

	Total	NDR	NDPR	PDR	*p*-Value
N	129	36	49	44	
Age (mean ± SD)	65.6 ± 8.9	67 ± 6.6	65.2 ± 7.1	63 ± 9.4	0.078 *
Males (n, %)	67 (51%)	16 (44.4%)	23 (46.9%)	26 (59%)	0.393 **
Duration of diabetes (yrs.)	8.9 ± 3.8	5.3 ± 2.4	9.36 ± 3.3	11 ± 3.1	<0.001 *
Associated DM complications:					
• Diabetic kidney disease	13 (10%)	3 (8.3%)	2 (4%)	8 (18.1%)	0.072 **
• Peripheral arterial disease	11 (8.5%)	2 (5.5%)	3 (6.1%)	6 (13.6%)	0.325 **
• Diabetic foot	4 (3.1%)	1 (2.7%)	1 (2%)	2 (4.5%)	0.778 **
Associated diseases (n, %):					
• Arterial hypertension	47	20	8	19	0.518 **
• Ischemic cardiopathy	4	2	1	1	0.563 **
• Glaucoma	5	1	2	2	0.916 **
FBG (mean ± SD)	164.6 ± 60.3	145.5 ± 41.4	173.6 ± 78.6	170.4 ± 46	0.078 *
HbA1C (mean ± SD)	7.6 ± 1.6	7.2 ± 1.1	7.5 ± 1.8	8.2 ± 1.8	0.113 *
Hb (mean ± SD)	13.5 ± 1.6	13.6 ± 1.3	13.5 ± 1.6	13.4 ± 1.2	0.646 *
RDW (mean ± SD)	13.9 ± 1.3	14 ±1.1	13.7 ± 1.4	14.2 ± 1.3	0.031 *
Neutrophils (mean ± SD)	4.9 ± 1.4	4.8 ± 1.4	4.6 ± 1.2	5.4 ± 1.5	0.007 *
Lymphocytes (mean ± SD)	2.1 ± 2.8	2.1 ± 0.7	2.2 ± 0.8	1.9 ± 0.7	0.285
Monocytes (mean ± SD)	0.5 ± 0.1	0.5 ± 0.1	0.5 ± 0.1	0.6 ± 0.1	0.066 *
Platelets (mean ± SD)	237.4 ± 53	224.3 ± 42.6	248.7 ± 46.3	235.3 ± 64.9	0.106 *
MPV mean ± SD)	9 ± 1.1	8.9 ± 1.1	8.7 ± 0.8	9.3 ± 1.2	0.02 *
TG (mean ± SD)	148.3 ± 78.2	136.3 ± 63.4	149 ± 64.2	152.9 ± 92.4	0.076 *
Cholesterol (mean ± SD)	163.19 ± 73.2	158.39 ± 61.7	162.14 ± 53.2	165.9 ± 70.5	0.341 *
Serum urea (mean ± SD)	51.8 ± 28.7	48.4 ± 31.2	49.1 ± 25.8	57.6 ± 29.3	0.264 *
Serum Creatinine	1.1 ± 0.6	0.9 ± 0.3	1 ± 0.5	1.3 ± 0.8	0.024 *
Creatinine > 1.2mg/dL (n, %):	29 (22.4%)	6 (16.6%)	6 (12.2%)	16 (36.3)	0.013 **
Urea > 60 mg/dL (n, %)	30 (23.2%)	7 (19.4%)	8 (16.3%)	15 (34%)	0.035 **

Footnote: * ANOVA; ** Chi-squared test; NDR; nodiabetic retinopathy; NPDR: non proliferative DR; PDR: proliferative DR; FBG: fasting blood glucose; RDW: red cells distribution width; MPV: mean platelet volume.

Serum creatinine was significantly different among the study groups ($p = 0.024$). The Tukey–Kramer test showed significantly higher values in PDR when compared to the NDR group. The proportion of patients with abnormal serum urea and creatinine values was significantly higher in the PDR group ($p = 0.035$; $p = 0.013$, respectively), with no differences between NDR and NPDR groups. These data suggest a higher prevalence of associated retinal and renal microvascular damages in the PDR group. The patients in the PDR group were also associated with more diabetic micro- and macrovascular complications, when compared to NDR and NPDR groups; however, the differences were not statistically significant.

When analyzing blood cell counts for different types of white cells, only neutrophils were significantly higher in the PDR group when compared to NDR and NPDR groups ($p = 0.007$). While platelet count was not significantly different among the three groups, MPV was higher in the PDR group when compared to NDR and NPDR groups, suggesting an increased platelet activation in the proliferative stage of DR.

Furthermore, white cell inflammatory biomarkers were comparatively analyzed (Table 2).

Table 2. NLR, PLR, SII, and LMR distribution in the study groups.

	Total	NDR	NDPR	PDR	*p*-Value
NLR (mean ± SD)	2.6 ± 1.3	2.4 ± 0.9	2.4 ± 1.1	3.2 ± 1.6	0.005 *
PLR (mean ± SD) × 10^9 cells/L	126 ± 54.1	115.4 ± 38.9	122.1 ± 35.4	138.9 ± 76.1	0.127 *
MLR (mean ± SD)	0.308 ± 0.157	0.269 ± 0.083	0.275 ± 0.111	0.376 ± 0.216	0.001 *
SII (mean ± SD) × 10^9 cells/L	624 ± 365.5	551.5 ± 215.1	560.3 ± 248.6	754.4 ± 514.4	0.013 *

Footnote: * ANOVA; NLR: neutrophil-lymphocyte ratio; PLR: platelet-lymphocyte ratio; LMR: lymphocyte to monocyte ratio; SII: systemic inflammatory index.

PLR values increased along with the severity of DR. However, the differences among groups were not statistically significant. NLR, MLR, and SII indices were significantly higher in the PDR group ($p = 0.005$; $p = 0.001$; $p = 0.013$, respectively) when compared to NDR and NDPR. The post hoc analysis shows no differences between NDR and NPDR groups.

3.1. ROC Curves and Predicting the Value of White Cell Inflammatory Biomarkers

For the systemic inflammatory indices that showed statistically significant different distribution among the study groups after post hoc analysis (MPV, NLR, SII, and LMR), the predictive power for DR and PDR was further assessed by ROC curves.

None of the studied biomarkers met the criteria of a minimum AUC ROC of 0.6 for predicting DR in the study group. When the prediction value for PDR was analyzed, NLR, LMR, and SII showed good discrimination power with high specificity and low sensitivity (Table 3).

Table 3. Sensitivity and specificity at the "cut-off" value predicting PDR.

	PDR Sensitivity (%)	PDR Specificity (5)	Cut-Off Value	AUC	*p*
NLR	40.0	86.9	>3.18	0.662	0.001
MLR	35.6	92.9	>0.364	0.643	0.006
SII	35.6	85.7	>763.8 (×10^9 cells/L)	0.627	0.015
MPV	55.6	63.1	>9.24	0.593	0.084
PLR	26.7	91.7	>168.8 (×10^9 cells/L)	0.536	0.518

3.2. Logistic Regression Approach

The binary logistic regression model was performed to analyze the correlation of PDR (1 = true or 0 = false) with each independent variable listed in Tables 1 and 2. The variables with a *p*-value of <0.05 and the minimal value of the confidence interval for each OR > 1 were selected as possible risk factors for PDR (Table 4), ensuring that each risk factor adds a 95%-significant extra hazard.

A higher correlation with PDR was encountered for creatinine (OR: 2.551), NLR (OR: 1.645), MPV (OR: 1.41), LMR (OR: 161.19), and duration of diabetes (OR: 1.301). Age, FBG, HbA1C, PLR, RDW, and urea were not significantly correlated with PDR.

Patients with diabetes present a diverse array of comorbidities and metabolic dysregulation features that may combine and accelerate the progression to proliferative diabetic retinopathy. A multivariate logistic regression analysis was further carried out to identify the predictive models with the best discrimination value based on the analyzed variables (Table 5).

We described three different models for predicting PDR based on different systemic inflammatory biomarkers. One model combines NLR and duration of diabetes, while the second combines PLR, MPV, serum creatinine, and duration of DM and the last MLR and duration of diabetes. All models proved a very good discrimination power for PDR, with an AUC ROC of 0.803, 0.830, and 0.809, respectively. The described models show a superior prediction when compared to each of the variables taken separately ($p < 0.05$), based on comparative ROC curve analysis (Figure 1).

Table 4. Logistic regression model for the dependent variable of PDR.

Risk	Estimated Co-Efficient	Standard Error	Wald	Degrees of Freedom	*p*-Value	OR	Lower	Upper
Duration of diabetes	0.263	0.061	18.55	1	<0.0001	1.301	1.154	1.467
MPV	0.348	0.174	3.984	1	0.045	1.41	1.006	1.994
NLR	0.498	0.165	9.062	1	0.002	1.645	1.189	2.275
MLR × 10	0.508	0.162	9.82	1	0.0017	1.662	1.209	2.284
SII	0.001	0.000	7.23	1	0.007	1.001	1	1.003
creatinine	0.936	0.414	5.11	1	0.02	2.551	1.132	5.746

Footnote: INPUT description: PDR (proliferative diabetic retinopathy) variable is binary (1 = present/0 = absent), and all risks are binary (1 = yes/0 = no)—OR, estimated odd ratio; CI, confidence interval.

Table 5. Logistic regression models based on multiple variables.

Model No.	Variable	Coefficient	Std.Error	Wald	*p*	OR	95% CI Lower	95% CI Upper
LOGREG_1	NLR	0.46364	0.17296	7.1857	0.0073	1.632	1.156	2.304
	Duration DM	0.28342	0.067991	17.3759	<0.0001	1.314	1.154	1.498
	Constant	−4.59859	0.89211	26.5713	<0.0001			
LOGREGR_2	MPV	0.46284	0.21874	4.4771	0.0344	1.5886	1.0347	2.439
	PLR	0.011012	0.0048070	5.2481	0.0220	1.0111	1.0016	1.0206
	creatinine	0.87851	0.42178	4.3384	0.0373	2.4073	1.0532	5.5025
	Duration of DM	0.30659	0.071052	18.6195	<0.0001	1.3588	1.1821	1.5618
	Constant	−10.14928	2.51771	16.2502	0.0001			
LOGREGR_3	MLR	0.5234	0.1747	8.9763	0.0027	1.6879	1.1984	2.3772
	Duration DM	0.2853	0.0706	16.3062	<0.0001	1.3302	1.1582	1.5278
	Constant	−4.97103	0.9542	27.1372	<0.0001			

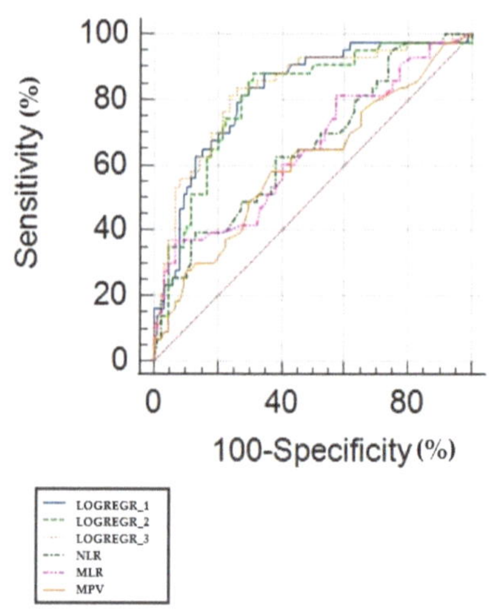

Figure 1. Comparative ROC curves for the three multivariate models described in Table 5, MLR, NLR, and MPV. The red line signifies a specificity and sensitivity of 50%.

Combining SII with different other parameters resulted in a logistic regression model with slightly lower AUC ROC, compared with NLR (0.784 vs. 0.803; $p = 0.23$) and MLR (0.784 vs. 0.830, $p = 0.17$).

4. Discussion

Monocytes, neutrophils, and platelets play a significant role in the pathology of diabetic retinopathy. While considered a metabolic disease initially, more and more evidence points out the role of inflammation in retinal damage. A large array of cytokines and chemokines, including monocyte chemotactic protein-1 (MCP1), interleukin-6 (IL-6), interleukin-8 (IL-8), tumor necrosis factor-α (TNF-α), and interferon-γ, were found to be elevated in serum and vitreous of DR patients [31,32]. Activated platelets are larger and display more enzymatic activities. Along with their key role in coagulation and thrombosis, platelets can release a large array of mediators of inflammation, regulating the leukocytes and endothelial cells' activity [33].

Several studies investigated the clinical value of the white cell fractions and ratios but with conflicting results. A large survey by Wan et al. [34], comparing 2709 patients with no retinopathy with 512 patients with DR, found no differences in neutrophils and platelets number but lower monocytes in early stages of DR. The lymphocyte-to-monocyte ratio (LMR) was considered a good indicator of the circulating immune status of the host in several oncologic diseases [35,36]. LMR was correlated with serum levels of interleukin-6 (IL-6), tumornecrosisfactor, IL-1β, and monocyte chemotactic protein 1, which was noticed in higher levels in vitreous and serum in patients with PDR [37]. Thus, MLR could be a cheaper biomarker to reflect the level of inflammatory changes in patients with DR.

There are only a few studies that analyzed the value of MLR in DR. Yue et al. [32] found a limited value for MLR in the diagnosis of DR and PDR, with NLR and PLR being much more relevant. However, other studies [37,38] proved a significant positive correlation between MLR values and PDR. Huang et al. [37] found a good predictive value for MLR in discriminating between T2DM patients with DR and those with no complication, with an AUC ROC of 0.868, which we could not find in the present study. One explanation may be the increased proportion of PDR cases included in the DR group (64%). We found similar MLR values in NDR groups with the study of Huang et al. [37] (0.27 ± 0.08 vs. 0.269 ± 0.083) but a slightly higher cut-off value for PDR (0.30 vs. 0.36).

The neutrophil-to-lymphocyte ratio (NLR) and platelet-to-lymphocyte ratio (PLR) are promising biomarkers for evaluating the inflammatory status in diabetic retinopathy (DR). However, to maximize their clinical utility, it is essential to establish the appropriate classification systems or cut-off values for NLR and PLR concerning the presence and severity of DR. The NLR values in the normal adult population vary between 0.43 and 3.53, in previous studies, with an average value of 1.56 [39], with slightly higher values in females versus males [40]. PLR is a newly explored biomarker which was correlated with cardiac diseases and all-cause mortality. An ample epidemiological study performed by Wu et al. [40] in a Chinese population found that PLR ranged between 36.63 and 149.13 in healthy males and between 43.36 and 172.68 in women. Moosazadeh et al. [41] found a mean PLR of 110.84 ± 56.53 and 123.12 ± 36.21 in Iranian healthy males and females, respectively. In a prospective cohort study including 8711 adults aged over 45 from the Rotterdam area, Fest and colleagues [42] described the mean value and 95% reference intervals for NLR, PLR, and SII of 1.76 (0.83–3.92), 120 (61–239), and 459 (189–1168), respectively (Fest). Previous studies showed a good prognostic value of PLR in oncologic and cardiovascular diseases [43]. In a large study on27,321individuals, the mean PLR was significantly higher in individuals who died during the follow-up period comparative with those who survived (145.7 vs. 133.0) [43].

Several studies have attempted to establish cut-off values for NLR and PLR to differentiate between patients with and without DR or to stratify patients according to DR severity. However, the proposed cut-off values have varied considerably across studies, reflecting differences in patient populations, study designs, and statistical methods [44,45].

Zeng et al. [31] found significant differences between T2DM with no DR and with DR for NLR and PLR, but no NLR or PLR was associated with the severity of retinopathy. However, in the study group only 24 patients with PDR vs. 124 with NPDR were included. Zeng et al. [31] found, however, that in a multivariate regression analysis, PLR was significantly correlated with the existence of DR, together with diabetic peripheral neuropathy, systolic blood pressure, and duration of diabetes.

In a study by Ilhan et al. [46], NLR was significantly higher in PDR versus healthy controls (2.67 ± 1.02 vs. 1.85 ± 0.49). An NLR value of 2.11 or more predicted PDR or severe NPDR with a sensitivity of 76% and a specificity of 80%.

Hu et al. analyzed the predictive value of NLR for the response to anti-VEGF therapy in DR patients and found that an NLR < 2.27 showed a better improvement in letter scores than those with NLR > 2.27 [47]. A recent study by Wang and colleagues [48] comparing T2DM patients with and without DR found that NLP is an independent risk factor for DR (OR: 1.37; 95%CI: 1.06, 1.78). Similar to our findings, although PLR was not independently associated with DR as a continuous variable, including PLR in a multivariate analysis improved the discrimination power of the statistical model [48].

In a systematic review by Luo et al. [33], MPV mean values ranged between 7.76 and 8.12 in NDR vs. between 8.18 and 10.76 in PDR groups, while the mean NLR had a range of 1.54–2.4 for NDR vs. 1.91–2.58 in PDR patients. The differences may be related to the methodology of research and classification of DR but also patients' age, BMI, comorbidities, and geographical and ethnic differences. Moreover, the relationship between NLR and DR appears to be non-linear [49], and some studies have suggested that changes in NLR and PLR may be more informative when analyzed as continuous variables rather than categorical variables based on arbitrary cut-off values [2,50,51]. T2DM patients often present with a unique but diverse series of risk factors, comorbidities, and complications, which may accumulate over time due to an inappropriate lifestyle [52]. For instance, a higher NLR or PLR may be associated with an increased risk of DR or more severe DR, but the precise risk may depend on other factors, such as glycemic control, duration of diabetes, and the presence of other microvascular or macrovascular complications [53,54].

While multiple studies proved that SII is an important indicator of systemic inflammation as well as for the risk of acute cardiovascular events, there is little evidence regarding the value range of SII in a normal population. A recent paper of Bai et al. [55] found a mean value of 374 (153, 832) in non-pregnant women.In a large epidemiological study of Luo et al. [56], the mean SII was 334, with a 95% CI of 142–804, with significantly higher values in females and at younger age.SII was investigated as a predictive biomarker for survival in various oncological and cardiovascular diseases. Feng et al. [57] found a cut-off value of 410×10^9 cells/L as a prognostic tool in estimating 5-year survival in patients with squamous cell carcinoma of the esophagus, while Hirara et al. [58] showed that low SII values <661.9 are corelated with overall survival in gastric cancer. A large study on 42,875 adults found that adults with SII levels of >655.56 had higher all-cause mortality and cardiovascular mortality than those with SII levels of <335.36 [59].Guo et al. [60] found a significantly higher SII in patients with T2DM with chronic kidney disease (634.14 ± 13.43 vs. 546.42 ± 10.13). There is also previous evidence that SII might be correlated with the level of retinal damage in T2DM. Eybeli et al. [61], in a recent study from 2022, found that SII and duration of diabetes may predict the incidence of diabetic macular edema (DME) in a group of patients with NPDR. He found a significantly higher value of SII in DME patients vs. the non-DME group (599.7 ± 279.2 and $464.\ 9 \pm 172.2$, respectively).

T2DM is associated with chronic inflammatory changes, higher levels of cytokines, and increased oxidative stress, which lead to neurodegeneration and destruction of blood vessels, causing damage to multiple organs [62]. This explains the higher values observed for all three study groups in our study, varying from $551.5 \pm 215.1 \times 10^9$ cells/L in non-DR patients to $754.4 \pm 514.4 \times 10^9$ cells/L for the proliferative DR group. This finding may be explained by the multiple micro- and macrovascular diabetes-related comorbidities often encountered in these very vulnerable patients. We also found a high cut-off value for SII in our study, of 763.8, but with a fair predictive value (AUC = 0.623) for PDR. We consider that, while not particularly specific for PDR, higher values of SII are useful in clinical management of DR patient, as an overall mirror of the level of inflammation and vascular damage.

In our study, we found that no statistical differences were noticed between NDR and NPDR groups for any of the white cell inflammatory biomarkers. However, significantly higher values for NLR, MLR, SII, and MPV were found in the PDR group when compared with NDR and NPDR groups. This finding may signify that the level of systemic inflammation is higher in the advanced stage of DR associated with neovascularization. These findings support the idea that diabetic retinopathy is a complex, multifactorial process. Systemic inflammation is one of the key elements that need to be further investigated. However, by combining multiple data points, the power of prediction increases and allows individualized management of each case.

The number of patients enrolled in this study is one of the main limitations of this study. This study did not evaluate the correlation of systemic inflammatory biomarkers with the level of macular edema, which can also be a significant cause of visual impairment in NPDR patients. This research is retrospective, based on the standard protocol of paraclinical exams used for patients admitted for cataract surgery. Data regarding the patients' medication were not available and were not taken into account in the statistical analysis. Several biological data were not evaluated, such as BMI, systolic and diastolic blood pressure, and the lipidic profile.

Systemic inflammatory biomarkers offer valuable information regarding the level of inflammation and vascular frailty in diabetic patients, and these two elements should be taken into account in evaluating the risk for developing PDR. The main shortcoming in the context of the current knowledge is the relatively wide range for normal values of these parameters. One future approach may be a dynamic evaluation of the changes in NLR, PLR, MPV, and SII analyzed with the correlation of DR progression. Further studies, including multiple risk factors evaluation, with diverse patient populations could identify better predictive models for clinical use. Such a model may provide a more accurate and personalized assessment of the inflammatory status and DR risk in individual patients with diabetes mellitus [63].

5. Conclusions

This paper brings new evidence that supports the role of chronic systemic inflammation in the pathology of diabetic retinopathy. Systemic white cell inflammatory biomarkers did not predict DR in our study group. However, they proved to be of clinical value in assessing PDR, reflecting better the changes associated with the proliferative diabetic retinopathy than each of the white cell count differential taken separately. There are cheap, inexpensive tools that can be valuable in clinical practice. Higher values of NLR, LMR, SII, PLR, and MPV are significantly correlated with PDR in T2DM patients. The best predictive value was obtained for NLR and MLR when combined with the duration of diabetes. Platelet-derived biomarkers (MPV and PLR) may be useful in evaluating the risk of PDR when correlated with other clinical and biological data.

Author Contributions: Conceptualization, A.M.D., D.T. and D.S. (Dragos Serban); methodology, A.G., G.V., V.A.N. and D.S. (Daniela Stana); software, A.M.D., C.T. and B.M.C.; validation, A.M.D., L.C.T., C.S. and M.S.T.; formal analysis, D.O.C. and D.G.B.; investigation, A.G.; resources, C.T. and M.S.T.; data curation, D.O.C. and B.M.C.; writing—original draft preparation, A.M.D., D.T., D.S. (Dragos Serban), A.G. and G.V.; writing—review and editing, D.O.C., D.G.B., V.A.N., L.C.T. and D.S. (Daniela Stana); visualization, A.M.D. and C.S.; supervision, D.S. (Dragos Serban); project administration, A.M.D. All authors have read and agreed to the published version of the manuscript.

Funding: This research received no external funding.

Institutional Review Board Statement: Ethical review and approval were waived for this study due to its retrospective nature.

Informed Consent Statement: Not applicable.

Data Availability Statement: The data presented in this study are available on request from the corresponding author. The data are not publicly available due to privacy.

Conflicts of Interest: The authors declare no conflict of interest.

References

1. Saeedi, P.; Petersohn, I.; Salpea, P.; Malanda, B.; Karuranga, S.; Unwin, N.; Colagiuri, S.; Guariguata, L.; Motala, A.A.; Ogurtsova, K.; et al. Global and regional diabetes prevalence estimates for 2019 and projections for 2030 and 2045: Results from the International Diabetes Federation Diabetes Atlas, 9th edition. *Diabetes Res. Clin. Pract.* **2019**, *157*, 107843. [CrossRef]
2. Serban, D.; Papanas, N.; Dascalu, A.M.; Kempler, P.; Raz, I.; Rizvi, A.A.; Rizzo, M.; Tudor, C.; Tudosie, M.; Tanasescu, D.; et al. Significance of Neutrophil to Lymphocyte Ratio (NLR) and Platelet Lymphocyte Ratio (PLR) in Diabetic Foot Ulcer and Potential New Therapeutic Targets. *Int. J. Low. Extrem. Wounds* **2021**, *18*, 15347346211057742. [CrossRef] [PubMed]
3. GBD Collaborator Group. Global, regional, and national burden of diabetes from 1990 to 2021, with projections of prevalence to 2050: A systematic analysis for the Global Burden of Disease Study 2021. *Lancet* **2023**, *402*, 203–234. [CrossRef] [PubMed]
4. Zegeye, A.F.; Temachu, Y.Z.; Mekonnen, C.K. Prevalence and factors associated with diabetes retinopathy among type 2 diabetic patients at Northwest Amhara Comprehensive Specialized Hospitals, Northwest Ethiopia 2021. *BMC Ophthalmol.* **2023**, *23*, 9. [CrossRef]
5. Yau, J.W.Y.; Rogers, S.L.; Kawasaki, R.; Lamoureux, E.L.; Kowalski, J.W.; Bek, T.; Chen, S.-J.; Dekker, J.M.; Fletcher, A.; Grauslund, J.; et al. Global prevalence and major risk factors of diabetic retinopathy. *Diabetes Care* **2012**, *35*, 556–564. [CrossRef]
6. Curran, K.; Piyasena, P.; Congdon, N.; Duke, L.; Malanda, B.; Peto, T. Inclusion of diabetic retinopathy screening strategies in national-level diabetes care planning in low- and middle-income countries: A scoping review. *Health Res. Policy Syst.* **2023**, *21*, 2. [CrossRef]
7. Rübsam, A.; Parikh, S.; Fort, P.E. Role of Inflammation in Diabetic Retinopathy. *Int. J. Mol. Sci.* **2018**, *19*, 942. [CrossRef]
8. Busnatu, S.-S.; Salmen, T.; Pana, M.-A.; Rizzo, M.; Stallone, T.; Papanas, N.; Popovic, D.; Tanasescu, D.; Serban, D.; Stoian, A.P. The Role of Fructose as a Cardiovascular Risk Factor: An Update. *Metabolites* **2022**, *12*, 67. [CrossRef]
9. Sharma, Y.; Saxena, S.; Mishra, A.; Saxena, A.; Natu, S.M. Advanced glycation end products and diabetic retinopathy. *J. Ocul. Biol. Dis. Inform.* **2013**, *5*, 63–69. [CrossRef] [PubMed]
10. Yue, S.; Zhang, J.; Wu, J.; Teng, W.; Liu, L.; Chen, L. Use of the monocyte-to-lymphocyte ratio to predict diabetic retinopathy. *Int. J. Environ. Res. Public Health* **2015**, *12*, 10009–10019. [CrossRef]
11. Benhar, I.; Reemst, K.; Kalchenko, V.; Schwartz, M. The retinal pigment epithelium as a gateway for monocyte trafficking into the eye. *EMBO J.* **2016**, *35*, 1219–1235. [CrossRef]
12. Kuo, C.Y.J.; Murphy, R.; Rupenthal, I.D.; Mugisho, O.O. Correlation between the progression of diabetic retinopathy and inflammasome biomarkers in vitreous and serum—A systematic review. *BMC Ophthalmol.* **2022**, *22*, 238. [CrossRef]
13. Gomułka, K.; Ruta, M. The Role of Inflammation and Therapeutic Concepts in Diabetic Retinopathy—A Short Review. *Int. J. Mol. Sci.* **2023**, *24*, 1024. [CrossRef]
14. Semeraro, F.; Cancarini, A.; dell'Omo, R.; Rezzola, S.; Romano, M.R.; Costagliola, C. Diabetic Retinopathy: Vascular and Inflammatory Disease. *J. Diabetes Res.* **2015**, *2015*, 582060. [CrossRef]
15. Yumnamcha, T.; Guerra, M.; Singh, L.P.; Ibrahim, A.S. Metabolic Dysregulation and Neurovascular Dysfunction in Diabetic Retinopathy. *Antioxidants* **2020**, *9*, 1244. [CrossRef]
16. Aldosari, D.I.; Malik, A.; Alhomida, A.S.; Ola, M.S. Implications of Diabetes-Induced Altered Metabolites on Retinal Neurodegeneration. *Front. Neurosci.* **2022**, *16*, 938029. [CrossRef] [PubMed]
17. Chittawar, S.; Dutta, D.; Qureshi, Z.; Surana, V.; Khandare, S.; Dubey, T.N. Neutrophil-lymphocyte Ratio is a Novel Reliable Predictor of Nephropathy, Retinopathy, and Coronary Artery Disease in Indians with Type-2 Diabetes. *Indian J. Endocrinol. Metab.* **2017**, *21*, 864–870. [CrossRef] [PubMed]
18. Branescu, C.; Serban, D.; Dascalu, A.M.; Oprescu, S.M.; Savlovschi, C. Interleukin 6 and lipopolysaccharide binding protein—Markers of inflammation in acute appendicitis. *Chirurgia* **2013**, *108*, 206–214. [PubMed]
19. Quevedo- Martínez, J.U.; Garfias, Y.; Jimenez, J.; Garcia, O.; Venegas, D.; de Lucio, V.M.B. Pro-inflammatory cytokine profile is present in the serum of Mexican patients with different stages of diabetic retinopathy secondary to type 2 diabetes. *BMJ Open Ophthalmol.* **2021**, *6*, e000717. [CrossRef]
20. Zhang, W.; Chen, S.; Liu, M.L. Pathogenic roles of microvesicles in diabetic retinopathy. *Acta Pharmacol. Sin.* **2018**, *39*, 1–11. [CrossRef]
21. Buch, A.; Kaur, S.; Nair, R.; Jain, A. Platelet volume indices as predictive biomarkers for diabetic complications in Type 2 diabetic patients. *J. Lab. Physicians* **2017**, *9*, 84–88. [CrossRef] [PubMed]
22. Jabeen, F.; Fawwad, A.; Rizvi, H.A.; Alvi, F. Role of platelet indices, glycemic control and hs-CRP in pathogenesis of vascular complications in type-2 diabetic patients. *Pak. J. Med. Sci.* **2013**, *29*, 152–156. [PubMed]
23. Guan, Y.; Zuo, W.; Jia, K.; Yu, C.; Liu, F.; Lv, Z.; Wang, D.; Shi, F.D.; Wang, X. Association of Red Blood Cell Distribution Width with Stroke Prognosis among Patients with Small Artery Occlusion: A Hospital-Based Prospective Follow-Up Study. *Int. J. Gen. Med.* **2022**, *15*, 7449–7457. [CrossRef]

24. Li, C.; Tian, W.; Zhao, F.; Li, M.; Ye, Q.; Wei, Y.; Li, T.; Xie, K. Systemic immune-inflammation index, SII, for prognosis of elderly patients with newly diagnosed tumors. *Oncotarget* **2018**, *9*, 35293–35299. [CrossRef]
25. Mercan, R.; Bitik, B.; Tufan, A.; Bozbulut, U.B.; Atas, N.; Ozturk, M.A.; Haznedaroglu, S.; Goker, B. The Association Between Neutrophil/Lymphocyte Ratio and Disease Activity in Rheumatoid Arthritis and Ankylosing Spondylitis. *J. Clin. Lab. Anal.* **2016**, *30*, 597–601. [CrossRef]
26. Khandare, S.A.; Chittawar, S.; Nahar, N.; Dubey, T.N.; Qureshi, Z. Study of Neutrophil-lymphocyte Ratio as Novel Marker for Diabetic Nephropathy in Type 2 Diabetes. *Indian J. Endocrinol. Metab.* **2017**, *21*, 387–392.
27. Wong, T.Y.; Sun, J.; Kawasaki, R.; Ruamviboonsuk, P.; Gupta, N.; Lansingh, V.C.; Maia, M.; Mathenge, W.; Moreker, S.; Muqit, M.M.K.; et al. Guidelines on Diabetic Eye Care: The International Council of Ophthalmology Recommendations for Screening, Follow-up, Referral, and Treatment Based on Resource Settings. *Ophthalmology* **2018**, *125*, 1608–1622. [CrossRef]
28. Wilkinson, C.P.; Ferris, F.L.; Klein, R.E.; Lee, P.P.; Agardh, C.D.; Davis, M.; Dills, D.; Kampik, A.; Pararajasegaram, R.; Verdaguer, J.T. Global Diabetic Retinopathy Project Group: Proposed international clinical diabetic retinopathy and diabetic macular edema disease severity scales. *Ophthalmology* **2003**, *110*, 1677–1682. [CrossRef]
29. Meng, X.; Chang, Q.; Liu, Y.; Chen, L.; Wei, G.; Yang, J.; Zheng, P.; He, F.; Wang, W.; Ming, L. Determinant roles of gender and age on SII, PLR, NLR, LMR and MLR and their reference intervals defining in Henan, China: A posteriori and big-data-based. *J. Clin. Lab. Anal.* **2018**, *32*, e22228. [CrossRef] [PubMed]
30. Yang, S.; Berdine, G. The receiver operating characteristic (ROC) curve. *Southwest Respir. Crit. Care Chron.* **2017**, *5*, 34–36. [CrossRef]
31. Zeng, J.; Chen, M.; Feng, Q.; Wan, H.; Wang, J.; Yang, F.; Cao, H. The Platelet-to-Lymphocyte Ratio Predicts Diabetic Retinopathy in Type 2 Diabetes Mellitus. *Diabetes Metab. Syndr. Obes.* **2022**, *15*, 3617–3626. [CrossRef]
32. Huang, L.; Li, L.; Wang, M.; Zhang, D.; Song, Y. Correlation between ultrawide-field fluorescence contrast results and white blood cell indexes in diabetic retinopathy. *BMC Ophthalmol.* **2022**, *22*, 231. [CrossRef]
33. Luo, W.J.; Zhang, W.F. The relationship of blood cell-associated inflammatory indices and diabetic retinopathy: A Meta-analysis and systematic review. *Int. J. Ophthalmol.* **2019**, *12*, 312–323.
34. Wan, H.; Cai, Y.; Wang, Y.; Fang, S.; Chen, C.; Chen, Y.; Xia, F.; Wang, N.; Guo, M.; Lu, Y. The unique association between the level of peripheral blood monocytes and the prevalence of diabetic retinopathy: A cross-sectional study. *J. Transl. Med.* **2020**, *18*, 248. [CrossRef]
35. Gawiński, C.; Michalski, W.; Mróz, A.; Wyrwicz, L. Correlation between Lymphocyte-to-Monocyte Ratio (LMR), Neutrophil-to-Lymphocyte Ratio (NLR), Platelet-to-Lymphocyte Ratio (PLR) and Tumor-Infiltrating Lymphocytes (TILs) in Left-Sided Colorectal Cancer Patients. *Biology* **2022**, *11*, 385. [CrossRef]
36. Huang, L.; Hu, Z.; Luo, R.; Li, H.; Yang, Z.; Qin, X.; Mo, Z. Predictive Values of the Selected Inflammatory Indexes in Colon Cancer. *Cancer Control J. Moffitt Cancer Cent.* **2022**, *29*, 10732748221091333. [CrossRef] [PubMed]
37. Huang, Q.; Wu, H.; Wo, M.; Ma, J.; Song, Y.; Fei, X. Clinical and predictive significance of Plasma Fibrinogen Concentrations combined Monocyte-lymphocyte ratio in patients with Diabetic Retinopathy. *Int. J. Med. Sci.* **2021**, *18*, 1390–1398. [CrossRef] [PubMed]
38. Wang, H.; Guo, Z.; Xu, Y. Association of monocyte-lymphocyte ratio and proliferative diabetic retinopathy in the US population with type 2 diabetes. *J. Transl. Med.* **2022**, *20*, 219. [CrossRef] [PubMed]
39. Forget, P.; Khalifa, C.; Defour, J.P.; Latinne, D.; Van Pel, M.C.; De Kock, M. What is the normal value of the neutrophil-to-lymphocyte ratio? *BMC Res. Notes* **2017**, *10*, 12. [CrossRef]
40. Wu, L.; Zou, S.; Wang, C.; Tan, X.; Yu, M. Neutrophil-to-lymphocyte and platelet-to-lymphocyte ratio in Chinese Han population from Chaoshan region in South China. *BMC Cardiovasc. Disord.* **2019**, *19*, 125. [CrossRef]
41. Moosazadeh, M.; Maleki, I.; Alizadeh-Navaei, R.; Kheradmand, M.; Hedayatizadeh-Omran, A.; Shamshirian, A.; Barzegar, A. Normal values of neutrophil-to-lymphocyte ratio, lymphocyte-to-monocyte ratio and platelet-to-lymphocyte ratio among Iranian population: Results of Tabari cohort. *Casp. J. Intern. Med.* **2019**, *10*, 320–325. [CrossRef]
42. Fest, J.; Ruiter, R.; Ikram, M.A.; Voortman, T.; van Eijck, C.H.J.; Stricker, B.H. Reference values for white blood-cell-based inflammatory markers in the Rotterdam Study: A population-based prospective cohort study. *Sci. Rep.* **2018**, *8*, 10566. [CrossRef]
43. Mathur, K.; Kurbanova, N.; Qayyum, R. Platelet-lymphocyte ratio (PLR) and all-cause mortality in general population: Insights from national health and nutrition education survey. *Platelets* **2019**, *30*, 1036–1041. [CrossRef] [PubMed]
44. Şahan, D.; Yıldırım, M.S.; Demir, M.; Çoban, N.F.; Şahan, L. Can neutrophil-to-lymphocyte ratio be used as a diagnostic tool in diabetic retinopathy? *Curr. Eye Res.* **2020**, *45*, 295–300.
45. Man, M.-A.; Davidescu, L.; Motoc, N.-S.; Rajnoveanu, R.-M.; Bondor, C.-I.; Pop, C.-M.; Toma, C. Diagnostic Value of the Neutrophil-to-Lymphocyte Ratio (NLR) and Platelet-to-Lymphocyte Ratio (PLR) in Various Respiratory Diseases: A Retrospective Analysis. *Diagnostics* **2022**, *12*, 81. [CrossRef]
46. Ilhan, C.; Citirik, M.; Uzel, M.M.; Tekin, K. The Optimal Cut-off Value of Neutrophil/Lymphocyte Ratio for Severe Grades of Diabetic Retinopathy. *Beyoglu Eye J.* **2019**, *4*, 76–81.
47. Hu, Y.; Cheng, Y.; Xu, X.; Yang, B.; Mei, F.; Zhou, Q.; Yan, L.; Wang, J.; Wu, X. Pretreatment neutrophil-to-lymphocyte ratio predicts prognosis in patients with diabetic macular edema treated with ranibizumab. *BMC Ophthalmol.* **2019**, *19*, 194. [CrossRef]

48. Wang, J.R.; Chen, Z.; Yang, K.; Yang, H.J.; Tao, W.Y.; Li, Y.P.; Jiang, Z.J.; Bai, C.F.; Yin, Y.C.; Duan, J.M.; et al. Association between neutrophil-to-lymphocyte ratio, platelet-to-lymphocyte ratio, and diabetic retinopathy among diabetic patients without a related family history. *Diabetol. Metab. Syndr.* **2020**, *12*, 55. [CrossRef] [PubMed]
49. He, X.; Qi, S.; Zhang, X.; Pan, J. The relationship between the neutrophil-to-lymphocyte ratio and diabetic retinopathy in adults from the United States: Results from the National Health and nutrition examination survey. *BMC Ophthalmol.* **2022**, *22*, 346. [CrossRef]
50. Chen, W.; Chen, K.; Xu, Z.; Hu, Y.; Liu, Y.; Liu, W.; Hu, X.; Ye, T.; Hong, J.; Zhu, H.; et al. Neutrophil-to-Lymphocyte Ratio and Platelet-to-Lymphocyte Ratio Predict Mortality in Patients with Diabetic Foot Ulcers Undergoing Amputations. *Diabetes Metab. Syndr. Obes.* **2021**, *14*, 821–829. [CrossRef]
51. Li, L.; Shen, Q.; Rao, S. Association of Neutrophil-to-Lymphocyte Ratio and Platelet-to-Lymphocyte Ratio with Diabetic Kidney Disease in Chinese Patients with Type 2 Diabetes: A Cross-Sectional Study. *Ther. Clin. Risk Manag.* **2022**, *18*, 1157–1166. [CrossRef] [PubMed]
52. Cho, Y.Y.; Cho, S.I. Treatment variation related to comorbidity and complications in type 2 diabetes: A real world analysis. *Medicine* **2018**, *97*, e12435. [CrossRef] [PubMed]
53. Demirtaş, L.; Degirmenci, H.; Aktaş, A.H.; Ozturk, C. Relationship between neutrophil-lymphocyte ratio and diabetic retinopathy. *Ann. Ophthalmol.* **2017**, *49*, 190–194.
54. Wang, X.; Li, W.; Zhang, Y.; Wan, X. Association of neutrophil-to-lymphocyte ratio and diabetic retinopathy in type 2 diabetes mellitus: A meta-analysis. *Ther. Adv. Endocrinol. Metab.* **2010**, *11*, 2042018820977801.
55. Bai, Y.Y.; Xi, Y.; Yin, B.B.; Zhang, J.H.; Chen, F.; Zhu, B. Reference intervals of systemic immune-inflammation index, neutrophil-to-lymphocyte ratio, lymphocyte-to-monocyte ratio, and platelet-to-lymphocyte ratio during normal pregnancy in China. *Eur. Rev. Med. Pharmacol. Sci.* **2023**, *27*, 1033–1044. [CrossRef]
56. Luo, H.; He, L.; Zhang, G.; Yu, J.; Chen, Y.; Yin, H.; Goyal, H.; Zhang, G.M.; Xiao, Y.; Gu, C.; et al. Normal Reference Intervals of Neutrophil-To-Lymphocyte Ratio, Platelet-To-Lymphocyte Ratio, Lymphocyte-To-Monocyte Ratio, and Systemic Immune Inflammation Index in Healthy Adults: A Large Multi-Center Study from Western China. *Clin. Lab.* **2019**, *65*, 255–265. [CrossRef]
57. Feng, J.F.; Chen, S.; Yang, X. Systemic immune-inflammation index (SII) is a useful prognostic indicator for patients with squamous cell carcinoma of the esophagus. *Medicine* **2017**, *96*, e5886. [CrossRef]
58. Hirahara, N.; Matsubara, T.; Fujii, Y.; Kaji, S.; Kawabata, Y.; Hyakudomi, R.; Yamamoto, T.; Taniura, T.; Tajima, Y. Comparison of the prognostic value of immunoinflammation-based biomarkers in patients with gastric cancer. *Oncotarget* **2020**, *11*, 2625–2635. [CrossRef]
59. Xia, Y.; Xia, C.; Wu, L.; Li, Z.; Li, H.; Zhang, J. Systemic Immune Inflammation Index (SII), System Inflammation Response Index (SIRI) and Risk of All-Cause Mortality and Cardiovascular Mortality: A 20-Year Follow-Up Cohort Study of 42,875 US Adults. *J. Clin. Med.* **2023**, *12*, 1128. [CrossRef]
60. Guo, W.; Song, Y.; Sun, Y.; Du, H.; Cai, Y.; You, Q.; Fu, H.; Shao, L. Systemic immune-inflammation index is associated with diabetic kidney disease in Type 2 diabetes mellitus patients: Evidence from NHANES 2011-2018. *Front. Endocrinol.* **2022**, *13*, 1071465. [CrossRef]
61. Elbeyli, A.; Kurtul, B.E.; Ozcan, S.C.; Ozarslan Ozcan, D. The diagnostic value of systemic immune-inflammation index in diabetic macular oedema. *Clin. Exp. Optom.* **2022**, *105*, 831–835. [CrossRef] [PubMed]
62. Sabbagh, F.; Muhamad, I.I.; Niazmand, R.; Dikshit, P.K.; Kim, B.S. Recent progress in polymeric non-invasive insulin delivery. *Int. J. Biol. Macromol.* **2022**, *203*, 222–243. [CrossRef] [PubMed]
63. Dincer, N.; Yildirim, A.; Yesil, N.; Sezer, S. The relationship between neutrophil-to-lymphocyte ratio and diabetic retinopathy: A systematic review and meta-analysis. *J. Clin. Lab. Anal.* **2021**, *35*, e23613.

biomedicines

MDPI

Article

Optical Coherence Tomography Angiography Changes in Macular Area in Patients with Proliferative Diabetic Retinopathy Treated with Panretinal Photocoagulation

Irini Chatziralli [1,*], Eleni Dimitriou [1], Chrysa Agapitou [1], Dimitrios Kazantzis [1], Petros Kapsis [1], Nick Morogiannis [1], Stylianos Kandarakis [2], George Theodossiadis [1] and Panagiotis Theodossiadis [1]

1 2nd Department of Ophthalmology, Attikon Hospital, National and Kapodistrian University of Athens, 12462 Athens, Greece
2 1st Department of Ophthalmology, National and Kapodistrian University of Athens, 12462 Athens, Greece
* Correspondence: eirchat@yahoo.gr

Abstract: Background: The purpose of this study was to investigate the changes in macular microvasculature using optical coherence tomography angiography (OCTA) in association with functional changes in patients with proliferative diabetic retinopathy (PDR) treated with panretinal photocoagulation (PRP) with a follow up of 12 months. Methods: The participants in this study were 28 patients with PDR and no macular oedema, who were eligible for PRP. All participants underwent best-corrected visual acuity (BCVA) measurement, optical coherence tomography (OCT), and OCT angiography (OCTA) at baseline (before treatment) and at months 1, 6, and 12 after the completion of PRP treatment. The comparison of OCTA parameters and BCVA between baseline and months 1, 6, and 12 after PRP was performed. Results: There was a statistically significant decrease in foveal avascular zone (FAZ) area at months 6 and 12 of the follow-up period compared to baseline ($p = 0.014$ and $p = 0.011$ for month 6 and 12, respectively). Of note is that FAZ became significantly more circular 6 months after PRP ($p = 0.009$), and remained so at month 12 ($p = 0.015$). There was a significant increase in the mean foveal and parafoveal vessel density (VD) at all quadrants at the superficial capillary plexus (SCP) at month 6 and month 12 after PRP compared to baseline. No difference was noticed in VD at the deep capillary plexus (DCP) at any time-point of the follow up. BCVA remained the same throughout the follow-up period. Conclusions: At months 6 and 12 after PRP, foveal and parafoveal VD at SCP significantly increased compared to baseline, while the FAZ area significantly decreased and FAZ became more circular.

Keywords: proliferative diabetic retinopathy; microvasculature; vessel density; OCTA; foveal avascular zone; panretinal photocoagulation

Citation: Chatziralli, I.; Dimitriou, E.; Agapitou, C.; Kazantzis, D.; Kapsis, P.; Morogiannis, N.; Kandarakis, S.; Theodossiadis, G.; Theodossiadis, P. Optical Coherence Tomography Angiography Changes in Macular Area in Patients with Proliferative Diabetic Retinopathy Treated with Panretinal Photocoagulation. *Biomedicines* **2023**, *11*, 3146. https://doi.org/10.3390/biomedicines11123146

Academic Editors: Ana Dascalu and Dragos Serban

Received: 17 September 2023
Revised: 12 November 2023
Accepted: 15 November 2023
Published: 26 November 2023

1. Introduction

Diabetes mellitus (DM) is a global epidemic which is estimated to affect about 785 million people in 2045 [1]. Diabetic retinopathy (DR) is a microvascular complication of DM and a leading cause of blindness worldwide, especially in the working-age population [2]. Diabetic retinopathy is classified as either non-proliferative (NPDR) and proliferative (PDR). NPDR is characterized by structural changes in retinal capillaries, including microaneurysms, haemorrhages, exudates, cotton wool spots, venous beading or looping, intraretinal microvascular abnormalities (IRMA), and retinal non-perfusion. The latter may trigger the development of neovascularization either on the disc (NVD) or elsewhere on the retina (NVE), leading to PDR [2].

Panretinal photocoagulation (PRP) using an argon laser is considered the "gold standard" for the treatment of PDR, since it has been proven to reduce the risk of severe visual loss or need for vitrectomy by over 50% in high-risk PDR patients at four-year follow up [3]. Regarding the mechanism by which PRP helps in the regression of neovascularization, there are several theories. Firstly, it was suggested that the destruction of the retina using laser thermal ablation can lead to the reduction in metabolic demand, or, in simple terms, the diseased retina could be debulked while also achieving the suppression of vascular endothelial growth factor (VEGF) production [4,5]. Additionally, the thinning of the retina due to photocoagulation would facilitate increased oxygen diffusion from the choroid to the vitreous and consequent improvement in inner retinal oxygenation via diffusion through the vitreous. Moreover, the stimulation of biological factors, such as heat shock proteins, might result in the improvement of the disease *per se* [5]. However, PRP may disrupt the blood–retinal barrier, leading to alterations in the retinal microvasculature and haemodynamics [6]. Specifically, a reduction in ocular blood flow velocities in the ophthalmic artery, central retinal artery, and central retinal vein has been reported after PRP in patients with DR using Doppler measurement or laser speckle flowgraphy [7–9].

On the other hand, advances in retinal imaging and the introduction of optical coherence tomography angiography (OCTA) allow the non-invasive visualization of the retina, also providing quantitative data, including measurements of the vessel density (VD) and the foveal avascular zone (FAZ) [10]. There have been a few studies reporting changes in circulatory distribution and vascular parameters in the macular area after PRP in patients with severe NPDR and PDR, with controversial results [11–14]. Russell et al. showed that retinal perfusion did not change significantly 3 months after PRP in 15 patients with PDR [11] in a prospective study using ultra-widefield fluorescein angiography (FA), the findings of which were in line with other investigators, who found similar results at months 3–6 after PRP using OCTA [12,13]. On the contrary, Fawzi et al. reported an increase in the flow metrics of all capillary layers in the macula using OCTA 3–6 months after PRP in 10 patients with PDR, suggesting an overall re-distribution of blood flow to the posterior pole following PRP in patients with PDR [14].

However, the as yet published studies have short follow ups of 1–6 months and small study sample, ranging from 6 to 15 patients, with an average age of patients between 46 and 63 years [11–14]. In light of the above, the purpose of this prospective study was to evaluate the changes in macular microvasculature using OCTA in association with visual changes in patients with PDR treated with PRP in a 12-month follow-up study.

2. Materials and Methods

Participants in this prospective study were 28 patients with type 2 DM and PDR. All patients were recruited, diagnosed, treated, and followed up at 2nd Department of Ophthalmology, National and Kapodistrian University of Athens, Athens, Greece between September 2020 and September 2022. The study was in adherence with the tenets of Helsinki Declaration and was approved by the Institutional Review Board of Attikon General Hospital (Ref: 291/2020). Written informed consent was obtained from all patients after explaining them the protocol of the study.

All participants were treatment-naïve patients with high-risk PDR, and did not have diabetic macular oedema. PDR was diagnosed through detection of retinal neovascularization on dilated fundoscopy, and confirmed with FA when there was uncertainty regarding its presence. In case of bilateral PDR, right eye was included to avoid selection bias. Eyes with other retinal diseases, glaucoma, intraocular pressure (IOP) $\geq$ 18 mmHg, ocular inflammation, spherical equivalent $\geq$ 6 diopters, axial length $\geq$ 26.5 mm, significant media opacity to preclude adequate imaging, vitreous haemorrhage, previous pars plana vitrectomy, previous cataract surgery within the last 6 months, trauma, and previous treatment for DME or DR were excluded. All patients were followed up for at least 12 months.

At baseline, we recorded patients' information, including age, gender, diabetes duration, presence of hypertension, dyslipidaemia, and glycated haemoglobin (HbA1c). In addition, at baseline, all participants underwent a thorough ophthalmic examination, including best-corrected visual acuity (BCVA) measurement by means of ETDRS charts, slit-lamp examination, IOP measurement, dilated fundoscopy, spectral domain optical coherence tomography (SD-OCT), OCTA, and fundus photography using HOCT-1F All-in-one (Huvitz, Korea). In cases where FA was needed, it was performed using Heidelberg Spectralis (Spectralis HRA+OCT, Heidelberg, Germany). BCVA assessment, slit-lamp examination, dilated fundoscopy, SD-OCT, and OCTA were repeated at each follow-up visit, while FA was performed at physician's discretion. BCVA was converted to LogMAR for statistical purposes.

All patients were treated with PRP, and were examined at month 1, 6, and 12 after treatment. Panretinal photocoagulation was carried out using Visulas Green Laser 532 nm (Carl Zeiss Meditec AG, Jena, Germany) in 3 sessions with an interval of 1 week in between, with a total number of about 2000–2500 burns. The covered area extended from optic disc by a diameter of about one disc, and to posteriorly just outside the arcades, reaching as far as possible into the periphery, while in the macula, the temporal demarcation border was defined as about 1.5 times the disc-to-fovea distance positioned temporal to the fovea. Laser settings were 200 μm spot size with pulse duration of 200 ms, and power of 200–250 mW, while the laser burns were spaced one laser spot size apart. No complications were reported during or after PRP in our study sample.

2.1. Optical Coherence Tomography (OCT) and OCT Angiography Protocol

The HOCT-1F All-in-one uses a diode beam source of 840 nm with a scanning speed of 68,000 axial scans per second, and has an intelligent fixation system that avoids artifacts from ocular movements. For SD-OCT, we used the macular radial 3D OCT scan pattern with 9 mm scan range, 512 A-scan points, and 96 B-scan lines, enabling "fine" OCT sensitivity. An area of 6 × 6 mm centred in the fovea was analysed in agreement with the Early Treatment Diabetic Retinopathy Study (ETDRS) map [15]. The macula was divided into nine regions and three concentric rings (the 1 mm diameter at foveal centre, the inner ring at 3 mm diameter, and the outer ring at 6 mm diameter), which were subdivided into four subfields (superior, temporal, inferior, and nasal) (Figure 1A). All measurements were obtained using the automated segmentation algorithms from the OCT unit. The central subfield thickness (CST) was defined as the mean thickness of the neurosensory retina in the central 1 mm diameter determined by the ETDRS map. In addition, the condition of ellipsoid zone and external limiting membrane was assessed at an area of 1500 μm radius around the foveola.

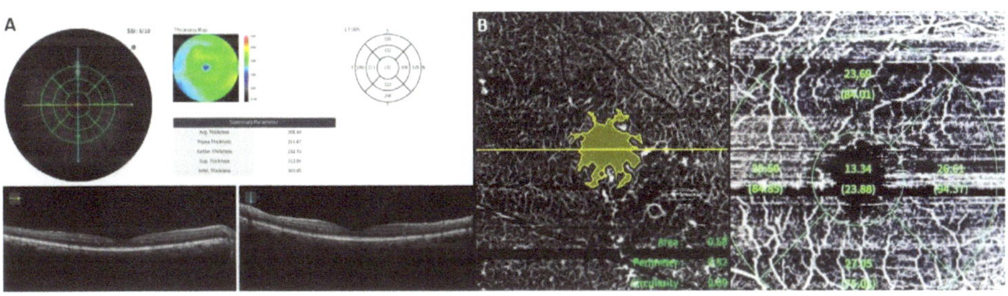

Figure 1. (**A**) Optical coherence tomography (OCT) of a patient with diabetes mellitus and no diabetic macular oedema, depicting the protocol of OCT imaging; (**B**) OCT–angiography of the same patient showing the foveal avascular zone (FAZ) area measurement, FAZ perimeter, and FAZ circularity, as well as vessel density measurement (**C**), according to the OCT–angiography imaging protocol.

Regarding OCTA, we used the "Macular Angio" protocol with 4.5 mm scan range, 384 A-scan points, and 384 B-scan lines. The OCTA analysis software calculated the foveal avascular zone (FAZ) area in mm^2, the FAZ perimeter in mm, and the FAZ circularity ratio (Figure 1B) automatically, as well as the vessel density (VD) in foveal and parafoveal region (Figure 1C) in both superficial capillary plexus (SCP) and deep capillary plexus (DCP). SCP was defined as from internal limiting membrane (ILM) to inner plexiform layer (IPL), while DCP was defined as from IPL to outer plexiform layer (OPL). The foveal region was defined as a 1 mm ring centred on the fovea, and the parafoveal region was defined as the zone between the 1 mm and 3 mm concentric rings centred on the fovea. FAZ represents a region absent of capillaries at the centre of the fovea.

2.2. Statistical Analysis

Patients' characteristics were presented using descriptive statistics. Mean ± standard deviation (SD) was used for continuous variables and counts for categorical variables. Normal distribution of data was analysed by the Kolmogorov–Smirnov test. Longitudinal comparisons for various parameters between baseline and months 1, 6, and 12 was performed, using ANOVA for repeated measures or McNemar test, as appropriate.

Statistical analysis was conducted using IBM SPSS (Version 22.0, Chicago, IL, USA). A p value ≤ 0.05 was considered as statistically significant. However, for longitudinal multiple comparisons, the Bonferroni correction was adopted, as appropriate. Specifically, since 3 comparisons were performed (baseline versus month 1, month 6, and month 12), statistical significance was set to 0.05/3 = 0.017.

3. Results

A total of 28 patients (28 eyes) with treatment-naïve PDR were enrolled in the study. Table 1 shows the demographic and clinical characteristics of the study sample. The mean age of patients was 57.3 ± 8.9 years. A total of 15 patients (53.6%) were male and 13 (46.4%) were female. The mean duration of DM was 12.4 ± 5.4 years. The mean HbA1c at baseline was 7.8 ± 1.3%. Regarding comorbidities, 18 out of 28 patients (64.3%) had hypertension, and 12 patients (42.9%) Had dyslipidaemia, while 11 out of 28 patients (39.2%) were current smokers. Furthermore, 6 out of 28 patients (21.4%) were pseudophakic.

Table 1. Demographic and clinical characteristics of the study sample.

Age (years, mean ± SD) (min–max)	57.3 ± 8.9 (48–68)
Gender (n, %)	
Male	15 (53.6%)
Female	13 (46.4%)
Duration of diabetes mellitus (years, mean ± SD) (min–max)	12.4 ± 5.4 (7–19)
HbA1c (%, mean ± SD) (min–max)	7.8 ± 1.3 (6–10.5)
Hypertension (n, %)	18 (64.3%)
Dyslipidemia (n, %)	12 (42.9%)
Smoking (n, %)	11 (39.2%)
Best-corrected visual acuity (logMAR, mean ± SD) (min–max)	0.41 ± 0.11 (0.3–0.6)
Intraocular pressure (mmHg, mean ± SD) (min–max)	15.7 ± 4.2 (11–18)
Lens status (n, %)	
Phakic	22 (78.6%)
Pseudophakic	6 (21.4%)
Refractive error (spherical equivalent in D, mean ± SD)	−1.25 ± 0.75
Central subfield thickness (μm, mean ± SD) (min–max)	262.4 ± 35.2 (223–309)
Ellipsoid zone condition (n, %)	
Intact	20 (71.4%)
Disrupted	8 (28.6%)
External limiting membrane condition (n, %)	
Intact	22 (78.6%)
Disrupted	6 (21.4%)

At baseline, the mean BCVA was 0.41 ± 0.11 logMAR. There was no statistically significant difference in BCVA at month 1 (0.41 ± 0.13 logMAR, p = 0.030), month 6

(0.40 ± 0.12 logMAR, $p = 0.103$), and month 12 after PRP (0.39 ± 0.12 logMAR, $p = 0.032$) compared to baseline, as it is shown in Table 2.

Table 2. Changes in visual acuity, central subfield thickness, ellipsoid zone, and external limiting membrane over time.

	Baseline	Month 1	Month 6	Month 12
Best-corrected visual acuity (logMAR, mean ± SD) (min–max)	0.41 ± 0.11 (0.3–0.6)	0.41 ± 0.13 (0.3–0.6) ($p = 0.030$)	0.40 ± 0.12 (0.3–0.6) ($p = 0.103$)	0.39 ± 0.12 (0.3–0.6) ($p = 0.032$)
Central subfield thickness (μm, mean ± SD) (min–max)	262.4 ± 35.2 (223–309)	299.1 ± 42.7 (242–350) ($p < 0.001$)	274.3 ± 37.7 (231–315) ($p = 0.228$)	270.1 ± 34.1 (232–311) ($p = 0.409$)
Ellipsoid zone integrity intact (n, %)	20/28 (71.4%)	21/28 (75%) ($p = 0.763$)	22/28 (78.6%) ($p = 0.537$)	22/28 (78.6%) ($p = 0.537$)
External limiting membrane integrity intact (n, %)	22/28 (78.6%)	22/28 (78.6%)	22/28 (78.6%)	22/28 (78.6%)

At baseline, the mean CST was 262.4 ± 35.2 μm. There was a statistically significant increase in CST at month 1 after PRP compared to baseline (299.1 ± 42.7 μm, $p < 0.001$), while CST did not differ at month 6 (274.3 ± 37.7 μm, $p = 0.228$) and month 12 (270.1 ± 34.1 μm, $p = 0.409$) after PRP compared to baseline, as it is shown in Table 2.

In addition, the ellipsoid zone was intact in 20 out of 28 patients (71.4%) at baseline. There was no statistically significant change in ellipsoid zone condition at month 1 ($p = 0.763$), month 6 ($p = 0.537$), and month 12 ($p = 0.537$) of the follow-up period, as it is illustrated in Table 2. Accordingly, at baseline, the external limiting membrane was intact in 22 out of 28 patients (78.6%), and remained the same throughout the whole follow-up period.

Table 3 shows the OCTA parameters in SCP and DCP at baseline and at the 1-month, 6-month and 12-month follow up after PRP. At baseline, the mean FAZ area was found to be 0.51 ± 0.18 mm^2, with a 6.43 ± 2.41 mm perimeter and 0.21 ± 0.09 circularity ratio. There was a statistically significant reduction in the FAZ area at month 6 (0.402 ± 0.13 mm^2, $p = 0.014$) and month 12 (0.406 ± 0.10 mm^2, $p = 0.011$) after PRP compared to baseline. Accordingly, at months 6 and 12, the FAZ perimeter was significantly lower compared to baseline ($p = 0.015$ and $p = 0.017$ for month 6 and 12, respectively). In addition, FAZ became more circular at month 6 (0.27 ± 0.08, $p = 0.009$) and month 12 (0.27 ± 0.10, $p = 0.015$) after PRP compared to baseline.

There was a statistically significant increase in the mean central VD at the SCP at month 6 ($p = 0.002$) and month 12 ($p = 0.003$) after PRP compared to baseline. In addition to the foveal area, the mean VD was significantly increased in the parafoveal area at all quadrants at the SCP at months 6 and 12 after PRP compared to baseline, as it is shown in Table 3. Regarding the DCP, no statistically significant difference was observed in foveal and parafoveal VD at any quadrant and at any time-point of the follow-up period compared to baseline, as it is depicted in Table 3.

Table 3. Optical coherence tomography angiography findings at baseline and at months 1, 6, and 12 after panretinal photocoagulation. Bold denotes statistical significance.

	Baseline	Month 1	Month 6	Month 12
FAZ area (mm^2, mean ± SD)	0.51 ± 0.18	0.51 ± 0.19 ($p = 0.952$)	0.40 ± 0.13 **($p = 0.014$)**	0.41 ± 0.10 **($p = 0.011$)**
FAZ perimeter (mm, mean ± SD)	6.43 ± 2.41	6.45 ± 2.03 ($p = 0.973$)	5.02 ± 1.73 **($p = 0.015$)**	5.01 ± 1.88 **($p = 0.017$)**

Table 3. *Cont.*

	Baseline	Month 1	Month 6	Month 12
FAZ circularity (mean ± SD)	0.21 ± 0.09	0.21 ± 0.10 (p > 0.999)	0.27 ± 0.08 **(p = 0.009)**	0.27 ± 0.10 **(p = 0.015)**
	Superficial capillary plexus (mean ± SD)			
Vessel density, central	9.07 ± 4.92	10.91 ± 4.01 (p = 0.102)	12.96 ± 4.06 **(p = 0.002)**	13.15 ± 4.89 **(p = 0.003)**
Vessel density, superior	26.05 ± 10.21	30.97 ± 7.17 (p = 0.028)	32.21 ± 7.56 **(p = 0.013)**	32.45 ± 7.74 **(p = 0.010)**
Vessel density, inferior	24.71 ± 11.63	26.95 ± 8.14 (p = 0.289)	31.99 ± 7.85 **(p = 0.008)**	32.12 ± 7.76 **(p = 0.007)**
Vessel density, temporal	19.52 ± 9.35	24.96 ± 9.55 (p = 0.036)	25.67 ± 9.40 **(p = 0.017)**	26.13 ± 9.42 **(p = 0.010)**
Vessel density, nasal	22.15 ± 10.61	27.13 ± 9.23 (p = 0.067)	28.77 ± 9.06 **(p = 0.015)**	28.93 ± 9.57 **(p = 0.015)**
	Deep capillary plexus (mean ± SD)			
Vessel density, central	5.01 ± 2.73	5.12 ± 2.96 (p = 0.886)	6.31 ± 2.57 (p = 0.059)	6.69 ± 3.17 (p = 0.038)
Vessel density, superior	12.99 ± 6.81	14.73 ± 6.62 (p = 0.337)	16.57 ± 7.59 (p = 0.069)	16.96 ± 6.53 (p = 0.030)
Vessel density, inferior	15.61 ± 5.53	16.06 ± 6.12 (p = 0.774)	17.89 ± 6.41 (p = 0.160)	18.15 ± 5.94 (p = 0.104)
Vessel density, temporal	16.11 ± 4.90	17.03 ± 4.51 (p = 0.468)	17.98 ± 4.99 (p = 0.163)	19.29 ± 5.23 (p = 0.023)
Vessel density, nasal	15.17 ± 3.78	15.88 ± 3.91 (p = 0.493)	16.91 ± 4.07 (p = 0.103)	17.52 ± 4.90 (p = 0.050)

4. Discussion

In the present study, we investigated the alterations in OCTA parameters after PRP in patients with PDR. The principal message that came from our study is that there was a statistically significant decrease in FAZ area 6 months after PRP for PDR compared to baseline, which remained until the end of our observation, at month 12 after PRP. Additionally, a significant increase in the foveal and parafoveal VD at SCP was found at month 6 and 12 of the follow-up period. However, although VD and FAZ area were improved, there was no statistically significant difference in BCVA over time compared to baseline.

Previous studies have investigated ocular blood flow, and tried to obtain quantitative data on retinal and choroidal circulation in eyes with DR after PRP, using various methods, such as colour Doppler imaging, laser interferometry, laser Doppler flowmetry and laser speckle flowgraphy [8,16–21]. Augsten et al. found that PRP improved the choroidal circulation in the macular area in patients with NPDR, using a reflection spectra method [19]. Takahashi et al. also reported a significant increase in subfoveal choroidal flow one month after PRP in patients without macular oedema, using laser doppler flowmetry [20]. They hypothesised that the redistribution of choroidal flow from obliterated peripheral capillaries to the macula and the inflammatory response to PRP were the main underlying mechanisms of macular choroidal flow increase [20]. These findings were in accordance with those of preclinical studies, since Flower et al. implied that PRP markedly increased the choriocapillaris blood flow in the macular area compared to the periphery in monkeys after the coagulation of the peripheral retina, using indocyanine green angiography [21].

On the other hand, several researchers tried to examine changes in VD after PRP in patients with PDR using OCTA, which can provide a non-invasive and more reliable assessment of retinal blood flow. Lorusso et al. did not find any significant change in the retinal haemodynamics indices at 6-month follow up after laser photocoagulation using a frequency-doubled Nd:YAG laser instead of an argon laser for PDR, and including patients with both type 1 and type 2 DM, which could have affected their results [12]. Similarly, Zhao et al. reported no change in OCTA parameters after PRP in eyes with PDR or severe NPDR, in a study with limited follow up, ranging from 1–3 months [22]. Moreover, Fawzi et al. did not find a significant change in macular VD following PRP in eyes with PDR, although the adjusted flow index (a self-created surrogate metric of blood flow) suggested an overall redistribution of blood flow to the posterior pole, with the improvement of capillary flow after PRP [14]. On the contrary, Kim et al. demonstrated a decrease in VD one month after PRP in eyes with PDR or severe NPDR compared to baseline findings, while they observed a significant increase in VD at both SCP and DCP 12 months after PRP [23]. Very similar results were found by Mirshahi et al., who showed a significant increase in VD in both SCP and DCP 3 months after PRP in 11 eyes with PDR or severe NPDR [13].

The exact mechanism of the early decrease in VD following PRP is still unknown. One possible explanation for this reduction is that the capillary flow following PRP might be associated with early PRP-induced inflammation and nitric oxide (NO) overproduction. After a sufficient time, the inflammation probably subsides, and subsequently, autoregulatory functions of retinal vasculature may occur, leading to an improvement in VD at the long-term follow up [23]. Furthermore, the main theory about the mechanism of action of PRP pertains to the fact that the retinal pigment epithelium (RPE) absorbs the laser energy, and, consequently, thermal energy is generated in the outer retina, destroying the peripheral RPE and adjacent photoreceptors to the reduced retinal oxygen consumption [13,14]. Due to this, it has been assumed that the redistribution of blood flow from the periphery to the macular region after PRP results in the re-organisation of capillary networks, resulting in the better oxygenation of retina and a consequent decrease in VEGF levels and increase in VD [23].

Regarding the FAZ alterations after PRP, the results of the existing studies are controversial. The enlargement and irregularity of the FAZ are important indications of diabetic macular ischemia [24–26]. Salz et al. have found that patients with PDR have a larger FAZ compared to normal eyes, and enlargement of the FAZ indicates an increase in non-perfusion, since the progression of DR may be associated with a more irregular FAZ due to capillary occlusion or altered blood flow [27]. Lorusso et al. found no change in FAZ six months after PRP treatment in PDR eyes [12], in line with the results of Misharhi et al., who also found no significant changes in the FAZ area three months following PRP [13]. On the other hand, Sabaner et al. reported a significant decrease in FAZ area after PRP in NPDR eyes [28], while Abdelhalim et al. presented a significant decrease in FAZ area 6 months after PRP in PDR [29]. Faghihi et al. also showed that the FAZ area constricted and became significantly more circular at the 6-month follow up after PRP [30], as we found in our study at the long-term follow up of 12 months. The latter could be explained by the fact that improved and more effective flow in the remaining capillaries of the macular area could result in the more effective perfusion of the posterior pole [30]. Of note is that the discrepancy between the results of various studies could be attributed to the different OCTA machines and software used, as well as the difference in population and follow up of each study.

Another important finding of our study was that the BCVA remained the same throughout the 12-month follow-up period, although there was an improvement in foveal and parafoveal VD at SCP. One could hypothesise that the functional improvement may need more time than the structural one, suggesting that improvement in visual acuity may occur a long time after the improvement in VD. In addition, there was no statistically significant difference in the VD at the DCP over time, and this could also have an impact on visual acuity. However, it should be noted that visual acuity is mainly affected by the ellipsoid zone condition [31,32].

A potential limitation of our study pertains to the lack of a control group, which prevents us from drawing a firm conclusion regarding the association between changes in OCTA parameters and PRP treatment. However, it is unethical not to treat patients with PDR, and thus the use of a control group is not feasible. In addition, the lack of considerations of other foveal area hemodynamic parameters, such as choroidal VD and fractal dimensions, could be added as a limitation of the study. Moreover, multivariate analysis taking into account HbA1c levels, hypertension, dyslipidaemia, and demographic data was not performed, since we lack of data at month 12, as they were not included in the protocol of the study. It is worth noting, however, that this is a prospective study with a relatively large sample size of 28 patients and a long-term follow up of 12 months, both considerable in comparison to previous studies, while we included only patients with high-risk PDR without diabetic macular oedema, so as to have a homogenous population. In addition, we conducted the first study in the literature on the ellipsoid zone condition in association with OCTA parameters in patients with PDR after PRP.

5. Conclusions

In conclusion, our study revealed that there was a significant improvement in foveal and parafoveal VD at the SCP at the 6- and 12-month follow up after PRP in patients with PDR without macular oedema. The increase in VD suggests that a redistribution of ocular blood flow may occur from the peripheral retina to the centre after PRP, leading to the better oxygenation of the macular area. This assumption may be supported by the fact that the FAZ area was restored 12 months after PRP, presenting better circularity and decreased FAZ area compared to the baseline condition. However, although VD at SCP and FAZ area were improved, the visual acuity was found to be the same throughout the follow up. Further studies with a larger sample size and better OCTA software are needed to corroborate our results.

Author Contributions: I.C. conceived and designed the study, collected, analysed and interpreted data, drafted the manuscript, and supervised the study; C.A., E.D., D.K. and P.K. collected data; N.M. interpreted data; S.K. performed statistical analysis; G.T. and P.T. interpreted data and critically revised the manuscript. All authors have read and agreed to the published version of the manuscript.

Funding: This research received no external funding.

Institutional Review Board Statement: The study was conducted in accordance with the Declaration of Helsinki and approved by the Institutional Review Board of Attikon University Hospital (Ref: 291/2020, 25 September 2020).

Informed Consent Statement: Informed consent was obtained from all subjects involved in the study.

Data Availability Statement: Data will be made available upon request.

Acknowledgments: This study was presented at the 18th Greek Vitreoretinal Society Congress (January 2023).

Conflicts of Interest: The authors declare no conflict of interest.

References

1. International Diabetes Federation. *IDF Diabetes Atlas*, 10th ed.; International Diabetes Federation: Brussels, Belgium, 2021; Available online: https://www.diabetesatlas.org (accessed on 10th November 2023).
2. Antonetti, D.A.; Klein, R.; Gardner, T.W. Diabetic retinopathy. *N. Engl. J. Med.* **2012**, *366*, 1227–1239. [CrossRef] [PubMed]
3. The Diabetic Retinopathy Study Research Group. Photocoagulation treatment of proliferative diabetic retinopathy. Clinical application of Diabetic Retinopathy Study (DRS) findings, DRS Report Number 8. *Ophthalmology* **1981**, *88*, 583–600.
4. Funatsu, H.; Hori, S.; Yamashita, H.; Kitano, S. Effective mechanisms of laser photocoagulation for neovascularization in diabetic retinopathy. *Nippon. Ganka Gakkai Zasshi* **1996**, *100*, 339–349. [PubMed]
5. Luttrull, J.K. Lasers in Medicine: The Changing Role of Therapeutic Laser-Induced Retinal Damage—From de rigeuer to Nevermore. *Photonics* **2023**, *10*, 999. [CrossRef]
6. Moriarty, A.P.; Spalton, D.J.; Shilling, J.S.; Ffytche, T.J.; Bulsara, M. Breakdown of the blood-aqueous barrier after argon laser panretinal photocoagulation for proliferative diabetic retinopathy. *Ophthalmology* **1996**, *103*, 833–838. [CrossRef] [PubMed]

7. Mendivil, A. Ocular blood flow velocities in patients with proliferative diabetic retinopathy after panretinal photocoagulation. *Surv. Ophthalmol.* **1997**, *42*, S89–S95. [CrossRef] [PubMed]
8. Yamada, Y.; Suzuma, K.; Onizuka, N.; Uematsu, M.; Mohamed, Y.H.; Kitaoka, T. Evaluation of retinal blood flow before and after panretinal photocoagulation using pattern scan laser for diabetic retinopathy. *Curr. Eye Res.* **2017**, *42*, 1707–1712. [CrossRef] [PubMed]
9. Iwase, T.; Kobayashi, M.; Yamamoto, K.; Ra, E.; Terasaki, H. Effects of photocoagulation on ocular blood flow in patients with severe non-proliferative diabetic retinopathy. *PLoS ONE* **2017**, *12*, e0174427. [CrossRef]
10. Cicinelli, M.V.; Cavalleri, M.; Brambati, M.; Lattanzio, R.; Bandello, F. New imaging systems in diabetic retinopathy. *Acta Diabetol.* **2019**, *56*, 981–994. [CrossRef]
11. Russell, J.F.; Al-Khersan, H.; Shi, Y.; Scott, N.L.; Hinkle, J.W.; Fan, K.C.; Lyu, C.; Feuer, W.J.; Gregori, G.; Rosenfeld, P.J. Retinal Nonperfusion in Proliferative Diabetic Retinopathy Before and After Panretinal Photocoagulation Assessed by Widefield OCT Angiography. *Am. J. Ophthalmol.* **2020**, *213*, 177–185. [CrossRef]
12. Lorusso, M.; Milano, V.; Nikolopoulou, E.; Ferrari, L.M.; Cicinelli, M.V.; Querques, G.; Ferrari, T.M. Panretinal Photocoagulation Does Not Change Macular Perfusion in Eyes With Proliferative Diabetic Retinopathy. *Ophthalmic Surg. Lasers Imaging Retin.* **2019**, *50*, 174–178. [CrossRef]
13. Mirshahi, A.; Ghassemi, F.; Fadakar, K.; Mirshahi, R.; Bazvand, F.; Riazi-Esfahani, H. Effects of panretinal photocoagulation on retinal vasculature and foveal avascular zone in diabetic retinopathy using optical coherence tomography angiography: A pilot study. *J. Curr. Ophthalmol.* **2019**, *31*, 287–291. [CrossRef] [PubMed]
14. Fawzi, A.A.; Fayed, A.E.; Linsenmeier, R.A.; Gao, J.; Yu, F. Improved Macular Capillary Flow on Optical Coherence Tomography Angiography after Panretinal Photocoagulation for Proliferative Diabetic Retinopathy. *Am. J. Ophthalmol.* **2019**, *206*, 217–227. [CrossRef] [PubMed]
15. Early Treatment Diabetic Retinopathy Study Research Group. Grading diabetic retinopathy from stereoscopic color fundus photographs—An extension of the modified Airlie House classification. ETDRS report number 10. *Ophthalmology* **1991**, *98*, 786–806. [CrossRef]
16. Lee, J.C.; Wong, B.J.; Tan, O.; Srinivas, S.; Sadda, S.R.; Huang, D.; Fawzi, A.A. Pilot study of Doppler optical coherence tomography of retinal blood flow following laser photocoagulation in poorly controlled diabetic patients. *Investig. Ophthalmol. Vis. Sci.* **2013**, *54*, 6104–6111. [CrossRef] [PubMed]
17. Grunwald, J.E.; Brucker, A.J.; Grunwald, S.E.; Riva, C.E. Retinal hemodynamics in proliferative diabetic retinopathy. A laser Doppler velocimetry study. *Investig. Ophthalmol. Vis. Sci.* **1993**, *34*, 66–71.
18. Feke, G.T.; Green, G.J.; Goger, D.G.; McMeel, J.W. Laser Doppler measurements of the effect of panretinal photocoagulation on retinal blood flow. *Ophthalmology* **1982**, *89*, 757–762. [CrossRef]
19. Augsten, R.; Königsdörffer, E.; Schweitzer, D.; Strobel, J. Nonproliferative diabetic retinopathy-reflection spectra of the macula before and after laser photocoagulation. *Ophthalmologica* **1998**, *212*, 105–111. [CrossRef]
20. Takahashi, A.; Nagaoka, T.; Sato, E.; Yoshida, A. Effect of panretinal photocoagulation on choroidal circulation in the foveal region in patients with severe diabetic retinopathy. *Br. J. Ophthalmol.* **2008**, *92*, 1369–1373. [CrossRef]
21. Flower, R.W.; Fryczkowski, A.W.; McLeod, D.S. Variability in choriocapillaris blood flow distribution. *Investig. Ophthalmol. Vis. Sci.* **1995**, *36*, 1247–1258.
22. Zhao, T.; Chen, Y.; Liu, D.; Stewart, J.M. Optical Coherence Tomography Angiography Assessment of Macular Choriocapillaris and Choroid Following Panretinal Photocoagulation in a Diverse Population With Advanced Diabetic Retinopathy. *Asia Pac. J. Ophthalmol.* **2020**, *10*, 203–207. [CrossRef]
23. Kim, K.; Kim, E.S.; Yu, S.Y. Longitudinal changes in retinal microvasculature after panretinal photocoagulation in diabetic retinopathy using swept-source OCT angiography. *Sci. Rep.* **2021**, *11*, 216. [CrossRef] [PubMed]
24. Takase, N.; Nozaki, M.; Kato, A.; Ozeki, H.; Yoshida, M.; Ogura, Y. Enlargement of foveal avascular zone in diabetic eyes evaluated by en face optical coherence tomography angiography. *Retina* **2015**, *35*, 2377–2383. [CrossRef]
25. Early Treatment Diabetic Retinopathy Study Research Group. Classification of diabetic retinopathy from fluorescein angiograms. ETDRS report number 11. *Ophthalmology* **1991**, *98*, 807–822. [CrossRef]
26. Krawitz, B.D.; Mo, S.; Geyman, L.S.; Agemy, S.A.; Scripsema, N.K.; Garcia, P.M.; Chui, T.Y.; Rosen, R.B. Acircularity index and axis ratio of the foveal avascular zone in diabetic eyes and healthy controls measured by optical coherence tomography angiography. *Vis. Res.* **2017**, *139*, 177–186. [CrossRef] [PubMed]
27. Salz, D.A.; de Carlo, T.E.; Adhi, M.; Moult, E.; Choi, W.; Baumal, C.R.; Witkin, A.J.; Duker, J.S.; Fujimoto, J.G.; Waheed, N.K. Select Features of Diabetic Retinopathy on Swept-Source Optical Coherence Tomographic Angiography Compared with Fluorescein Angiography and Normal Eyes. *JAMA Ophthalmol.* **2016**, *134*, 644–650. [CrossRef] [PubMed]
28. Sabaner, M.C.; Dogan, M.; Akdogan, M.; Şimşek, M. Panretinal laser photocoagulation decreases large foveal avascular zone area in non-proliferative diabetic retinopathy: A prospective OCTA study. *Photodiagnosis Photodyn. Ther.* **2021**, *34*, 102298. [CrossRef]
29. Abdelhalim, A.S.; Abdelkader, M.F.S.O.; Mahmoud, M.S.E.-D.; Mohamed Mohamed, A.A. Macular vessel density before and after panretinal photocoagulation in patients with proliferative diabetic retinopathy. *Int. J. Retin. Vitr.* **2022**, *8*, 21. [CrossRef]
30. Faghihi, H.; Riazi-Esfahani, H.; Khodabande, A.; Khalili Pour, E.; Mirshahi, A.; Ghassemi, F.; Mirshahi, R.; Khojasteh, H.; Bazvand, F.; Hashemi, A.; et al. Effect of panretinal photocoagulation on macular vasculature using optical coherence tomography angiography. *Eur. J. Ophthalmol.* **2021**, *31*, 1877–1884. [CrossRef] [PubMed]

31. Theodossiadis, P.G.; Theodossiadis, G.P.; Charonis, A.; Emfietzoglou, I.; Grigoropoulos, V.G.; Liarakos, V.S. The photoreceptor layer as a prognostic factor for visual acuity in the secondary epiretinal membrane after retinal detachment surgery: Imaging analysis by spectral-domain optical coherence tomography. *Am. J. Ophthalmol.* **2011**, *151*, 973–980. [CrossRef]
32. Chatziralli, I.; Theodossiadis, G.; Dimitriou, E.; Kazantzis, D.; Theodossiadis, P. Association between the patterns of diabetic macular edema and photoreceptors' response after intravitreal ranibizumab treatment: A spectral-domain optical coherence tomography study. *Int. Ophthalmol.* **2020**, *40*, 2441–2448. [CrossRef] [PubMed]

biomedicines

MDPI

Article

Retinal Functional Impairment in Diabetic Retinopathy

Cornelia Andreea Tănasie [1,†], Alexandra Oltea Dan [1,*], Oana Maria Ică [2,*], Maria Filoftea Mercuț [3,†], George Mitroi [4,†], Citto-Iulian Taisescu [1], Veronica Sfredel [1], Ramona Ingrid Corbeanu [1], Carmen Luminița Mocanu [3] and Ciprian Danielescu [5]

1 Department of Physiology, University of Medicine and Pharmacy of Craiova, 200349 Craiova, Romania; andreea.tanasie@umfcv.ro (C.A.T.); citto.taisescu@umfcv.ro (C.-I.T.); veronica.sfredel@umfcv.ro (V.S.); ramonna.ingrid@yahoo.com (R.I.C.)
2 Department of Dermatology, University of Medicine and Pharmacy of Craiova, 200349 Craiova, Romania
3 Department of Ophthalmology, University of Medicine and Pharmacy of Craiova, 200349 Craiova, Romania; maria.mercut@umfcv.ro (M.F.M.); carmen.mocanu@umfcv.ro (C.L.M.)
4 Department of Urology, University of Medicine and Pharmacy of Craiova, 200349 Craiova, Romania; gmitroi@yahoo.com
5 Department of Ophthalmology, University of Medicine and Pharmacy "Grigore T. Popa", 700111 Iasi, Romania; ciprian.danielescu@umfiasi.ro
* Correspondence: puiu.alexandra.oltea@gmail.com (A.O.D.); oana.maria.corici@gmail.com (O.M.I.)
† These authors contributed equally to this work.

Abstract: Background: Diabetic retinopathy (DR) is a neurodegenerative disease of the retina. The aim of our study was to analyze latency changes in a full-field electroretinogram (ERG) in patients with type 2 diabetes. Material: This prospective study included 15 diabetic patients without DR, 16 diabetic patients with non-proliferative DR, 14 patients with pre-proliferative DR, 15 patients with proliferative DR, and 14 age-matched controls. All the participants underwent ophthalmologic examination and full-field ERGs. The ERGs were recorded with the Metrovision MonPackOne system. The latencies were analyzed for "a"- and "b"-waves in the dark-adapted (DA) 0.01 ERG, DA 3.0 ERG, DA oscillatory potentials, light-adapted (LA) 3.0 ERG, and 30 Hz flicker ERG. Results: The delayed responses of healthy subjects compared to diabetic patients without DR were the DA oscillatory potentials (25.45 ± 1.04 ms vs. 26.15 ± 0.96 ms, $p = 0.027$). When comparing diabetic patients without DR and with non-proliferative DR, we did not obtain statistically significant delays. Significant delays in the DA 0.01 "b"-wave (61.91 ± 5.52 ms vs. 66.36 ± 8.12 ms, $p = 0.029$), DA 3.0 "b"-wave (41.01 ± 2.50 ms vs. 44.16 ± 3.78 ms, $p = 0.035$), and LA 3.0 "a"-wave (16.21 ± 0.91 ms vs. 16.99 ± 1.16 ms, $p = 0.045$) were found between non-proliferative DR and pre-proliferative DR. When comparing the groups of patients with pre-proliferative DR and proliferative DR, the LA 3.0 ERG "b"-wave (32. 63 ± 2.53 ms vs. 36.19 ± 3.21 ms, $p < 0.0001$), LA 30 Hz flicker ERG "a"-wave (19.56 ± 3.59 vs. 21.75 ± 4.74 ms, p= 0.025), and "b"-wave (32.23 ± 4.02 vs. 36.68 ± 3.48 ms, $p = 0.017$) were delayed. Conclusions: the electrophysiological findings from our study indicate that there is a substantial dysfunction of the neural retina in all stages of DR.

Keywords: full-field ERG; diabetes mellitus; diabetic retinopathy

Citation: Tănasie, C.A.; Dan, A.O.; Ică, O.M.; Mercuț, M.F.; Mitroi, G.; Taisescu, C.-I.; Sfredel, V.; Corbeanu, R.I.; Mocanu, C.L.; Danielescu, C. Retinal Functional Impairment in Diabetic Retinopathy. *Biomedicines* **2024**, *12*, 44. https://doi.org/10.3390/biomedicines12010044

Academic Editors: Ana Dascalu and Dragos Serban

Received: 19 October 2023
Revised: 4 December 2023
Accepted: 6 December 2023
Published: 22 December 2023

1. Introduction

Diabetic retinopathy (DR) is one of the most common causes of vision impairment worldwide, affecting one-third of people with diabetes mellitus [1]. The risk of developing DR is correlated with the duration of diabetes and, especially, with blood glucose levels [2]. The risk of blindness in patients with diabetes is 5.2 times higher than in individuals without diabetes [3]. DR affects the retina gradually and can remain unnoticed until the patient experiences severe vision reduction. This is why, once diabetes mellitus is diagnosed, screening for retinal complications is started in order to initiate the correct treatment on time, which consists of maintaining good glycemic control, administering laser photocoagulation, and administering an intravitreal injection with steroids and/or antibodies against a vascular endothelial growth factor [4]. Because it was thought that DR was only a microangiopathy, screening usually consists of morphological investigations, such as fundus photography, optical coherence tomography, and fluorescein angiography, which detect the presence of retinal lesions and aid in disease staging. But some new studies have shown that DR is a chronic neurodegenerative disease of the retina, as DR manifests a significant reduction in the thickness of the inner plexiform layer, inner nuclear layer, and overall retina [5]; increased apoptosis in retinal cells [6]; and glial cell activation [7].

An objective functional examination of the retina is obtained using electroretinography. A full-field electroretinogram is a non-invasive tool for the objective assessment of not only the retinal function but also the individual functionality of photoreceptors, bipolar cells, amacrine cells, and retinal ganglion cells [8,9]. Since clinical changes may not be correlated with the degree of functional retinal damage, the use of functional investigations, such as electroretinography, as a routine investigation, along with the morphological investigations, can provide a better understanding of disease progression. Also, the use of electroretinography could offer effective data for studies that target neural preservation in DR. The aim of our study was to analyze and compare ERG latency changes in patients with diabetes but without DR, patients with non-proliferative DR, patients with pre-proliferative DR, and patients with proliferative DR.

2. Materials and Methods

2.1. Subjects

A total of 60 type 2 diabetic patients and 14 age-matched controls were enrolled in our study. The patients were divided into 4 subgroups, according to their stage of retinopathy: 15 diabetic patients without DR, 16 diabetic patients with non-proliferative DR, 14 patients with pre-proliferative DR, and 15 patients with proliferative DR, according to the Wilkinson classification [10]. All the diabetic patients had a clinical diagnosis of type 2 diabetes mellitus of at least 10 years.

Subgroup 1 (control) included 28 eyes from 14 healthy subjects. The inclusion criteria were ages between 50 and 80, no diagnosis of diabetes mellitus, no history and no present signs of ophthalmological diseases, and clear optic media. The exclusion criteria were ages lower than 50 and higher than 80, diabetes mellitus, neurological diseases, and any ophthalmological disease.

Subgroup 2 included 30 eyes from 15 patients with type 2 diabetes mellitus without clinical signs of DR (DR-). The inclusion criteria were ages between 50 and 80, a clinical diagnosis of type 2 diabetes mellitus, no history of other ophthalmological diseases, normal findings from ophthalmological investigations, clear optic media, and no signs of DR from an ophthalmoscopic examination. The exclusion criteria were ages lower than 50 and higher than 80, neurological diseases, and ophthalmological diseases.

Subgroup 3 included 32 eyes from 16 patients with type 2 diabetes mellitus with non-proliferative diabetic retinopathy (NDR). The inclusion criteria were ages between 50 and 80; a clinical diagnosis of type 2 diabetes; no history of other ophthalmological diseases; signs of NDR: microaneurysms, retinal hemorrhages, venous loops, exudate or cotton wool spots, and venous beading; and no other ophthalmological diseases. The exclusion criteria were ages lower than 50 and higher than 80, neurological diseases, and ophthalmological diseases, except for NDR.

Subgroup 4 included 28 eyes from 14 patients with type 2 diabetes mellitus with pre-proliferative diabetic retinopathy (PPDR). The inclusion criteria were ages between 50 and 80; a clinical diagnosis of type 2 diabetes; no history of other ophthalmological diseases; clear optic media; signs of PPDR: microaneurysms, retinal hemorrhages, venous loops, exudate or cotton wool spots, venous beading, intraretinal hemorrhages, and intraretinal microvascular abnormalities; and no other ophthalmological diseases. The exclusion criteria were ages lower than 50 and higher than 80, neurological diseases, and ophthalmological diseases, except for PPDR.

Subgroup 5 included 30 eyes from 15 patients with type 2 diabetes mellitus with proliferative diabetic retinopathy (PDR). The inclusion criteria were ages between 50 and 80; clinical diagnosis of type 2 diabetes; no history of other ophthalmological diseases; clear optic media; signs of PDR: microaneurysms, retinal hemorrhages, venous loops, exudate or cotton wool spots, venous beading, intraretinal microvascular abnormalities, and new vessels on disc, new vessels elsewhere, preretinal or vitreous hemorrhages, preretinal fibrosis, and no other ophthalmological disease. The exclusion criteria were ages lower than 50 and higher than 80, neurological diseases, and ophthalmological diseases, except for PDR.

2.2. ERG Measurements

All subjects were recruited from the Ophthalmology Outpatient Clinic of Emergency County Hospital of Craiova, Romania, and the "Ocularius" Ophthalmological Research Centre in Craiova, Romania, where the ophthalmological examinations were performed. We recorded the ERGs at Ophthalmological Research Centre "Ocularius", Craiova, with MonPackOne System (Metrovision, Perenchies, France), according to the ISCEV standards [11]. We used HK loop ("Hawlina-Konec loop") electrodes as active electrodes and the Ag-AgCl cup type as reference and ground electrodes.

For each participant, the recording protocol consisted of the following:

A. Ophthalmological investigation: visual acuity, refraction, anterior segment examination by biomicroscopy, intraocular pressure, fundus examination, and color vision in order to diagnose the stage of diabetic retinopathy and to exclude other ophthalmological diseases.
B. ERG recording according to the ISCEV protocol:
    1. Mydriasis using 2.5% phenylephrine eye drops and 1% tropicamide;
    2. Skin preparation for the placement of the electrodes;
    3. Oxybuprocaine chlorhydrate in the lower conjunctival bag;
    4. Electrode settlement: we placed the active electrodes on the lower eyelid bilaterally, the reference electrodes on each orbital rim, and the ground electrode on the vertex;
    5. Twenty minutes of scotopic adaptation;
    6. Scotopic ERG recording: rod response: we used a dim blue flash of 0.01 cd.s.m^{-2}, with a 2 s interval between flashes; combined rod–cone response: we used a white flash of 3.0 cd.s.m^{-2}, with a 10 s interval between flashes; oscillatory potentials: we used a white flash of 2.0 cd.s.m^{-2}, with a 165 s interval between flashes;
    7. Ten minutes of photopic adaptation;
    8. Light-adapted ERG recording: single-flash cone response: we used a 3.0 cd.s.m^{-2} stimulus, with a background luminance of 30 cd.s.m^{-2} and a 0.5 s interval between flashes; 30 Hz flicker: 30 stimuli of 3.0 cd.s.m^{-2} per second.

We analyzed the implicit time for a- and b-waves for the DA 0.01, DA 3.0, LA 3.0, and LA 30 Hz flicker ERGs and implicit time for N1 and N2 for DA 3.0 oscillatory potentials (Figure 1). N1 and N2 waves are low-amplitude and high-frequency waves that are superimposed on the rising phase of the b-wave and show the function of the inner retina, especially for the amacrine and horizontal cells [11].

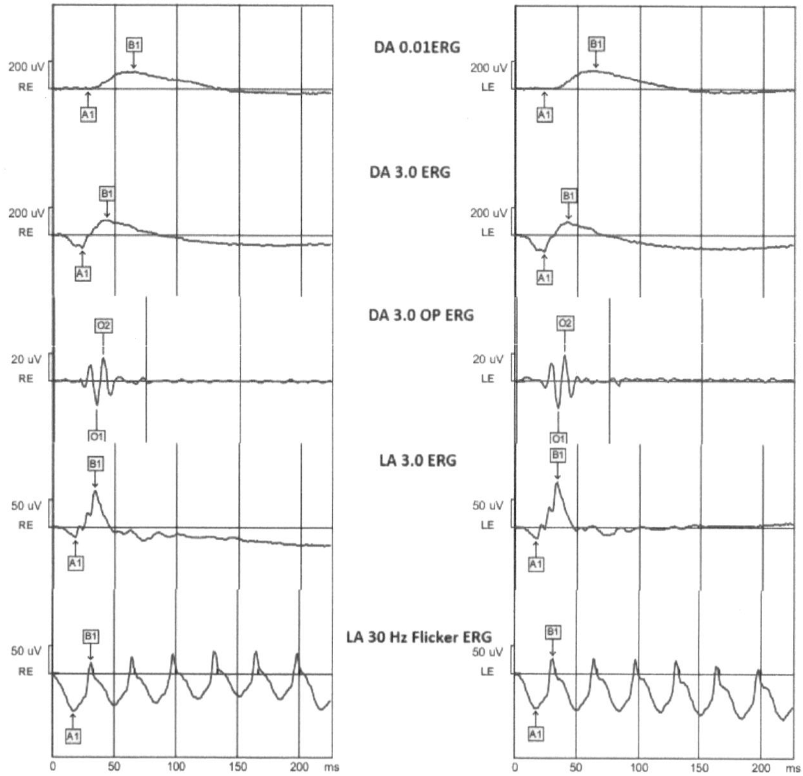

Figure 1. Standard ERG in a healthy individual. DA—dark adapted; LA—light adapted; RE—right eye; LE—left eye; A1—a-wave; B1—b-wave; O1—OP N1; O2—OP N2.

2.3. Data Analysis

The obtained data were processed using the Microsoft Excel program (Microsoft Corp., Redmond, WA, USA), with XLSTAT suite for MS Excel (Addinsoft SARL, Paris, France) and the Vision Monitor software Mon2016J integrated in the Metrovision MonPackOne system. The ERG responses were recorded using the MonPackOne program and embedded in the Metrovision MonPackOne acquisition system, which also performed the initial analysis of the signals, the automatic marking, and the manual correction of the landmarks necessary for the quantification of the parameters of the standard electroretinogram waves. All records were stored in the system database, from which they were exported as Excel files for each subject. The secondary processing of the data, the calculation of the fundamental statistical parameters (the mean and the standard deviation of their ratio—the coefficient of variation), and the comparison of the data were carried out by means of ANOVA test, post hoc Fisher LSD test, and Tukey's test, which is the most used multiple comparison test. Statistical significance was considered at $p < 0.05$.

3. Results

For each wave of all standard ERG responses, the mean implicit times and standard deviations were analyzed and compared (Table 1). In our research, we did not obtain statistically significant gender differences, so we analyzed both genders together, for each group.

Table 1. Mean implicit times and standard deviations for healthy subjects (control), patients with diabetes mellitus without DR (DR-), patients with diabetes mellitus and NDR (NDR), patients with diabetes mellitus and PPDR (PPDR), and patients with diabetes mellitus and PDR (PDR). a—a-wave; b—b-wave.

Parameter	CONTROL	DR-	NDR	PPDR	PDR
DA 0.01 ERG a	29.65 ± 2.29	30.34 ± 2.57	26.48 ± 3.36	27.16 ± 5.06	28.46 ± 3.70
DA 0.01 ERG b	61.23 ± 2.99	62.87 ± 2.99	61.91 ± 5.52	66.36 ± 8.12	68.62 ± 8.01
DA 3.0 ERG a	21.06 ± 2.92	20.49 ± 2.30	20.78 ± 2.60	22.06 ± 3.35	20.89 ± 3.04
DA 3.0 ERG b	42.84 ± 4.80	43.96 ± 6.19	41.01 ± 2.50	44.16 ± 3.78	46.53 ± 6.11
OP N1	21.96 ± 0.89	22.19 ± 1.02	22.48 ± 0.58	23.09 ± 1.06	22.98 ± 1.28
OP N2	25.45 ± 1.04	26.15 ± 0.96	26.08 ± 0.54	26.65 ± 1.05	26.56 ± 1.04
LA 3.0 ERG a	15.85 ± 1.26	15.81 ± 0.69	16.21 ± 0.91	16.99 ± 1.16	16.79 ±1.73
LA 3.0 ERG b	31.62 ± 1.75	32.73 ± 1.40	32.75 ± 3.76	32.63 ±2.53	36.19 ± 3.21
LA 30 Hz Flicker a	18.54 ± 2.50	18.36 ± 2.16	19.22 ± 4.85	19.56 ± 3.59	21.75 ± 4.74
LA 30 HZ Flicker b	30.99 ± 2.87	31.97 ± 2.28	33.63 ± 10.37	32.23 ± 4.02	36.68 ± 3.48

Due to the small sample size included in our study, we used two multiple comparison tests, Fisher's post hoc LSD test and Tukey's test, in order to avoid the errors that could occur and to obtain valid results.

For the DA 0.01 ERG a-wave, performing the ANOVA test revealed statistically significant differences (p ANOVA = 0.000) globally between the five analyzed subgroups. Consequently, Fisher's LSD post hoc test was used to detect the pairs of subgroups between which these differences were manifested (Table 2). Statistically significant differences were found between the control subgroup and diabetic patients with the NDR (p = 0.005) and the control subgroup and patients with PPDR (p = 0.022) (Figure 2).

Table 2. Fisher LSD post hoc test p values for the DA 0.01 a- and b-waves, DA 3.0 a- and b-waves, OP N1 and N2 waves, LA 3 a- and b-waves, LA 30 Hz flicker a- and b-waves for the comparisons between normal subjects (control), patients with diabetes mellitus without DR (DR-), patients with diabetes mellitus and NDR (NDR), patients with diabetes mellitus and PPDR (PPDR), and patients with diabetes mellitus and PDR (PDR). a—a-wave; b—b-wave.

Contrast	DA 0.01 a	DA 0.01 b	DA 3.0 a	DA 3.0 b	DA 3.0 OP N1	DA 3.0 OP N2	LA 3.0 a	LA 3.0 b	LA 30 a	LA 30 b
Control vs. NDR	0.005	0.663	0.746	0.183	0.064	0.045	0.272	0.126	0.756	0.076
Control vs. PPDR	0.022	0.006	0.231	0.357	0.0001	0.0001	0.002	0.163	0.557	0.269
Control vs. DR-	0.214	0.592	0.442	0.431	0.400	0.027	0.877	0.146	0.542	0.681
Control vs. PDR	0.255	<0.0001	0.831	0.009	0.002	0.001	0.009	<0.0001	0.004	0.0005
PDR vs. NDR	0.094	0.001	0.916	0.0003	0.143	0.152	0.142	<0.0001	0.013	0.074
PDR vs. PPDR	0.241	0.334	0.183	0.114	0.674	0.770	0.582	<0.0001	0.025	0.017
PDR vs. DR-	0.922	0.001	0.599	0.083	0.016	0.160	0.009	<0.0001	0.002	0.005
DR- vs. NDR	0.114	0.935	0.674	0.046	0.312	0.753	0.237	0.941	0.392	0.226
DR- vs. PPDR	0.279	0.0032	0.066	0.891	0.002	0.065	0.002	0.954	0.270	0.547
PPDR vs. NDR	0.629	0.029	0.152	0.035	0.043	0.053	0.045	0.895	0.793	0.519

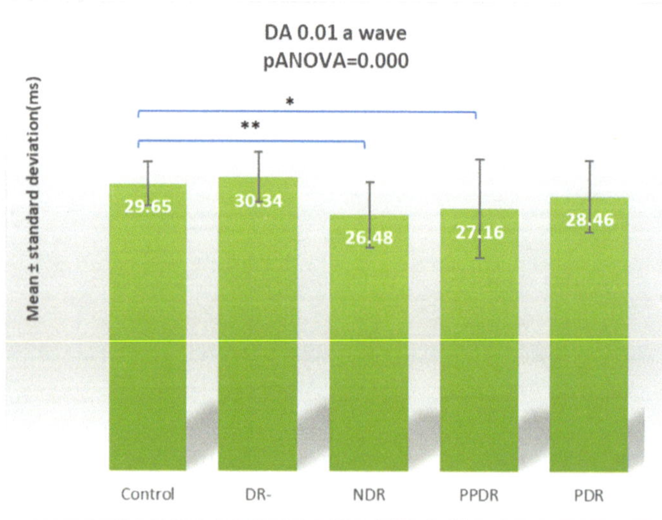

Figure 2. Comparison of mean values of l DA 0.01 a-wave latencies between normal subjects (control) and the subgroups of patients with diabetes mellitus without DR (DR-), patients with non-proliferative diabetic retinopathy (NDR), patients with pre-proliferative diabetic retinopathy (PPDR), and patients with proliferative diabetic retinopathy (PDR). * $p < 0.05$; ** $p < 0.01$.

For the DA 0.01 ERG b-wave, the ANOVA test revealed statistically significant differences (p ANOVA = 0.000) globally among the five subgroups. Consequently, Fisher's LSD post hoc test was used to detect the pairs of subgroups between which these differences were manifested (Table 2). Statistically significant differences were found between patients with PPDR and the control ($p = 0.006$), DR- ($p = 0.0032$), and NDR ($p = 0.029$) subgroups and between patients with PDR and the control ($p < 0.0001$), DR- ($p = 0.001$), and NDR ($p = 0.001$) subgroups (Figure 3).

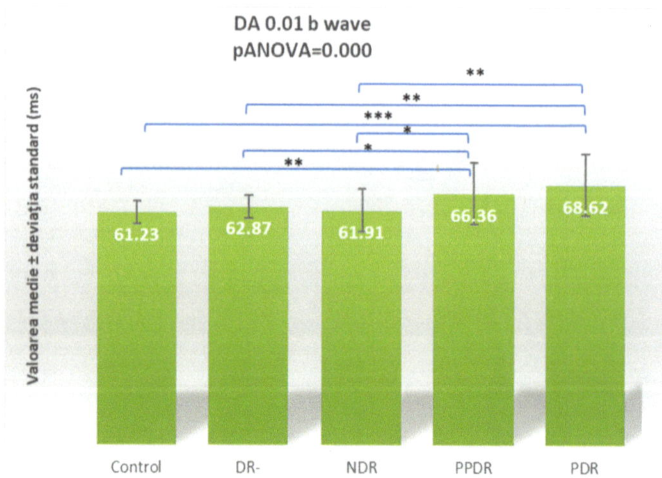

Figure 3. Comparison of mean values of DA 0.01 b-wave latencies between normal subjects (control) and the groups of patients with diabetes mellitus without DR (DR-), patients with non-proliferative diabetic retinopathy (NDR), patients with pre-proliferative diabetic retinopathy (PPDR), and patients with proliferative diabetic retinopathy (PDR). * $p < 0.05$; ** $p < 0.01$; *** $p < 0.001$.

According to Tukey's test, for the DA 0.01 a-wave, the statistically significant delays are between healthy subjects and patients with NDR ($p = 0.0405$). For the DA 0.01 b-wave, the significant delays are between healthy subjects and PPDR ($p = 0.0477$) and PDR ($p = 0.0009$) groups and between patients with PDR and DR- ($p = 0.0108$) and NDR ($p = 0.0108$) groups (Table 3).

Table 3. Tukey's test *p* values for the DA 0.01 a- and b-waves, DA 0.01 a- and b-waves, OP N1 and N2 waves, LA 3 a- and b-waves, and LA 30 Hz flicker a- and b-waves for the comparisons between normal subjects (control), patients with diabetes mellitus without DR (DR-), patients with diabetes mellitus and NDR (NDR), patients with diabetes mellitus and PPDR (PPDR), and patients with diabetes mellitus and PDR (PDR). a—a-wave; b—b-wave.

Contrast	DA 0.01 a	DA 0.01 b	DA 3.0 a	DA 3.0 b	DA 3.0 OP N1	DA 3.0 OP N2	LA 3.0 a	LA 3.0 b	LA 30 a	LA 30 b
Control vs. DR-	0.7223	0.9832	0.9386	0.9329	0.9151	0.9127	0.9999	0.1017	0.9728	0.9938
Control vs. NDR	0.0405	0.9923	0.9976	0.6663	0.3338	0.2596	0.8042	0.0858	0.9979	0.3778
Control vs. PPDR	0.1463	0.0477	0.7491	0.8868	0.0009	0.0008	0.0130	0.1155	0.9762	0.7986
Control vs. PDR	0.7820	0.0009	0.9995	0.0700	0.0145	0.0096	0.0654	<0.0001	0.0347	0.0041
DR- vs. NDR	0.5034	>0.9999	0.9933	0.2663	0.8468	0.7777	0.7573	>0.9999	0.9099	0.7380
DR- vs. PPDR	0.8128	0.1977	0.3459	>0.9999	0.0195	0.0179	0.0143	>0.9999	0.7999	0.9738
DR- vs. PDR	>0.9999	0.0108	0.9845	0.4089	0.1322	0.0956	0.0655	0.0012	0.0155	0.0371
NDR vs. PPDR	0.9886	0.1864	0.6018	0.2161	0.2475	0.2963	0.2592	>0.9999	0.9989	0.9663
NDR vs. PDR	0.4451	0.0108	>0.9999	0.0025	0.6388	0.6147	0.5790	0.0015	0.1029	0.3832
PPDR vs. PDR	0.7625	0.8677	0.6680	0.5061	0.9812	0.9952	0.9816	0.0010	0.1778	0.1185

For the DA 3.0 ERG a-wave, the ANOVA test revealed no statistically significant differences globally between the five analyzed subgroups ($p = 0.434$) (Table 2, Figure 4).

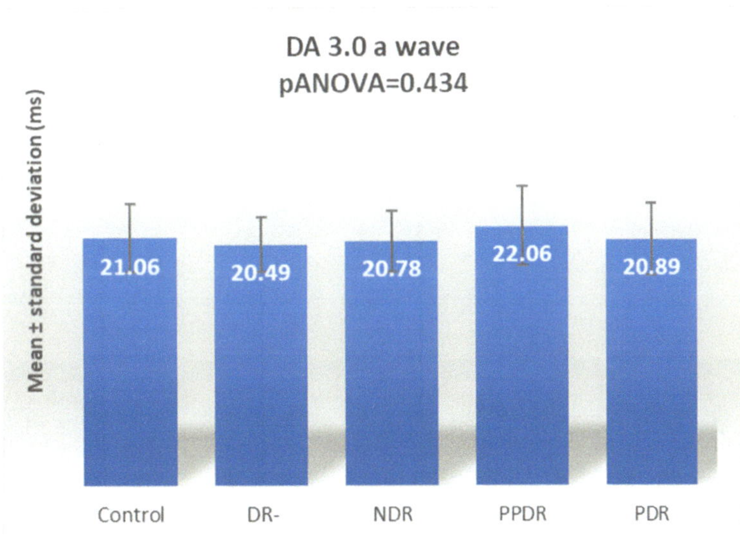

Figure 4. Comparison of mean values of DA 3.0 a-wave latencies between healthy subjects (control) and the subgroups of patients with diabetes mellitus without DR (DR-), patients with non-proliferative diabetic retinopathy (NDR), patients with pre-proliferative diabetic retinopathy (PPDR), and patients with proliferative diabetic retinopathy (PDR).

The ANOVA test for the DA 3.0 b-wave revealed statistically significant differences (p ANOVA = 0.006) globally between the five analyzed subgroups. By using the Fisher LSD post hoc test, statistically significant differences were found between patients with PDR and subjects from the control (p = 0.009) and NDR (p = 0.0003) subgroups and between patients with NDR and patients with PPDR (p = 0.035) (Table 2) (Figure 5).

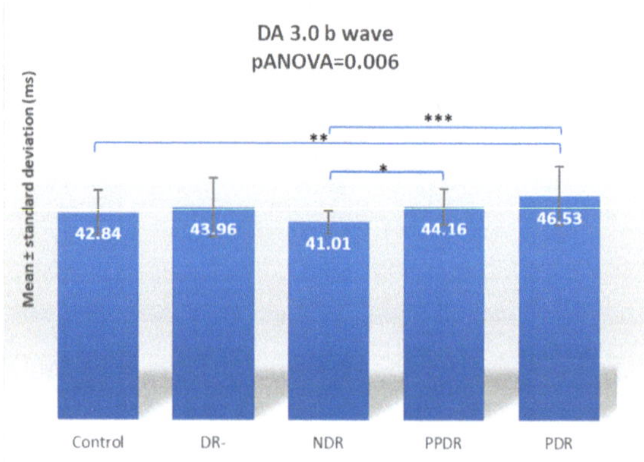

Figure 5. Comparison of mean values of DA 3.0 b-wave latencies between healthy subjects (control) and the groups of patients with diabetes mellitus without DR (DR-), patients with non-proliferative diabetic retinopathy (NDR), patients with pre-proliferative diabetic retinopathy (PPDR), and patients with proliferative diabetic retinopathy (PDR). * $p < 0.05$; ** $p < 0.01$; *** $p < 0.001$.

According to Tukey's test, for the DA 3.0 a-wave, there is no statistically significant delay between groups. For the DA 3 b-wave, the significant delay is between the NDR and PDR groups (p = 0.0025) (Table 3).

The oscillatory potentials (OPs) are characterized by a highly significant overall increase in latency for both the N1 and N2 waves, with ANOVA p values of 0.000 (Figures 6 and 7). The analysis of the five subgroups, using the Fisher LSD post hoc test, showed that for the N1 wave, significant differences were found between the PPDR and the control subgroups (p = 0.0001) and the DR- (p = 0.002) and NDR (p = 0.043) subgroups, as well as between the PDR and the control (p = 0.002) subgroups and DR- (p = 0.016) subgroups (Table 2).

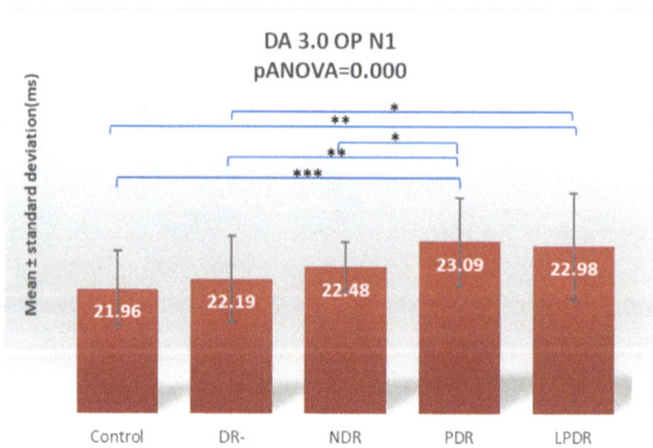

Figure 6. Comparison of mean values of DA 3.0 OP N1 wave latencies between healthy subjects (control) and the groups of patients with diabetes mellitus without DR (DR-), patients with non-proliferative diabetic retinopathy (NDR), patients with pre-proliferative diabetic retinopathy (PPDR), and patients with proliferative diabetic retinopathy (PDR). * $p < 0.05$; ** $p < 0.01$; *** $p < 0.001$.

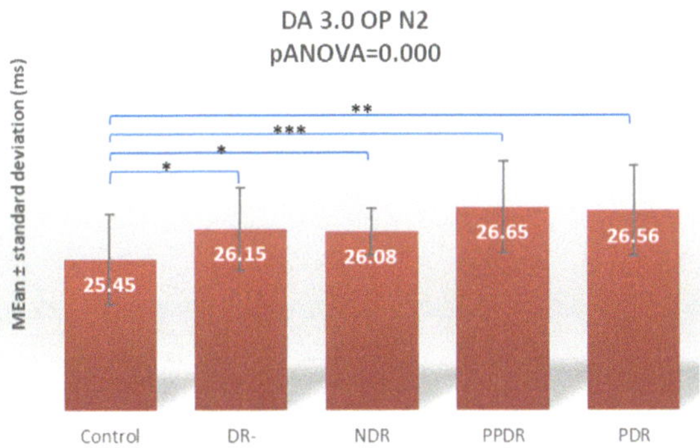

Figure 7. Comparison of mean values of DA 3.0 OP N2 wave latencies between healthy subjects (control) and the groups of patients with diabetes mellitus without DR (DR-), patients with non-proliferative diabetic retinopathy (NDR), patients with pre-proliferative diabetic retinopathy (PPDR), and patients with proliferative diabetic retinopathy (PDR). * $p < 0.05$; ** $p < 0.01$; *** $p < 0.001$.

For the N2 wave, the Fisher LSD post hoc test showed significant changes between the control subgroup and all four other groups: DR- ($p = 0.027$), NDR ($p = 0.045$), PPDR ($p = 0.0001$), and PDR ($p = 0.001$) (Table 2).

Performing Tukey's test for both N1 and N2 waves of DA 3OP, the statistically significantly increased latencies are between healthy subjects and PPDR ($p = 0.009$ and 0.008, respectively) and PDR ($p = 0.0145$ and 0.0096, respectively) groups and between DR- and PPDR groups ($p = 0.0195$ and 0.0179, respectively) (Table 3).

Analyzing the a- and b-waves in the standard LA 3.0 ERG recording, a significant change in latency for both the a-wave (p ANOVA = 0.002) and b-wave (p ANOVA = 0.000) were found (Figures 8 and 9). For the a-wave, this overall difference was due to significant increases in latency between the subgroup of subjects with PPDR versus healthy subjects ($p = 0.002$) and the subgroup with type 2 diabetes mellitus but without diabetic retinopathy ($p = 0.002$) and subjects with PDR versus subjects in the control ($p = 0.009$) and DR- ($p = 0.009$) subgroups (Figure 8). For the b-wave, the difference is due to the major increase in latency in the subgroup of subjects with PDR compared to the other four groups included in the study: control ($p < 0.0001$), DR- ($p < 0.0001$), NDR ($p < 0.0001$), and PPDR ($p < 0.0001$) (Table 2) (Figure 9).

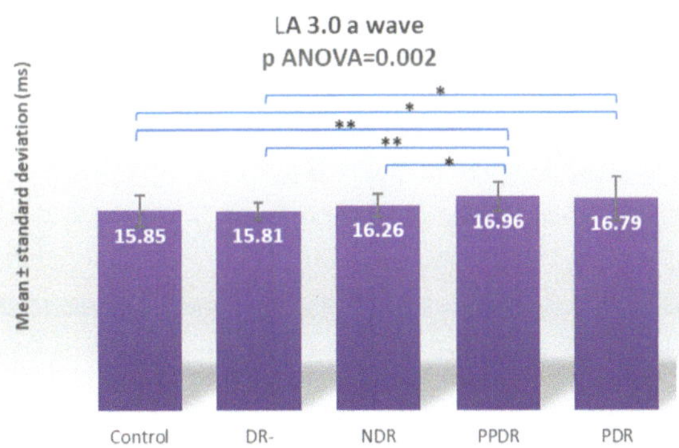

Figure 8. Comparison of mean values of LA 3.0 a-wave latencies between healthy subjects (control) and the groups of patients with diabetes mellitus without DR (DR-), patients with non-proliferative diabetic retinopathy (NDR), patients with pre-proliferative diabetic retinopathy (PPDR), and patients with proliferative diabetic retinopathy (PDR). * $p < 0.05$; ** $p < 0.01$.

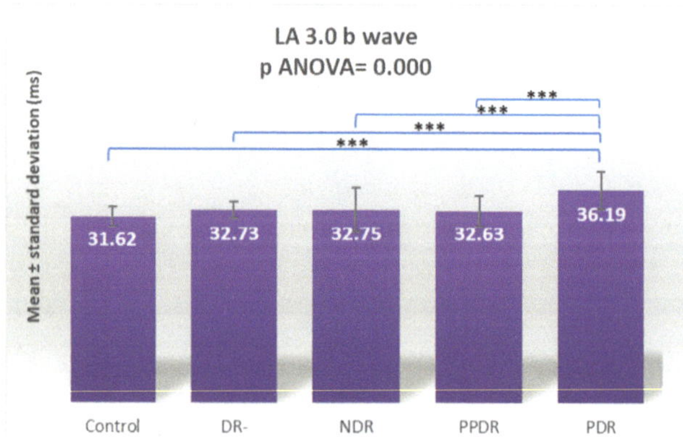

Figure 9. Comparison of mean values of LA 3.0 b-wave latencies between healthy subjects (control) and the groups of patients with diabetes mellitus without DR (DR-), patients with non-proliferative diabetic retinopathy (NDR), patients with pre-proliferative diabetic retinopathy (PPDR), and patients with proliferative diabetic retinopathy (PDR). *** $p < 0.001$.

Performing Tukey's test for the LA 3 a-wave, the significant delays are between PPDR patients and the control ($p = 0.0130$) and DR- ($p = 0.0143$) subjects. For the LA 3 b-wave, the delays are between PDR and all the other four groups: control ($p < 0.0001$), DR- ($p = 0.012$), NDR ($p = 0.0015$), and PPDR ($p = 0.0010$) (Table 3).

By analyzing the a-wave from the LA 30 Hz flicker ERG, we found a globally significant difference between the five study subgroups (p ANOVA = 0.005). The Fisher LSD post hoc test showed that this overall difference was due to increased latency in the subgroup of subjects with PDR compared to the control ($p = 0.004$), DR- ($p = 0.002$), NDR ($p = 0.013$), and PPDR ($p = 0.025$) (Table 2) (Figure 10).

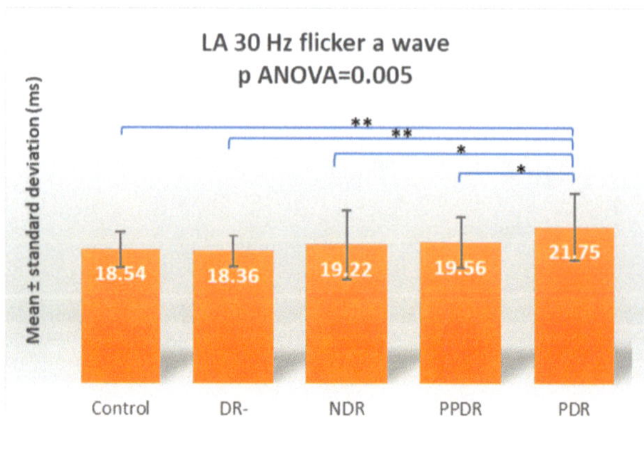

Figure 10. Comparison of mean values of LA 30 Hz flicker a-wave latencies between healthy subjects (control) and the groups of patients with diabetes mellitus without DR (DR-), patients with non-proliferative diabetic retinopathy (NDR), patients with pre-proliferative diabetic retinopathy (PPDR), and patients with proliferative diabetic retinopathy (PDR). * $p < 0.05$; ** $p < 0.01$.

For the b-wave, the globally significant difference (p ANOVA = 0.002) was due to the increased implicit time between subjects from the PDR subgroup and those in the control ($p = 0.0005$), DR- ($p = 0.005$), and PPDR ($p = 0.017$) subgroups (Table 2) (Figure 11).

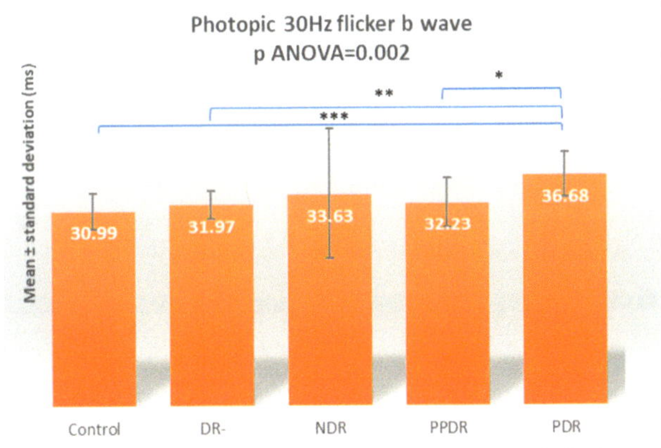

Figure 11. Comparison of mean values of LA 30 Hz flicker b-wave latencies between healthy subjects (control) and the groups of patients with diabetes mellitus without DR (DR-), patients with non-proliferative diabetic retinopathy (NDR), patients with pre-proliferative diabetic retinopathy (PPDR), and patients with proliferative diabetic retinopathy (PDR). * $p < 0.05$; ** $p < 0.01$; *** $p < 0.001$.

According to Tukey's test, for 30 Hz flicker, both a- and b-waves have statistically significant delays between PDR and control ($p = 0.0347$ and 0.0041, respectively) and DR- ($p = 0.0155$ and 0.0371, respectively) groups (Table 3).

4. Discussion

The objectives of this study were to observe the electroretinographic evolution of DR and to obtain a comparison between the various stages of DR regarding retinal functional aspects, while excluding changes due to age, by comparing the diabetic patients with a subgroup of subjects similar in age.

All parameters analyzed in our study showed a statistically significant global delay, except for the a-wave from the Scotopic 3.0 recording of the standard electroretinogram, which assesses the mixed rod–cone response.

The first electroretinographic change identified is the increase in the implicit time of the second wave of OP in patients with type 2 diabetes without diabetic retinopathy compared to the healthy subjects. In our study between healthy subjects and those with diabetes without diabetic retinopathy, we did not find other statistically significant changes.

The results of our research correlate with the majority of studies that claim that OPs represent the most sensitive electroretinographic indicator in diabetic eye disease [12–16], demonstrating an early impairment of neuronal synaptic activity of amacrine and horizontal retinal cells.

Yonemura was the first to show that OPs are impaired in more than 50% of patients with diabetes without DR [17]. Holopigian showed that the latency of OP was not significantly altered in diabetics compared to healthy subjects; only the amplitude was decreased, indicating that the OPs are the first parameter impaired in the ERG [18]. In a study on rats, in which diabetes was induced by an injection of streptozotocin, Hancock demonstrated a relationship between OP and the appearance and evolution of DR lesions [19]. More recently, Gualtieri also showed OP abnormalities in 82.95% of subjects with diabetes mellitus without DR [20].

In subjects with NDR, we did not obtain statistically significant delays, but we observed a slight increase in the a-wave of the rod–cone response and a-wave of the cone response. As is known, the scotopic a-wave reflects the activity of the photoreceptors, in our case both rods and cones, but also has postreceptoral contributions [21]. Hanock showed a delay in DA 3.0 a-wave in diabetic patients without DR in his study [19]. In our research, this change appeared later in DR evolution in patients with NDR, showing a slightly decreased activity of rods [4,21]. We also obtained a discrete delay of the response of cone cells. Usually, cone damage occurs in the more advanced stages of diabetic retinopathy; however, Gualtieri et al. have found evidence of damage to cone cells in the early stages [20]. These small differences in ERG parameters between DR- and NDR could be due to the small sample size included in our study but also due to variability in patient populations.

Our results are consistent with other studies that found dark-adapted ERG delays in the early stages of DR [22,23]. The inferred pathway for this ERG change is the increased metabolic demand and high oxidative stress of the diabetic retina under scotopic conditions [24,25].

In the group of subjects with PPDR, the main increases in latency were the b-wave of the rod cell response and the mixed rod–cone response and the "a"-wave of the cone cell response. Thus, as the ischemia progressed, we observed a marked alteration of the scotopic system with damage to the internal retinal layers but also damage to the photoreceptors in the photopic system.

Comparing our results with those of the study published by Luu et al., which included 18 patients with diabetes without DR and 10 patients with NDR, both studies show that the photopic system is not affected in the early stages of DR. On the other hand, Luu's study showed the impairment of the b-wave in the scotopic system before the onset of DR and NDR; in our research, the b-wave shows a significant increase in latency, with the appearance of a pre-proliferative stage [24]. Both studies have a small number of participants and the same ERG recording protocol, so this difference could be due to patients' variability and statistical methods.

Overall, with the progress of ischemia, our research showed minimal cone cell damage, in addition to marked alteration of the scotopic system. This fact is in agreement with other studies that reported cone pathway abnormalities in subjects with advanced DR [26–28].

In the group of subjects with PDR, the affected waves were the LA 3.0 b-wave and LA 30 Hz flicker a- and b-waves, showing a severe impairment of the photopic system.

The flicker stimulation assesses neurovascular coupling, that is, the ability to adjust blood flow in the microcirculation of the neuronal retina in response to neuronal activity. The pathway involved in flicker stimulation changes is that an increased frequency of the light stimulus induces greater activity in nervous cells, with increased release of nitric oxide and other vasodilatory substances, increasing the microvascular response [29,30].

In fact, two mechanisms contribute to the basis of this altered response: firstly, the alteration of the function of photoreceptors and retinal neurons and, secondly, the impairment of the retinal microcirculation, creating a vicious circle [31–33].

In our research, the flicker response was delayed in subjects with PDR, a fact that is in agreement with other studies that report an increased implicit time in diabetics with advanced stages of DR [34–38] and are normal or minimally affected in early stages [39–43].

By using Tukey's test, we obtained only a significant delay of the scotopic a-wave for patients with NDR compared to healthy subjects, also showing a slight impairment of rods in early stages of DR. With the disease progression to a pre-proliferative stage, we also obtained a delay of the b-wave within the rod pathway and the a-wave within the cone system. We obtained a different result regarding oscillatory potentials—these being delayed in patients with PPDR. For the PDR group, we also obtained a severe impairment of cone pathway, through a significant delay of the LA 3 a-wave and LA 30 Hz flicker a- and b-waves.

Research over the past 20 years has shown that normal or slightly delayed a- and b-waves are consistently presented during the early stages of DR. However, studies examining more advanced DR stages or combinations of subjects across multiple DR stages have reported delayed b-waves. There is a lack of large-scale clinical studies of electroretinography conducted in diabetic patients able to clearly define the disease stages. Further research is necessary to clarify questions regarding the extent to which the a- and b-waves are abnormal in diabetic individuals who have different disease severity.

The correspondence in clinical practice of the ERG for diabetic patients is related to quantifying neural dysfunction prior to the appearance of the first signs of retinopathy, which would help set individual goals for better glycemic control. Further research is necessary to determine whether diabetic patients who have ERG abnormalities are more likely to progress in a more advanced stage and therefore adjust the optimal individualized follow-up intervals for diabetic patients without clinically apparent retinopathy. Also, patients with flicker ERG abnormalities may show a risk of developing pre-proliferative or proliferative DR. This marker could indicate the need to decrease the duration between the usual check-ups and initiate laser therapy as soon as the neovascularization appears. Thus, electroretinography can strengthen the relationship between the ophthalmologist and the diabetologist in the sense of individualized therapy in correlation not only with glycemic values but also with preclinical retinal complications.

Considering that electroretinography is a technique not very often used in diabetic ophthalmological pathology and that most studies [3,44] evaluated electroretinographic changes, especially in the early stages, we can say that this research is one of the few studies to have followed the evolution of retinal function impairment in one of the most common complications of diabetes. Another important feature of our study is the highlighting of the different effects on rod cells versus cone cells as the disease progresses.

Based on the ERG changes obtained in our research, as the DR progresses, the following elements are affected: amacrine and horizontal cells, photoreceptors from the scotopic system followed by the inner retinal layers of the rods pathway, photoreceptors from the photopic system, and ON and OFF bipolar cells within the cone pathway. The early ERG sign that we have signaled in patients with type 2 diabetes without DR is the increase in the implicit time of the second wave of OP. In subjects with NDR, we did not obtain statistically significant delays, but we observed a slight increase in the "a"-wave of the rod–cone response and the "a"-wave of the cone response. There was a significant increase in the latency of the dark-adapted 0.01 "b"-wave and 3.0 "b"-wave from the early to the late stages of DR. In the group of subjects with PDR, the affected waves were the LA 3.0 b-wave and LA 30 Hz flicker "a"- and "b"-waves, showing a severe impairment of the photopic system.

A limitation of the current study could be the small sample size for each group: 14 participants for the DR- group, 15 for the NDR group, and 16 and 15 patients for the PPDR and PDR groups, respectively. It is well known that studies with a small sample size may not be able to detect a proper difference between groups due to inaccurate statistics leading to erroneous conclusions [45]. Large sample size studies could offer more accurate data regarding inferential statistics. We tried to neutralize the errors that could appear due to small sample size by using two types of multiple comparison tests, and we obtained similar results. Also, we analyzed a parameter, the implicit time of all standard ERG waves, which is stable compared to the amplitude of the waves, which has a higher variability. It is considered that studies with a small sample size are required to evaluate stable parameters [45]. Even if the small number of subjects could be a limitation of our study, due to the fact that we investigated a stable parameter and we used two multiple comparison tests, we consider our results and conclusions accurate.

Another limitation is that the participants were not monitored over a longer period of time through electroretinography so that we could evaluate the retinal progression of the disease. Although this investigation is non-invasive, the patients show some reluctance to repeat this investigation due to the increased examination time, approximately one hour, as well as slight eye discomfort due to the induced mydriasis and to the introduction of the electrodes into the conjunctival sac.

5. Conclusions

The standard ERG is an objective test able to assess global retinal dysfunction, giving us important insights into the neural effects of diabetes mellitus on the retinal cells that may not be similar to the clinically visible lesions. Our study showed that, as DR progresses, the following elements are affected: amacrine and horizontal cells and the rod pathway followed by the cone pathway. The obtained results support the idea that electroretinography is a useful technique that can be used not only in research but also in current practice and could help to set individual goals for better control of the disease.

Author Contributions: Conceptualization, C.A.T. and C.D.; methodology, C.A.T. and M.F.M.; software, O.M.I. and M.F.M.; validation, C.L.M., G.M. and V.S.; investigation, C.A.T. and A.O.D.; resources, C.-I.T. and C.A.T.; data curation, C.A.T., G.M. and O.M.I.; writing—original draft preparation, C.A.T. and M.F.M.; writing—review and editing, R.I.C. and C.D.; visualization, C.A.T. and O.M.I.; supervision, C.D and C.L.M.; project administration, C.A.T. and C.D. All authors have read and agreed to the published version of the manuscript.

Funding: This research received no external funding.

Institutional Review Board Statement: This study was conducted in accordance with the Declaration of Helsinki and approved in 9 May 2022, by the Ethics Committee of the University of Medicine and Pharmacy of Craiova, No. 81/09.05.2022.

Informed Consent Statement: Informed consent was obtained from all subjects involved in this study.

Data Availability Statement: The authors declare that the data for this research are available from the corresponding authors upon reasonable request.

Conflicts of Interest: The authors declare no conflict of interest.

References

1. Flaxman, S.R.; Bourne, R.R.A.; Resnikoff, S.; Ackland, P.; Braithwaite, T.; Cicinelli, M.V.; Das, A.; Jonas, J.B.; Keeffe, J.; Kempen, J.H.; et al. Global causes of blindness and distance vision impairment 1990–2020: A systematic review and meta-analysis. *Lancet Glob. Health* **2017**, *5*, 1221–1234. [CrossRef] [PubMed]
2. Engelgau, M.M.; Geiss, L.S.; Saaddine, J.B.; Boyle, J.P.; Benjamin, S.M.; Gregg, E.W.; Tierney, E.F.; Rios-Burrows, N.; Mokdad, A.H.; Ford, E.S.; et al. The Evolving diabetes burden in the United States. *Ann. Intern. Med.* **2004**, *140*, 945–950. [CrossRef] [PubMed]
3. Tyrberg, M.; Lindblad, U.; Melander, A.; Lövestam-Adrian, M.; Ponjavic, V.; Andréasson, S. Electrophysiological studies in newly onset type 2 diabetes without visible vascular retinopathy. *Doc. Ophthalmol.* **2011**, *123*, 193–198. [CrossRef] [PubMed]
4. Tonadea, D.; Kern, T.S. Photoreceptor cells and RPE contribute to the development of diabetic retinopathy. *Prog. Retin. Eye Res.* **2021**, *83*, 1009–1019. [CrossRef] [PubMed]
5. Barber, A.J.; Antonetti, D.A.; Kern, T.S.; Reiter, C.E.N.; Soans, R.S.; Krady, J.K.; Levison, S.W.; Gardner, T.W.; Bronson, S.K. The Ins2Akita mouse as a model of early retinal com-plications in diabetes. *Investig. Ophthalmol. Vis. Sci.* **2005**, *46*, 2210–2218. [CrossRef] [PubMed]
6. Barber, A.J.; Gardner, T.W.; Abcouwer, S.F. The significance of vascular and neural apoptosis to the pathology of diabetic retinopathy. *Investig. Ophthalmol. Vis. Sci.* **2011**, *52*, 1156–1163. [CrossRef] [PubMed]
7. Villarroel, M.; Ciudin, A.; Hernández, C.; Simó, R. Neurodegeneration: An early event of diabetic retinopathy. *World J. Diabetes* **2010**, *15*, 257–264. [CrossRef]
8. McAnany, J.J.; Persidina, O.S.; Park, J.C. Clinical electroretinography in diabetic retinopathy: A review. *Surv. Ophthalmol.* **2021**, *67*, 712–722. [CrossRef]
9. Ştefănescu-Dima, A.; Corîci, C.A.; Mănescu, M.R.; Sas, T.N.; Iancău, M.; Mocanu, C.L. Posterior vitreous detachment and macular anatomical changes—A tomographic-electroretinographic study. *Rom. J. Morphol. Embryol.* **2016**, *57* (Suppl. 2), 751–758.
10. Wilkinson, C.P.; Ferris FL 3rd Klein, R.E.; Lee, P.P.; Agardh, C.D.; Davis, M.; Dills, D.; Kampik, A.; Pararajasegaram, R.; Verdaguer, J.T. Proposed international clinical diabetic retinopathy and diabetic macular edema disease severity scales. *Ophthalmology* **2003**, *110*, 1677–1682. [CrossRef]
11. McCulloch, D.L.; Marmor, M.F.; Brigell, M.G.; Hamilton, R.; Holder, G.E.; Tzekov, R.; Bach, M. ISCEV Standard for full-field clinical electroretinography (2015 update). *Doc. Ophthalmol.* **2014**, *130*, 1–12. [CrossRef]
12. Bresnick, G.H.; Palta, M. Predicting progression to severe proliferative diabetic retinopathy. *Arch. Ophthalmol.* **1987**, *105*, 810–814. [CrossRef] [PubMed]
13. Juen, S.; Kieselbach, G.F. Electrophysiological changes in juvenile diabetics without retinopathy. *Arch. Ophthalmol.* **1990**, *108*, 372–375. [CrossRef] [PubMed]
14. Yoshida, A.; Kojima, M.; Ogasawara, H.; Ishiko, S. Oscillatory potentials and permeability of the blood-retinal barrier in noninsulin-dependent diabetic patients without retinopathy. *Ophthalmology* **1991**, *98*, 1266–1271. [CrossRef] [PubMed]
15. Vadala, M.; Anastasi, M.; Lodato, G.; Cillino, S. Electroretinographic oscillatory potentials in insulin-dependent diabetes pa-tients: A long-term follow-up. *Acta Ophthalmol. Scand.* **2002**, *80*, 305–309. [CrossRef] [PubMed]
16. Bresnick, G.H.; Palta, M. Oscillatory Potential Amplitudes. *Arch. Ophthalmol.* **1987**, *105*, 929–933. [CrossRef] [PubMed]

17. Yonemura, D.; Aoki, T.; Tsuzuki, K. Electroretinogram in Diabetic Retinopathy. *Arch. Ophthalmol.* **1962**, *68*, 19–24. [CrossRef] [PubMed]
18. Holopigian, K.; Seiple, W.; Lorenzo, M.; Carr, R. A comparison of photopic and scotopic electroretinographic changes in early diabetic retinopathy. *Investig. Ophthalmol. Vis. Sci.* **1992**, *33*, 2773–2780.
19. Hancock, H.A.; Kraft, T.W. Oscillatory potential analysis and ergs of normal and diabetic rats. *Investig. Ophthalmol. Vis. Sci.* **2004**, *45*, 1002–1008. [CrossRef]
20. Gualtieri, M.; Feitosa-Santana, C.; Lago, M.; Nishi, M.; Ventura, D.F. Early visual changes in diabetic patients with no retinopathy measured by color discrimination and electroretinography. *Psychol. Neurosci.* **2013**, *6*, 227–234. [CrossRef]
21. Jamison, J.; Bush, R.; Lei, B.; Sieving, P. Characterization of the rod photoresponse isolated from the dark-adapted primate ERG. *Vis. Neurosci.* **2001**, *18*, 445–455. [CrossRef] [PubMed]
22. Kim, M.; Kim, R.; Park, W.; Kim, I.; Park, Y. Electroretinography and retinal microvascular changes in type 2 diabetes. *Acta Ophthalmol.* **2020**, *98*, E807–E813. [CrossRef] [PubMed]
23. Luu, C.D.; Szental, J.A.; Lee, S.-Y.; Lavanya, R.; Wong, T.Y. Correlation between retinal oscillatory potentials and retinal vascular caliber in type 2 diabetes. *Investig. Ophthalmol. Vis. Sci.* **2010**, *51*, 482–486. [CrossRef] [PubMed]
24. Arden, G.B.; Sidman, R.L.; Arap, W.; Schlingemann, R.O. Spare the rod and spoil the eye. *Br. J. Ophthalmol.* **2005**, *89*, 764–769. [CrossRef] [PubMed]
25. Arden, G.B.; Sivaprasad, S. Hypoxia and oxidative stress in the causation of diabetic retinopathy. *Curr. Diabetes Rev.* **2011**, *7*, 291–304. [CrossRef] [PubMed]
26. Chen, H.; Zhang, M.; Huang, S.; Wu, D. The photopic negative response of flash ERG in nonproliferative diabetic retinopathy. *Doc. Ophthalmol.* **2008**, *117*, 129–135. [CrossRef] [PubMed]
27. Jansson, R.W.; Raeder, M.B.; Krohn, J. Photopic full-field electroretinography and optical coherence tomography in type 1 diabetic retinopathy. *Graefes Arch. Clin. Exp. Ophthalmol.* **2015**, *253*, 989–997. [CrossRef] [PubMed]
28. Dan, A.O.; Ștefănescu-Dima, A.; Bălășoiu, A.T.; Puiu, I.; Mocanu, C.L.; Ionescu, M.; Tănasie, A.C.; Târtea, A.E.; Sfredel, V. Early Retinal Microvascular Alterations in Young Type 1 Diabetic Patients without Clinical Retinopathy. *Diagnostics* **2023**, *13*, 1648. [CrossRef]
29. Mandecka, A.; Dawczynski Blum, J.M. Influence of flickering light on the retinal vessels in diabetic patients. *Diabetes Care* **2007**, *30*, 3048–3052. [CrossRef]
30. Lecleire-Collet, A.; Audo, I.; Aout, M.; Girmens, J.-F.; Sofroni, R.; Erginay, A.; Le Gargasson, J.-F.; Mohand-Saïd, S.; Meas, T.; Guillausseau, P.-J.; et al. Evaluation of retinal function and flicker light-induced retinal vascular response in normotensive patients with diabetes without retinopathy. *Investig. Ophthalmol. Vis. Sci.* **2011**, *52*, 2861–2867. [CrossRef]
31. Fukuo, M.; Kondo, M.; Hirose, A.; Fukushima, H.; Ikesugi, K.; Sugimoto, M.; Kato, K.; Uchigata, Y.; Kitano, S. Screening for diabetic retinopathy using new mydriasis-free, full-field flicker ERG recording device. *Sci. Rep.* **2016**, *6*, 36591. [CrossRef] [PubMed]
32. Garhofer, G.; Zawinka, C.; Resch, H.; Kothy, P.; Schmetterer, L.; Dorner, G.T. Reduced response of retinal vessel diameters to flicker stimulation in patients with diabetes. *Br. J. Ophthalmol.* **2004**, *88*, 887–891. [CrossRef] [PubMed]
33. Scimone, C.; Bramanti, P.; Ruggeri, A.; Donato, L.; Alafaci, C.; Crisafulli, C.; Mucciardi, M.; Rinaldi, C.; Sidoti, A.D.; Angelo, R. CCM3/SERPINI1 bidirectional promoter variants in patients with cerebral cavernous malformations: A molecular and func-tional study. *BMC Med. Genet.* **2016**, *17*, 74. [CrossRef] [PubMed]
34. Maa, A.Y.; Feuer, W.J.; Davis, C.Q.; Pillow, E.K.; Brown, T.D.; Caywood, R.M.; Chasan, J.E.; Fransen, S.R. A novel device for accurate and efficient testing for vision-threatening diabetic retinopathy. *J. Diabetes Its Complicat.* **2015**, *30*, 524–532. [CrossRef] [PubMed]
35. Zeng, Y.; Cao, D.; Yang, D.; Zhuang, X.; Hu, Y.; He, M.; Yu, H.; Wang, J.; Yang, C.; Zhang, L. Retinal vasculature–function correlation in non-proliferative diabetic retinopathy. *Doc. Ophthalmol.* **2019**, *140*, 129–138. [CrossRef] [PubMed]
36. Danielescu, C.; Moraru, A.D.; Anton, N.; Bilha, M.I.; Donica, V.C.; Darabus, D.M.; Munteanu, M.; Stefanescu-Dima, A.S. The Learning Curve of Surgery of Diabetic Tractional Retinal Detachment-A Retrospective, Comparative Study. *Medicina* **2022**, *59*, 73. [CrossRef] [PubMed]
37. Zeng, Y.; Cao, D.; Yang, D.; Zhuang, X.; Yu, H.; Hu, Y.; Zhang, Y.; Yang, C.; He, M.; Zhang, L. Screening for diabetic retinopathy in diabetic patients with a mydriasis-free, full-field flicker electroretinogram recording device. *Doc. Ophthalmol.* **2019**, *140*, 211–220. [CrossRef]
38. Scimone, C.; Bramanti, P.; Ruggeri, A.; Katsarou, Z.; Donato, L.; Sidoti, A.; D'angelo, R. Detection of Novel Mutation in Ccm3 Causes Familial Cerebral Cavernous Malformations. *J. Mol. Neurosci.* **2015**, *57*, 400–403. [CrossRef]
39. McAnany, J.J.; Park, J.C.; Chau, F.Y.; Leiderman, Y.I.; Lim, J.I.; Blair, N.P. Amplitude loss of the high-frequency flicker electroretinogram in early diabetic retinopathy. *Retina* **2019**, *39*, 2032–2039. [CrossRef]
40. McAnany, J.J.; Chen, Y.F.; Liu, K.; Park, J.C. Nonlinearities in the flicker electroretinogram: A tool for studying retinal dys-function applied to early-stage diabetic retinopathy. *Vision. Res.* **2019**, *161*, 1–11. [CrossRef]
41. McAnany, J.J.; Park, J.C. Temporal Frequency Abnormalities in Early-Stage Diabetic Retinopathy Assessed by Electroretino-graphy. *Investig. Ophthalmol. Vis. Sci.* **2018**, *59*, 4871–4879. [CrossRef]
42. McAnany, J.J.; Park, J.C. Cone photoreceptor dysfunction in early-stage diabetic retinopathy: Association between the acti-vation phase of cone phototransduction and the flicker electroretinogram. *Investig. Ophthalmol. Vis. Sci.* **2019**, *60*, 64–72. [CrossRef]

43. Scimone, C.; Donato, L.; Alafaci, C.; Granata, F.; Rinaldi, C.; Longo, M.; D'angelo, R.; Sidoti, A. High-Throughput Sequencing to Detect Novel Likely Gene-Disrupting Variants in Pathogenesis of Sporadic Brain Arteriovenous Malformations. *Front. Genet.* **2020**, *11*, 146. [CrossRef]
44. Arden, G.B.; Sivaprasad, S. The pathogenesis of early retinal changes of diabetic retinopathy. *Doc. Ophthalmol.* **2012**, *124*, 15–26. [CrossRef]
45. Indrayan, A.; Mishra, A. The importance of small samples in medical research. *J. Postgrad. Med.* **2021**, *67*, 219–223. [CrossRef]

biomedicines

MDPI

Article

The Multi-Kinase Inhibitor RepSox Enforces Barrier Function in the Face of Both VEGF and Cytokines

Lina Lietuvninkas [1], Basma Baccouche [1] and Andrius Kazlauskas [1,2,*]

[1] Department of Ophthalmology and Visual Sciences, University of Illinois at Chicago, Chicago, IL 60612, USA; lel2@uic.edu (L.L.)
[2] Department of Physiology and Biophysics, University of Illinois at Chicago, Chicago, IL 60612, USA
* Correspondence: ak20@uic.edu

Abstract: The therapeutic benefit provided by anti-vascular endothelial growth factor (VEGF) for patients with vision-threatening conditions such as diabetic retinopathy (DR) demonstrates the important role of VEGF in this affliction. Cytokines, which can be elevated in the vitreous of patients with DR, promote leakage of retinal blood vessels, and may also contribute to pathology, especially in those patients for whom anti-VEGF does not provide adequate benefit. In this in vitro study using primary human retinal endothelial cells, we compared anti-VEGF with the (transforming growth factor beta) TGFβ receptor inhibitor RepSox (RS) for their ability to enforce barrier function in the face of VEGF, cytokines, and the combination of both. RS was superior to anti-VEGF because it prevented permeability in response to VEGF, cytokines, and their combination, whereas anti-VEGF was effective against VEGF alone. The inhibitory effect of RS was associated with suppression of both agonist-induced pore formation and disorganization of adherens junctions. RS-mediated inhibition of the TGFβ pathway and increased expression of claudin-5 did not adequately explain how RS stabilized the endothelial cell barrier. Finally, RS not only prevented barrier relaxation, but also completely or partially reclosed a barrier relaxed with tumor necrosis factor α (TNF α) or VEGF, respectively. These studies demonstrate that RS stabilized the endothelial barrier in the face of both cytokines and VEGF, and thereby identify RS as a therapeutic that has the potential to overcome permeability driven by multiple agonists that play a role in the pathology of DR.

Keywords: RepSox; endothelial cells; permeability; diabetes; retinopathy; anti-VEGF; cytokines

Citation: Lietuvninkas, L.; Baccouche, B.; Kazlauskas, A. The Multi-Kinase Inhibitor RepSox Enforces Barrier Function in the Face of Both VEGF and Cytokines. *Biomedicines* **2023**, *11*, 2431. https://doi.org/10.3390/biomedicines11092431

Academic Editors: Ana Dascalu and Dragos Serban

Received: 30 June 2023
Revised: 11 August 2023
Accepted: 17 August 2023
Published: 31 August 2023

1. Introduction

Diabetic retinopathy (DR) is a complication that can develop in the eyes of patients with diabetes mellitus (DM) [1]. The diabetic milieu within the circulation damages the retinal vasculature and thereby increases the vitreal concentration of growth factors (such as vascular endothelial growth factor (VEGF)) and cytokines (such as tumor necrosis factor α (TNFα) and interleukin-1 beta (IL-1β)). Elevated levels of such agents drive blood–retinal barrier (BRB) breakdown and neovascularization—diagnostic features of advanced DR [2].

While intensive blood glucose management may delay the onset of DR, most patients with DM eventually develop DR [3]. Reducing the vitreal level of VEGF by intravitreal administration of anti-VEGFs has become the standard of care for patients with DR [4]. Not all patients experience an adequate therapeutic benefit from anti-VEGF treatment, perhaps because of elevated levels of cytokines, which are unaffected by anti-VEGFs [2,3]. Antagonizing both VEGF and cytokines is expected to be effective for a greater percentage of patients with DR.

RepSox (RS) is a small-molecule multi-kinase inhibitor that prevents VEGF-induced permeability [5]. Because one of the RS targets is TGF-βR1 [6], and the TGFβ family governs vascular homeostasis [7], it is possible that TGF-βR1 is an essential mediator of VEGF-induced permeability. However, not all TGF-βR1 inhibitors prevent VEGF-driven barrier relaxation [5], which suggests that the relevant RS target is a kinase other than TGF-βR1.

In this report we investigated the therapeutic potential of RS. We considered whether RS prevents both cytokine- and VEGF-induced permeability of primary human retinal endothelial cells (HRECs), its mechanism of action, and its ability to reclose a relaxed barrier. Finally, we performed a head-to head comparison of RS and an anti-VEGF (aflibercept).

2. Materials and Methods

2.1. Materials

Primary human retinal endothelial cells (HRECs) were purchased from Cell Systems (Kirkland, WA, USA). They were cultured in EBM™-2 Basal Medium (CC-3156) supplemented with EGM™-2 MV Microvascular Endothelial Cell Growth Medium SingleQuots™ (CC-4147) purchased from Lonza Biosciences (Verviers, Belgium). D-(+)-glucose (G7021) and 2% gelatin solution (G1393) were purchased from Sigma Aldrich (St. Louis, MO, USA). Corning™ Penicillin–Streptomycin Solution (MT30002CI) was purchased from ThermoFisher Scientific (Waltham, MA, USA).

RS was purchased from both Selleckchem (S7223) and Cayman (14794). Recombinant human VEGF 165 (100-20), recombinant human TNFα (300-01A), recombinant human IL-1β (200-01B), and recombinant human TGFβ1 (100-21) were from Peprotech, Inc. (Cranbury, NJ, USA). Aflibercept (Eylea) was from Regeneron Pharmaceuticals Inc. (Tarrytown, NY, USA). ON-TARGETplus human SMAD4 siRNA (L-003902-00-0005), ON-TARGETplus Non-targeting Control Pool (D-001810-10-20), and DharmaFECT 1 Transfection Reagent (T-2001-03) were purchased from Horizon Discovery, Ltd. (Waterbeach, UK). Opti-MEM™ reduced serum medium (31985070) was purchased from ThermoFisher. The 8-well chamber slides for ECIS (8W10E+) were purchased from Applied Biophysics, Inc. (Troy, NY, USA).

Dulbecco's Phosphate-Buffered Salt Solution 1× (MT21030CM), EZ-Link™ NHS-LC-LC-Biotin (21343), Pierce™ 16% Formaldehyde (28908), and Streptavidin—Alexa Fluor™ 594 Conjugate (S32356) were purchased from ThermoFisher Scientific. Bovine serum albumin (BSA (A2153-50G)), and DAPI (D9542-10MG) were purchased from Sigma Aldrich. Human VE-cadherin antibody (MAB9381-SP) was purchased from R&D Systems (Minneapolis, MN, USA). Goat Anti-Mouse IgG H&L (Alexa Fluor® 488) (ab150113) and normal goat serum (ab7481) were purchased from Abcam (Boston, MA, USA). DAPI Fluoromount-G® (0100-20) was purchased from SouthernBiotech (Birmingham, AL, USA).

SMAD4 polyclonal antibody (PA5-34806) was purchased from ThermoFisher Scientific. Phospho-SMAD3 (Ser423/425) antibody (9520S) and anti-rabbit IgG HRP-linked antibody (7074) were purchased from Cell Signaling, Inc. (Danvers, MA, USA). RasGAP antibody was an antiserum that was raised as described previously [8]. Claudin-5 monoclonal antibody (35–2500) was purchased from ThermoFisher. IRDye® 800CW Donkey anti-Mouse IgG secondary antibody (926–32,212) and Goat anti-Rabbit IgG secondary antibody (926–32,211) were purchased from LI-COR, Inc. (Lincoln, NE, USA).

Precast 7.5%, 10%, and 12% Mini-PROTEAN® TGX™ gels (4561034) were from Bio-Rad Laboratories, Inc. (Hercules, CA, USA). Pierce™ ECL Western Blotting Substrate (PI32106) and iBlot™ 2 Transfer Stacks were purchased from ThermoFisher Scientific.

2.2. Cell Culture

HRECs were seeded onto standard polystyrene plates coated with a 0.04% gelatin solution and cultured in Lonza EBM basal media supplemented with EGM SingleQuots and 5% penicillin–streptomycin solution. Cells were cultured in either 5 mM D-glucose-containing media, which we denoted as normal glucose (NG) media, or media supplemented with additional glucose to a final concentration of 30 mM, which we denoted as high-glucose (HG) media. Our in vitro system emulating diabetic retinopathy requires HRECs to be treated with HG for at least 10 days with daily media changes. Experiments commenced after this 10-day treatment with HG.

2.3. Electrical Cell-Substrate Impedance Sensing (ECIS)

The transendothelial electrical resistance (TEER) of HRECs was measured using an ECIS ZTheta instrument (Applied Biophysics, Troy, NY, USA) housed within a standard tissue culture incubator, as described previously [9]. First, 17.6×10^6 cells were plated onto gelatin-coated 8-well chamber slides containing gold-plated microelectrodes (Applied Biophysics catalogue #8W10E+, New York, NY, USA). The following day, each well was inspected under a phase-contrast microscope to verify that the monolayer was complete and overtly normal. At the start of the experiment (typically 24–48 h after plating), the impedance was typically 1500 ohms at 16 kHz. Four wells of the TEER chamber were used for a single experimental condition.

After placing the chamber in the TEER instrument, the impedance was monitored until it stabilized (typically at least 30 min). The electric current passing through the endothelial monolayers was measured independently in each well. TEER was measured continuously and in real time before, during and after the treatment of cells. When the medium was changed during a measurement period, the recording was paused until the medium change-induced noise subsided.

TEER data are presented in ohms without normalization. The starting impedance was typically very similar for all wells; variants, wells in which the impedance was ±15% of the average of all of the wells, were not included. The TEER tracing shows the mean ± SEM of the four wells of a given experimental condition. The area under the curve (AUC) of data sets was quantified using ImageJ (version: 2.0.0-rc-69/1.52v; build: 269a0ad53f; date: 2018-12-04; open source image processing software; https://imagej.net/software/fiji; Bethesda, MD, USA). Unless indicated otherwise, the TEER tracing is of a single, representative experiment, whereas the AUC data are the compilation of all independent experiments within a series. The AUC data, determined from the TEER tracing for the same time period for all experimental conditions, were expressed as a ratio of the agent-treated/vehicle-treated samples. The bar graph shows the mean ± SEM of the agent-treated/vehicle-treated ratio for all of the independent experiments within a series.

For experiments testing the ability of RS to prevent agonist-induced barrier relaxation, 10 μM RS was added, followed by the immediate addition of 100 ng/mL VEGF and/or 50 ng/mL TNFα or 50 ng/mL IL-1β. For RS and aflibercept reclosure experiments, cells were exposed to 100 ng/mL VEGF and/or 50 ng/mL TNFα for 8 h before the addition of either 10 μM RS or 143.75 μg/mL aflibercept for the remainder of the time course. To reflect that clinical scenario, aflibercept was added at a 500-fold molar excess of VEGF (1250 nM aflibercept and 2.5 nM VEGF). The composition of the aflibercept vehicle was 10 mM sodium phosphate pH 6.2, 40 mM sodium chloride, 0.03% polysorbate 20, and 5% sucrose. In this experimental system, bevacizumab, which neutralizes VEGF-A, had the same effect as aflibercept, which neutralizes VEGF-B and PIGF in addition to VEGF-A [9].

2.4. Gelatin Trapping Assay (GTA)

The formulation of biotinylated gelatin and the gelatin trapping assay (GTA) protocol was as previously described [10]. Briefly, cells were seeded at 100% confluency on a 35 mm glass bottom dish (P35G-1.5-14-C) from MatTek Corporation (Ashland, MA, USA) that had been coated with 100 μL of 0.25 μg/mL biotinylated gelatin overnight at 4 °C. The medium was refreshed every 24 h; 48 h post-plating, 10 μM RS or vehicle, 100 ng/mL VEGF, and 50 ng/mL TNFα or vehicle were added and incubated for 5 h. Once stimulation was complete, cells were treated briefly with a 1:2000 dilution of streptavidin in DPBS. The remaining protocol was completed in the dark using 1000 μL of treatment solution in every step. Cells were washed three times with room-temperature DPBS and treated with 4% PFA in DPBS for 10 min at room temperature. After three 5 min washes with ice-cold DPBS, cells were treated with permeabilization solution (0.25% TritonX100 in DPBS) for 30 min at room temperature. Once cells were permeabilized, they were washed again with DPBS three times for 5 min per wash. Plates were blocked with AD buffer (10% goat serum, 1% BSA, 0.05% Tween20 in DPBS) for 1 h at room temperature. After aspirating the AD buffer, cells were incubated overnight at 4 °C with a 1:200 dilution of VE-cadherin primary antibody in AD buffer. The next day, the cells were washed with DPBS containing 0.05% Tween 20 (PBST) three times for 5 min per wash, and then incubated for 1 h at room temperature with a 1:1000 dilution of fluorescently tagged secondary antibody in AD buffer. Once incubation with the secondary antibody was complete, the cells were washed two times in PBST and once in PBS for 5 min per wash, and then counterstained with DAPI diluted in DPBS 1:1000 for one minute. After one final wash with DPBS, 12 mm glass coverslips were mounted onto the glass bottom using a drop of antifade mounting medium.

Plates containing fixed cells were subjected to imaging with a Zeiss confocal microscope (Jena, Germany). At least two images per plate were collected at 20× objective using 405 nm (DAPI), 488 nm (VE-cadherin), and 594 nm (streptavidin) lasers. Quantification of pore intensity was accomplished using ImageJ.

2.5. Small Interfering RNA (siRNA)

HG HRECs were plated onto 60 mm gelatin-coated plates and grown to confluency in complete Lonza medium without antibiotic supplements. Cells were transfected with 2 mL of 10 nM non-targeting control or siSMAD4 complexed with Dharmacon transfecting reagent (1:2 ratio) in Opti-mem medium. The medium was replaced with complete Lonza medium 24 h post-transfection and incubated for another 24 h (total of 48 h of knockdown) before sample collection for Western blot analysis.

For ECIS analysis, cells were plated to confluency on two 100 mm gelatin-coated dishes cultured in Lonza medium without antibiotic supplements. Cells were transfected with 4 mL of 10 nM siSMAD4 or 10 nM non-targeting control complexed with Dharmacon transfecting reagent (1:2 ratio) in Opti-mem medium for 24 h. After this incubation, cells were plated on gelatin-coated 8-well chamber slides for ECIS (8W10E+); 24 h later (after 48 h of knockdown), the ECIS assay began in which 100 ng/mL VEGF or vehicle was added to the wells. TEER tracings were recorded for at least 20 h.

2.6. Western Blotting

HG HRECs were plated onto 60 mm or 35 mm gelatin-coated plates and grown to confluency before treatment with 10 µM RS or vehicle for 30 min, followed by 50 ng/mL TGFβ or vehicle for an additional 30 min. Cells were then washed with ice-cold PBS and lysed in SDS-βMe (0.3% sodium dodecyl sulfate, 1% β-mercaptoethanol, and 50 mM Tris-HCl pH 7.5). Lysates were then treated with DNase/RNase (1 mg/mL DNase I, 500 µg/mL RNaseA, 100 mM Tris, pH 7.5, 25 mM MgCl2, and 5 mM CaCl2) and 5× sample buffer (10 mM EDTA, 2% sodium dodecyl sulfate, 0.2 M 2-mercaptoethanol, 20% glycerol, 20 mM Tris-HCl, pH 6.8, and 0.2% bromophenol blue) before being heated at 95 °C for 5 min to denature the proteins. Samples were then cooled on ice (or stored at −20 °C) before being resolved on a 7.5% SDS–polyacrylamide gel and subjected to Western blot analysis. siRNA-transfected cells and RS (10 µM)-stimulated cells were collected in the same way and resolved on 10% and 12% SDS–polyacrylamide gels, respectively.

After transferring the gel using iBlot stacks, membranes were incubated on an orbital shaker for 1 h in TBST (10 mM Tris, pH 7.5, 150 mM NaCl, and 0.05% Tween20) containing 5% bovine serum albumin (BSA) at room temperature. Blots were subsequently incubated with primary antibodies (SMAD4 (1:1000), pSMAD3 (1.3:1000) claudin-5 (1:1000), and Rasgap (1:2000)) at room temperature for 2–3 h or overnight at 4 °C before washing and probing with secondary antibody for 1 h at room temperature. An HRP-linked secondary antibody was used for the SMAD4 and pSMAD3 blots, and IRDye 800CW secondary antibodies were used for the claudin-5 blots. ECL substrate kits were used to visualize the SMAD4 and pSMAD3 blots. The claudin-5 blots were visualized with the Chemidoc MP imager. The intensity of the bands was quantified using ImageJ.

2.7. Statistical Analysis

Unless indicated otherwise, TEER data are mean ± SEM; all other data are mean ± SD. Statistical significance of differences between means of two experimental groups was assessed using the *t*-test; ANOVA was used when comparisons involved groups larger than two. Significance was defined as $p < 0.05$.

3. Results

3.1. RS Prevented Barrier Relaxation by Distinct Agonists

We investigated the effect of RS on both basal and agonist-induced permeability in primary human retinal endothelial cells (HRECs) using electric cell-substrate impedance sensing (ECIS), which measures the electrical resistance across a monolayer of cells. Addition of RS rapidly improved basal barrier function and completely prevented VEGF from relaxing the barrier (Figure 1A). Dose–response experiments indicated that 10 μM was the minimum concentration of RS that had a maximal effect; 1 and 0.1 μM had a partial effect and no effect, respectively. These observations resonate with a previous publication investigating the influence of a panel of TGFβ receptor antagonists on basal and VEGF-induced permeability [5]. We extended this line of investigation by considering the effect of RS on cytokine-induced permeability. The effect was even more pronounced. RS not only prevented TNFα or IL-1β permeability, but it also converted TNFα from a barrier relaxer to a modest barrier enforcer (Figure 1B,D). Finally, RS prevented relaxation in response to the combination of TNFα and VEGF (Figure 1C). These studies demonstrate that RS enforces basal barrier function and prevents relaxation in response to multiple types of agonists.

The experiments in Figure 1 were conducted with HRECs cultured in high glucose (30 mM). We recently reported that these experimental conditions induced hyperglycemia-induced adaptation (HIMA) [11]. To test if the effects of RS described above were unique to this in vitro model of diabetic retinopathy, we repeated key experiments with HRECs cultured in normal glucose (5 mM). The effect of RS was the same in cells that had not undergone HIMA (Figure S1). Thus, HG-induced changes to gene expression, osmolality, and metabolism did not influence the response of cells to RS. We conclude that acquisition of HIMA was not required for the RS effect, which is consistent with the report of Roudnicky et al. using cells cultured in normal glucose [5].

3.2. Antagonizing TGFβ Signaling Was Insufficient for the RS Effect

In light of the well-established ability of RS to inhibit the kinase activity of TGFβ receptors [5,12], we sought to test the hypothesis that RS enforced basal barrier function and prevented agonist-induced relaxation by antagonizing constitutive TGFβ signaling. We found that while RS inhibited TGFβ-induced signaling (Figure 2A), constitutive TGFβ signaling (phosphorylation of SMAD3) was low (Figure 2A; see also the uncropped blot in Figure S2. Furthermore, antagonizing the endogenous TGFβ pathway by suppressing the expression of SMAD4 did not phenocopy the RS effect (Figure 2B,C; see also the uncropped blot in Figure S2). In addition, exposing cells to exogenous TGFβ for either 18 or 48 h had no effect on basal or VEGF-induced permeability (Figure 3 and [13]). These data show that antagonizing or activating the TGFβ pathway did not influence basal or VEGF-mediated barrier relaxation. Moreover, although RS prevented TGFβ signaling, this did not appear to be how it enforced barrier function. This concept is supported by Roudnicky et al., showing that TGFβ receptor inhibitors other than RS (e.g., SB-431542) had no effect on barrier function. This unique feature of RS appears to require more than suppression of TGFβ-dependent signaling.

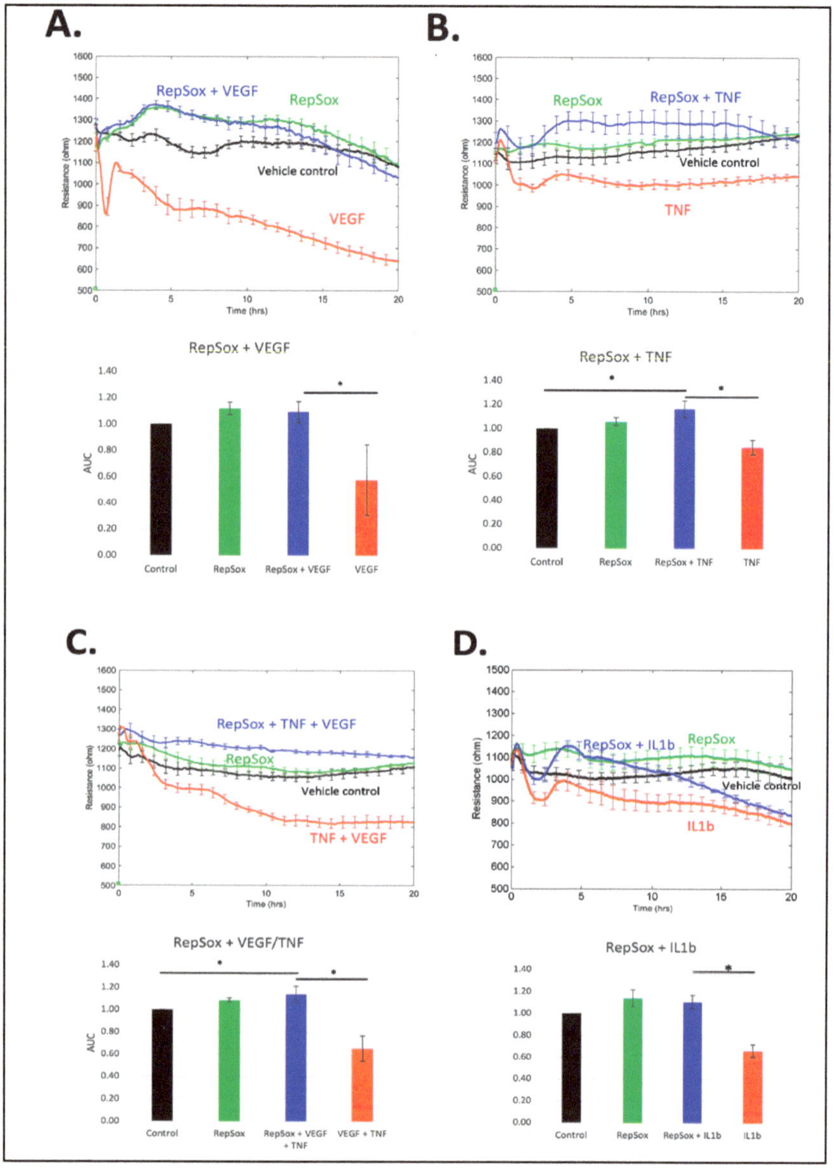

Figure 1. RS prevented barrier permeability by distinct agonists. TEER tracings of cells treated with either DMSO or 10 μM RS that was added concurrently with agonist vehicle (0.1% BSA in water) or (**A**) 100 ng/mL VEGF, (**B**) 50 ng/mL TNFα, (**C**) 100 ng/mL VEGF and 50 ng/mL TNFα, or (**D**) 50 ng/mL IL-1β. The TEER tracings from a single, representative experiment are shown; the data are the mean ± SEM of 4 wells that were used for a single experimental condition. The bar graphs show the average area under the curve (AUC) from three independent experiments. The AUC was calculated for the time interval from 0 to 15 h. Differences between the indicated pairs were determined using the Student's *t*-test; * $p < 0.05$.

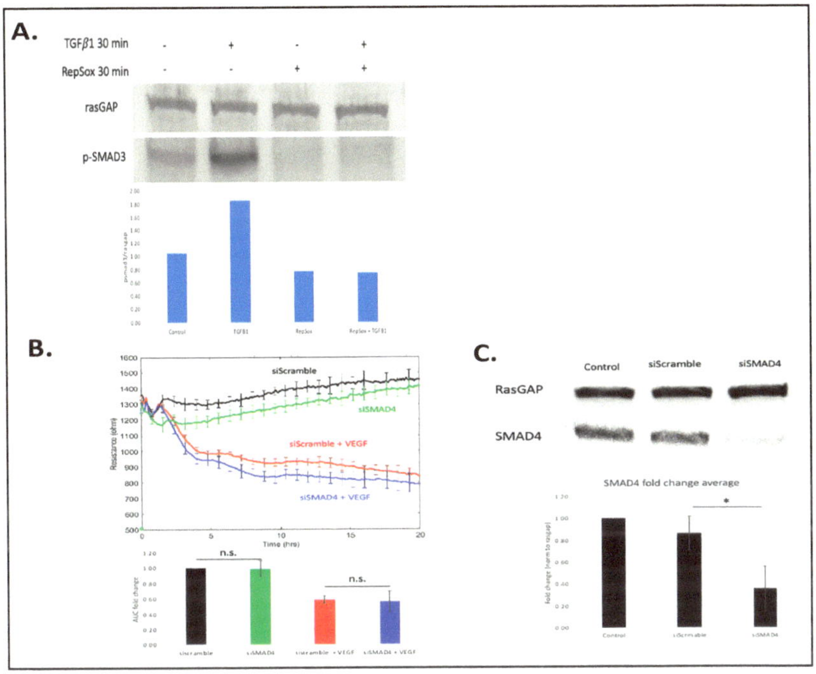

Figure 2. RS's inhibition of TGFβ-induced signaling was not responsible for its effect on agonist-induced permeability. (**A**) Cells were pretreated with RS vehicle (DMSO) or 10 μM RS for 30 min before adding TGFβ vehicle (0.1% BSA in water) or 50 ng/mL TGFβ for an additional 30 min, after which time the cells were lysed and subjected to Western blot analysis with the indicated antibodies. Band intensities were quantified using ImageJ. The pSMAD3/Rasgap ratio for the blot shown is presented in the bar graph. Similar results (efficient RS-mediated suppression of the TGFβ pathway) were observed on at least three independent occasions. (**B**) Same as Figure 1A, except the cells were transfected with either siScramble or siSMAD4. The TEER tracing is of a single, representative experiment; the bar graph is the compilation of at least three independent experiments. (**C**) The cells used in panel (**B**) were subjected to Western blot analysis using the indicated antibodies. The blot is of a single, representative experiment, while the bar graph shows the fold change in the intensity of SMAD4 in three independent experiments. * $p < 0.05$; n.s. $p > 0.05$.

3.3. *RS Prevented Agonist-Induced Pore Formation and VE-Cadherin Disorganization*

To better understand how RS enforced barrier function, we investigated its influence on pore formation and the organization of adherens junctions, which are components of the barrier that are affected by many types of agonists [14]. To observe pore formation, we used a previously described gelatin trapping assay (GTA) [10] in which cells are plated on biotinylated gelatin and stained with fluorescently labeled streptavidin. Pores permit interaction between biotin and streptavidin, which can be detected via fluorescent microscopy and quantified. The organization of adherens junctions was assessed by co-staining the monolayers with a fluorescently labeled anti-VE-cadherin antibody. Control experiments demonstrated the specificity of the signals that were observed in this series of experiments (Figure S3).

Exposing cells to the combination of VEGF and TNFα for 5 h increased permeability (Figure 1C) induced pore formation and disorganized the adherens junctions (Figure 4). A high-magnification image shows that pores preferentially formed in areas where the adherens junctions had disorganized (Figure 4). RS prevented all of these changes (Figures 1C and 4). While RS improved the electrical resistance of the barrier in unstimulated cells (Figure 1), RS did not detectably affect pores or the organization of the adherens junctions under such experimental conditions (Figure 4). These data indicate that a plausible mechanism by which RS suppresses agonist-induced permeability is by preventing changes in the barrier that are required for its relaxation.

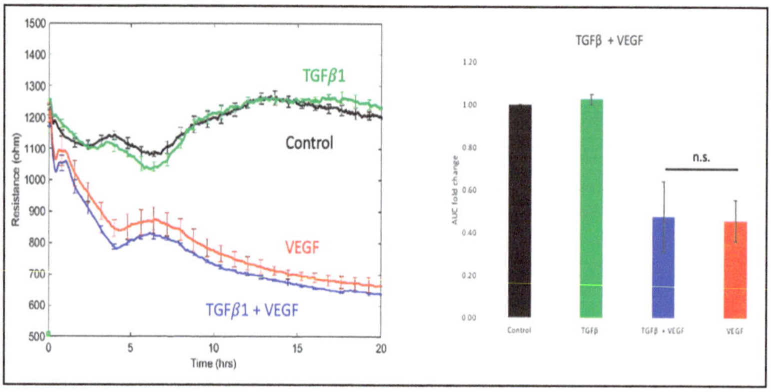

Figure 3. Overnight treatment of TGFβ had no effect on VEGF-induced permeability. Cells were pretreated overnight with 50 ng/mL TGFβ or TGFβ vehicle (0.1% BSA in water), and then 100 ng/mL VEGF or PBS (control) was added, and the TEER tracing was recorded for the indicated duration. The TEER tracing from a single, representative experiment is shown; the data are the mean ± SEM of the 4 wells that were used for a single experimental condition. The bar graphs show the mean ± SD of the area under the curve (AUC) from three independent experiments. The AUC was calculated for the time interval from 0 to 20 h. Differences between the indicated pairs were determined using the Student's *t*-test; n.s. $p > 0.05$.

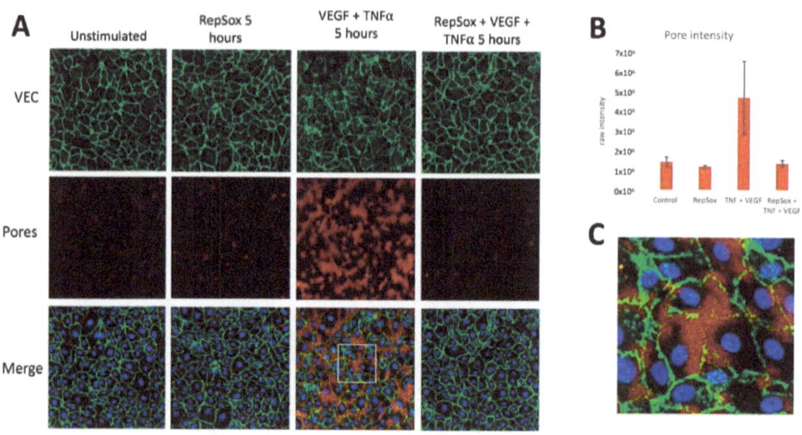

Figure 4. RS prevented agonist-induced pore formation and VEC disorganization. (**A**) Cells that had been treated with the indicated agents were subject to the GTA assay as described in the Materials and Methods section. The top row of this panel shows staining with an anti-VE-cadherin antibody, which reveals the nature of the adherens junction. (**B**) Quantification of pore intensity (red signal). The bars are the median of the intensity from two arbitrarily selected images; error bars show the range. In 3 independent experiments, RS inhibited VEGF/TNFα-induced pore formation. (**C**) High magnification of the boxed region within the merged VEGF and TNFα image (the third panel from the right in the bottom row). This panel demonstrates that pores formed preferentially in regions of VEC disorganization.

We also considered whether the RS effect was associated with increased expression of claudin-5, a component of tight junctions that contributes to endothelial barrier stability [15]. As expected, and as in [5], expression of claudin-5 was markedly increased following exposure to RS for 24 h (Figure S4; see also the uncropped blot in Figure S5). However, the level of claudin-5 was unchanged in cells incubated with RS for 1 or 6 h (Figure S4; see also the uncropped blot in Figure S5). Because the RS effect was observed very quickly, well before the increase in expression of claudin-5, we conclude that upregulation of claudin-5 is unlikely to be responsible for how RS stabilized the barrier in our experimental setting.

3.4. RS Reclosed the Barrier Relaxed by Distinct Agonists

In addition to preventing agonist-induced permeability (Figure 1), we considered whether RS was able to reclose a relaxed barrier. The design of these experiments was to first relax the barrier for 8 h with VEGF, TNFα, or their combination before adding RS and continuing to record the permeability of the monolayers. Thus, we tested if RS could reclose the barrier in the continued presence of the agonist that relaxed it.

RS partially reclosed a barrier relaxed by VEGF, and the maximum effect took approximately 4 h (Figure 5A). In contrast, RS completely and quickly overcame TNFα-induced permeability (Figure 5B). Cells treated with both VEGF and TNFα responded to the addition of RS in a manner similar to cells treated with VEGF alone (Figure 5C). It seems likely that the combination of RS and anti-VEGF treatment would completely reclose the barrier because anti-VEGF is more effective than RS at reclosing a VEGF-relaxed barrier (ref. [9] and Figure 6 below). These results indicate that RS induced reclosure of the barrier in the continued presence of the agonist which relaxed it. Furthermore, the extent of this effect was determined by the agonist.

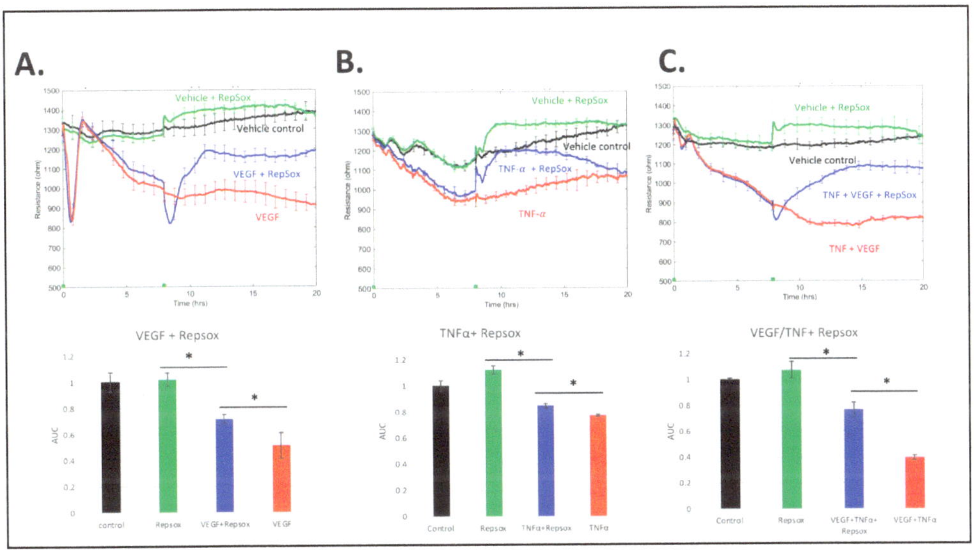

Figure 5. RS reclosed barriers that were relaxed by distinct agonists. Barriers were relaxed with either (**A**) 100 ng/mL VEGF, (**B**) 50 ng/mL TNFα, or (**C**) 100 ng/mL VEGF and 50 ng/mL TNFα for 8 h before the addition of either RS vehicle (DMSO) or 10 μM RS. Each TEER tracing is from a single, representative experiment along with the bar graph showing the area under the curve (AUC) for that experiment, which was calculated for the time interval from 15 to 20 h. The error bars show the mean ± SEM for quadruplicate wells that were used for each experimental condition. The Student's *t*-test was used to determine statistical significance between the indicated pairs. * $p < 0.05$ Three independent experiments showed similar results. Repeat experiments can be found in Figures S6–S8.

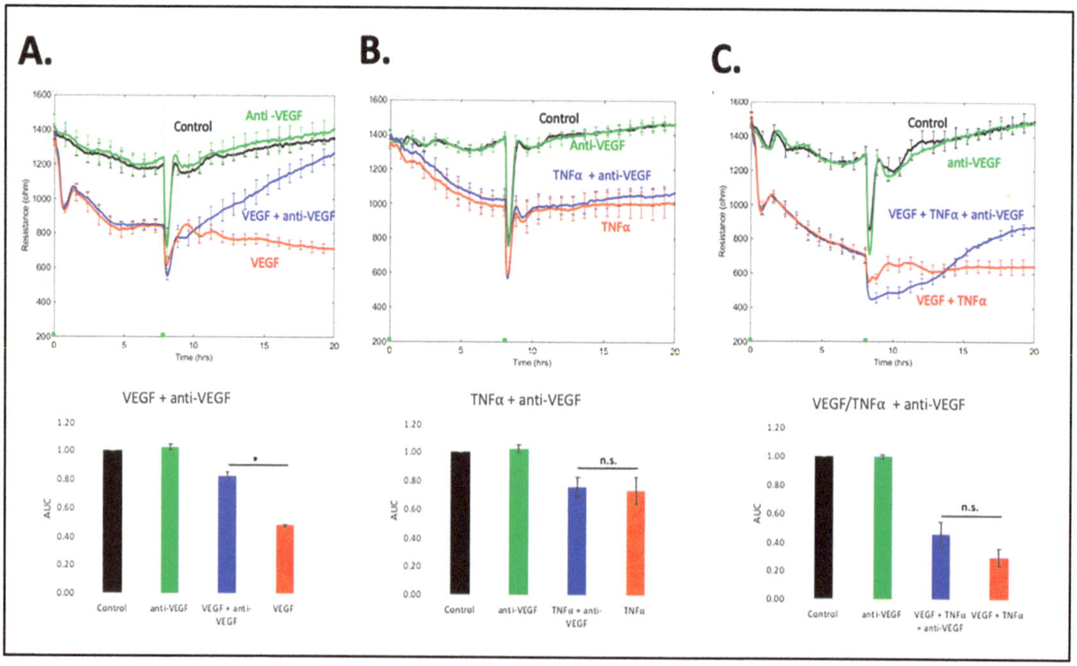

Figure 6. Anti-VEGF reverses only VEGF-mediated barrier relaxation. Same as Figure 5, except aflibercept vehicle or aflibercept was used instead of RS. Barriers were relaxed with either (**A**) 100 ng/mL VEGF, (**B**) 50 ng/mL TNFα, or (**C**) 100 ng/mL VEGF and 50 ng/mL TNFα for 8 h before the addition of either aflibercept vehicle or aflibercept (anti-VEGF). TEER tracings are from a single, representative experiment; the data are the mean ± SEM of the 4 wells that were used for each experimental condition. The bar graphs under the TEER tracings show the mean ± SD of the area under the curve (AUC) for the TEER tracing shown above. The AUC was calculated for the time interval from 15 to 20 h. Statistically significant differences between the indicated pairs were determined using the Student's *t*-test; * $p < 0.05$; n.s. $p > 0.05$. Three independent experiments showed similar results.

3.5. Comparison of RS with Anti-VEGF

In order to compare RS with anti-VEGF, we repeated the experiments described in Figure 5 using an anti-VEGF (aflibercept) instead of RS. Anti-VEGF reclosed the barrier relaxed with VEGF (Figure 6A) to a greater extent than RS (Figure 5A). Anti-VEGF had no effect on the TNFα -relaxed barrier (Figure 6B), whereas RS completely reclosed it (Figure 5B). Permeability driven by the combination of TNFα and VEGF was largely insensitive to anti-VEGF (Figure 6C), whereas RS partially reclosed a barrier that had been relaxed by this agonist combination (Figure 5C). We conclude that RS either completely or partially overcame permeability driven by different types of agonists. In contrast, anti-VEGF was effective only when VEGF was the sole agonist; under such conditions, it was more effective than RS.

4. Discussion

We observed that the therapeutic potential of RS extended beyond the previously reported inhibition of VEGF-induced permeability [5]. RS completely suppressed cytokine (TNFα or IL-1β)-induced and even VEGF/TNFα-induced barrier relaxation. Furthermore, RS either partially (VEGF) or completely (TNFα) reclosed barriers, despite the continued presence of the relaxation-promoting agent. Finally, a head-to head comparison with anti-VEGF indicated that RS stabilized barriers in the face of multiple types of agonists (VEGF and cytokines), whereas anti-VEGF was effective when VEGF was the only agonist present.

The results described herein, together with those published by other groups [12,16–18], raise the question of how RS enforces barrier stability. While we observed the expected RS-mediated inhibition of the TGFβ pathway, this did not appear to suffice in our experimental system. Activating or suppressing the TGFβ pathway did not influence either basal or agonist-driven permeability [13] (Figures 2 and 3). Furthermore, other TGFβ pathway inhibitors (SB-431542) do not prevent VEGF-induced permeability [5]. While these observations do not rule out a contribution of the TGFβ pathway, they show that additional targets play an essential role in the effect of RS on barrier function.

Given that RS prevented agonists that engage distinct signaling pathways, it seems most likely that the relevant RS target lies within steps controlling the barrier that are common to multiple agonists (e.g., VEGF, TNFα, and IL1β) [19]. Indeed, RS prevented pore formation and adherens junction disorganization driven by the combination of VEGF and TNFα (Figure 4). While these data are consistent with the RS target being involved with control of components of the barrier, they do not rule out the possibility that the target is proximal to the receptor. For instance, candidate RS targets exist among the kinases that are involved with clathrin-mediated internalization [5], a process that both VEGF and TNFα receptors require for proper signal transduction [20,21]. It is plausible that RS simultaneously antagonizes many kinases, and that inhibition of more than one of them is required for the RS effect.

Is it likely that the relevant RS-regulated kinases govern expression of claudin-5? Claudin-5 is a key component of tight junctions, and its expression declines in response to agents that induce permeability [22]. Furthermore, RS increases the level of claudin-5, and this change correlates with RS-mediated barrier stabilization [23]. However, while RS increased expression of claudin-5 in HRECs, this occurred well after the barrier had been stabilized (Figure S4). Thus, regulation of claudin-5 expression does not adequately explain how RS stabilizes the barrier of HRECs. Additional studies are necessary to identify the RS targets (kinases and pathways that they regulate) responsible for enforcing barrier function.

Unlike anti-VEGF, which physically binds and thereby neutralizes the agonist responsible for destabilizing the barrier, RS prevented cells from responding to the agonist, despite its continued presence. Similarly, RS reclosed a relaxed barrier without neutralizing the agent instructing cells to keep the barrier open. This was especially dramatic with TNFα-treated cells, in which case RS completely reclosed the barrier within minutes (Figure 5B). The rapid and complete reclosure of the barrier raises the intriguing possibility that RS engages an as yet-undescribed pathway that instructs cells to close a relaxed barrier.

The next steps to develop RS as a therapeutic include determining the best way to deliver it (eye drops, slow-release depots implanted into the eye, etc.), its pharmacokinetics, and its safety/toxicity. Investigators focused on outcomes other than permeability have reported the feasibility of treating experimental animals with RS [16,18,24]. Given that RS not only prevents agonist-induced permeability but also tightens the basal barrier, it will be particularly important to assess the effect of RS on kidney function, which depends on the relative high permeability of the glomerular vasculature. Additional considerations include the spectrum of RS targets, especially within the TGFβ family.

5. Conclusions

RS stabilized the endothelial barrier in the face of both cytokines and VEGF. Consequently, RS is a candidate therapeutic because of its capability of overcoming permeability driven by multiple agonists that play a role in the pathology of DR.

Supplementary Materials: The following supporting information can be downloaded at: https://www.mdpi.com/article/10.3390/biomedicines11092431/s1. Figure S1: The effect of RS was independent of HIMA acquisition. Figure S2: Uncropped western blot images from Figure 2A with molecular weight standards. Figure S3. Controls for the GTA shown in Figure 4. Figure S4. Claudin-5 expression increases after 24-h of RS treatment. Figure S5. Uncropped western blot images from Figure S3 with molecular weight standards. Figure S6. Experimental repeats for Figure 5A. Figure S7. Experimental repeats for Figure 5B. Figure S8. Experimental repeats for Figure 5C.

Author Contributions: The experimental design was prepared by A.K., L.L. and B.B. Experiments were performed by L.L. and B.B. Data were analyzed by A.K., L.L. and B.B., A.K. and L.L. wrote the manuscript. A.K., L.L. and B.B. reviewed and edited the manuscript. All authors have read and agreed to the published version of the manuscript.

Funding: This research was supported by the Illinois Society to Prevent Blindness, the National Institute of Health (EY031350; EY001792), and an unrestricted grant from the Research to Prevent Blindness Foundation.

Institutional Review Board Statement: Not applicable.

Informed Consent Statement: Not applicable.

Data Availability Statement: The data provided herein are available upon request from the corresponding author.

Acknowledgments: We would like to thank Tung Nguyen and Monica Lee for help in establishing the GTA assay.

Conflicts of Interest: We have no conflict of interest to declare.

References

1. Kuo, C.Y.J.; Murphy, R.; Rupenthal, I.D.; Mugisho, O.O. Correlation between the progression of diabetic retinopathy and inflammasome biomarkers in vitreous and serum—A systematic review. *BMC Ophthalmol.* **2022**, *22*, 238. [CrossRef] [PubMed]
2. Stitt, A.W.; Curtis, T.M.; Chen, M.; Medina, R.J.; McKay, G.J.; Jenkins, A.; Gardiner, T.A.; Lyons, T.J.; Hammes, H.P.; Simo, R.; et al. The progress in understanding and treatment of diabetic retinopathy. *Prog. Retin. Eye Res.* **2016**, *51*, 156–186. [CrossRef] [PubMed]
3. Duh, E.J.; Sun, J.K.; Stitt, A.W. Diabetic retinopathy: Current understanding, mechanisms, and treatment strategies. *JCI Insight* **2017**, *2*, e93751. [CrossRef]
4. Ellis, M.P.; Lent-Schochet, D.; Lo, T.; Yiu, G. Emerging Concepts in the Treatment of Diabetic Retinopathy. *Curr. Diabetes Rep.* **2019**, *19*, 137. [CrossRef] [PubMed]
5. Roudnicky, F.; Zhang, J.D.; Kim, B.K.; Pandya, N.J.; Lan, Y.; Sach-Peltason, L.; Ragelle, H.; Strassburger, P.; Gruener, S.; Lazendic, M.; et al. Inducers of the endothelial cell barrier identified through chemogenomic screening in genome-edited hPSC-endothelial cells. *Proc. Natl. Acad. Sci. USA* **2020**, *117*, 19854–19865. [CrossRef]
6. Gellibert, F.; Woolven, J.; Fouchet, M.H.; Mathews, N.; Goodland, H.; Lovegrove, V.; Laroze, A.; Nguyen, V.L.; Sautet, S.; Wang, R.; et al. Identification of 1,5-naphthyridine derivatives as a novel series of potent and selective TGF-beta type I receptor inhibitors. *J. Med. Chem.* **2004**, *47*, 4494–4506. [CrossRef]
7. Goumans, M.J.; Liu, Z.; ten Dijke, P. TGF-beta signaling in vascular biology and dysfunction. *Cell Res.* **2009**, *19*, 116–127. [CrossRef]
8. Valius, M.; Bazenet, C.; Kazlauskas, A. Tyrosines 1021 and 1009 are phosphorylation sites in the carboxy terminus of the platelet-derived growth factor receptor b subunit and are required for binding of phospholipase Cg and a 64-kilodalton protein, respectively. *Mol. Cell. Biol.* **1993**, *13*, 133–143.
9. Li, Y.; Yan, Z.; Chaudhry, K.; Kazlauskas, A. The Renin-Angiotensin-Aldosterone System (RAAS) Is One of the Effectors by Which Vascular Endothelial Growth Factor (VEGF)/Anti-VEGF Controls the Endothelial Cell Barrier. *Am. J. Pathol.* **2020**, *190*, 1971–1981. [CrossRef]
10. Dubrovskyi, O.; Birukova, A.A.; Birukov, K.G. Measurement of local permeability at subcellular level in cell models of agonist- and ventilator-induced lung injury. *Lab. Investig.* **2013**, *93*, 254–263. [CrossRef]
11. Serikbaeva, A.; Li, Y.; Ganesh, B.; Zelkha, R.; Kazlauskas, A. Hyperglycemia Promotes Mitophagy and Thereby Mitigates Hyperglycemia-Induced Damage. *Am. J. Pathol.* **2022**, *192*, 1779–1794. [CrossRef]
12. He, D.; Gao, J.; Zheng, L.; Liu, S.; Ye, L.; Lai, H.; Pan, B.; Pan, W.; Lou, C.; Chen, Z.; et al. TGF-beta inhibitor RepSox suppresses osteosarcoma via the JNK/Smad3 signaling pathway. *Int. J. Oncol.* **2021**, *59*, 84. [CrossRef] [PubMed]
13. Baccouche, B.; Lietuvninkas, L.; Kazlauskas, A. Activin A Limits VEGF-Induced Permeability via VE-PTP. *Int. J. Mol. Sci.* **2023**, *24*, 8698. [CrossRef] [PubMed]
14. Komarova, Y.; Malik, A.B. Regulation of endothelial permeability via paracellular and transcellular transport pathways. *Annu. Rev. Physiol.* **2010**, *72*, 463–493. [CrossRef]
15. Claesson-Welsh, L.; Dejana, E.; McDonald, D.M. Permeability of the Endothelial Barrier: Identifying and Reconciling Controversies. *Trends Mol. Med.* **2021**, *27*, 314–331. [CrossRef] [PubMed]
16. Guo, Y.; Zhu, H.; Li, X.; Ma, C.; Li, Y.; Sun, T.; Wang, Y.; Wang, C.; Guan, W.; Liu, C. RepSox effectively promotes the induced differentiation of sheep fibroblasts into adipocytes via the inhibition of the TGF-beta1/Smad pathway. *Int. J. Mol. Med.* **2021**, *48*, 148. [CrossRef]
17. Shi, C.J.; Lian, J.J.; Zhang, B.W.; Cha, J.X.; Hua, Q.H.; Pi, X.P.; Hou, Y.J.; Xie, X.; Zhang, R. TGFbetaR-1/ALK5 inhibitor RepSox induces enteric glia-to-neuron transition and influences gastrointestinal mobility in adult mice. *Acta Pharmacol. Sin.* **2023**, *44*, 92–104. [CrossRef]

18. Ichida, J.K.; Blanchard, J.; Lam, K.; Son, E.Y.; Chung, J.E.; Egli, D.; Loh, K.M.; Carter, A.C.; Di Giorgio, F.P.; Koszka, K.; et al. A small-molecule inhibitor of tgf-Beta signaling replaces sox2 in reprogramming by inducing nanog. *Cell Stem Cell* **2009**, *5*, 491–503. [CrossRef]
19. Dejana, E.; Orsenigo, F.; Lampugnani, M.G. The role of adherens junctions and VE-cadherin in the control of vascular permeability. *J. Cell Sci.* **2008**, *121*, 2115–2122. [CrossRef]
20. Simons, M. An inside view: VEGF receptor trafficking and signaling. *Physiology* **2012**, *27*, 213–222. [CrossRef]
21. Schneider-Brachert, W.; Tchikov, V.; Neumeyer, J.; Jakob, M.; Winoto-Morbach, S.; Held-Feindt, J.; Heinrich, M.; Merkel, O.; Ehrenschwender, M.; Adam, D.; et al. Compartmentalization of TNF receptor 1 signaling: Internalized TNF receptosomes as death signaling vesicles. *Immunity* **2004**, *21*, 415–428. [CrossRef] [PubMed]
22. Greene, C.; Hanley, N.; Campbell, M. Claudin-5: Gatekeeper of neurological function. *Fluids Barriers CNS* **2019**, *16*, 3. [CrossRef] [PubMed]
23. Zihni, C.; Mills, C.; Matter, K.; Balda, M.S. Tight junctions: From simple barriers to multifunctional molecular gates. *Nat. Rev. Mol. Cell Biol.* **2016**, *17*, 564–580. [CrossRef] [PubMed]
24. Greene, C.; Hanley, N.; Reschke, C.R.; Reddy, A.; Mae, M.A.; Connolly, R.; Behan, C.; O'Keeffe, E.; Bolger, I.; Hudson, N.; et al. Microvascular stabilization via blood-brain barrier regulation prevents seizure activity. *Nat. Commun.* **2022**, *13*, 2003. [CrossRef] [PubMed]

MDPI AG
Grosspeteranlage 5
4052 Basel
Switzerland
Tel.: +41 61 683 77 34

Biomedicines Editorial Office
E-mail: biomedicines@mdpi.com
www.mdpi.com/journal/biomedicines

www.ingramcontent.com/pod-product-compliance
Lightning Source LLC
Chambersburg PA
CBHW042011260926
51850CB00020B/3050
* 9 7 8 3 7 2 5 8 2 4 9 9 1 *